Signals and Communication Technology

This series is devoted to fundamentals and applications of modern methods of signal processing and cutting-edge communication technologies. The main topics are information and signal theory, acoustical signal processing, image processing and multimedia systems, mobile and wireless communications, and computer and communication networks. Volumes in the series address researchers in academia and industrial R&D departments. The series is application-oriented. The level of presentation of each individual volume, however, depends on the subject and can range from practical to scientific.

Indexing: All books in "Signals and Communication Technology" are indexed by Scopus and zbMATH

For general information about this book series, comments or suggestions, please contact Mary James at mary.james@springer.com or Ramesh Nath Premnath at ramesh.premnath@springer.com.

Yesenia Cevallos
Paola G. Vinueza-Naranjo • Deysi Inca
Elvis Pilligua • Luis Tello-Oquendo
Marcelo León • Cristian Vacacela Gómez
Talía Tene • Jiří Svozilík • Elba Bodero-Poveda
Albert Espinal

Markup Languages and Simulators for Molecular Communications

Springer

Yesenia Cevallos
IDeTIC, University of Las Palmas de Gran Canaria
Las Palmas de Gran Canaria, Spain

College of Engineering
Universidad Nacional de Chimborazo
Riobamba, Ecuador

Deysi Inca
College of Engineering
Universidad Nacional de Chimborazo
Riobamba, Ecuador

Luis Tello-Oquendo
College of Engineering
Universidad Nacional de Chimborazo
Riobamba, Ecuador

Cristian Vacacela Gómez
Universidad Tecnológica ECOTEC
Guayaquil, Ecuador

INFN-Laboratori Nazionali di Frascati
Frascati, Italy

Jiří Svozilík
Facultad de Ciencias
Escuela Superior Politécncia de Chimborazo
Riobamba, Ecuador

Albert Espinal
Faculty of Electrical and Computer Engineering
Escuela Superior Politécnica del Litoral
Guayaquil, Ecuador

Paola G. Vinueza-Naranjo
IDeTIC, University of Las Palmas de Gran Canaria
Las Palmas de Gran Canaria, Spain

College of Engineering
Universidad Nacional de Chimborazo
Riobamba, Ecuador

Elvis Pilligua
IDeTIC, University of Las Palmas de Gran Canaria
Las Palmas de Gran Canaria, Spain

Marcelo León
Universidad Tecnológica ECOTEC
Guayaquil, Ecuador

Talía Tene
Department of Chemistry
Universidad Técnica Particular de Loja
Loja, Ecuador

Elba Bodero-Poveda
College of Engineering
Universidad Nacional de Chimborazo
Riobamba, Ecuador

ISSN 1860-4862 ISSN 1860-4870 (electronic)
Signals and Communication Technology
ISBN 978-3-032-33039-0 ISBN 978-3-032-33040-6 (eBook)
https://doi.org/10.1007/978-3-032-33040-6

This Springer imprint is published by the registered company Springer Nature Switzerland AG
The registered company address is: Gewerbestrasse 11, 6330 Cham, Switzerland

To God, for supporting our lives in every moment and in everything…
and to our loved ones, for their constant love and support…

Preface

Technological advances in engineering and biology have allowed these two scientific areas to join their efforts to produce efficient models, using inspirational paradigms based on the behavior of nature and its diverse characteristics. Industries, medicine, biochemistry, biotechnology, and computational sciences, among other disciplines, are precisely using such models. For example, bio-nanotechnology is a branch of nanotechnology that uses biological materials and employs biologically based designs for applications primarily in medicine. Molecular communications (MCs) propagate biochemical signals, representing information between various natural and artificial components in biological transmission systems that communicate at the nanoscale. Recently, other areas where MCs are gaining strength are food production, materials science, environmental sciences, and military applications, among others.

Currently, achieving increasingly efficient treatments for diseases is one of the main objectives of the medical sciences, particularly genetics and the various sciences that support it. The demand for genomic signal processing has dramatically increased in recent years, leading science and technology to amalgamate various specialties, among which MCs stand out. These MCs are supported by the information and communication theories of engineering, primarily for medical purposes. MCs can be used for many purposes, such as smart drug delivery, which allows for precise targeting with fewer side effects, and health monitoring through intrabody nanoscale sensors, which adds new dimensions to collecting real-time data and diagnosing diseases.

Essentially, MCs analyze the information exchanged between cells, which are the intelligent components of our bodies, and the transfer of information primarily through molecules and their various bonds in physicochemical mediums of diffusion. Contrary to traditional communication systems, MCs use molecules as information carriers, and these in turn constitute the information itself. Biological emitters, known as nanomachines (or bio-nanomachines if they contain a biological component), encode information into molecules, also known as informational signals, and then release these signals into the environment. These molecules then move through a transmission medium (which is generally a fluid) until they reach the receiving nanomachine (or bio-nanomachine if it has a biological component). This communication end destination generates a biochemical reaction with the arriving molecules in order to decode (decipher) the content of the information. This information plays a vital role for the human body in the target cell, tissue, or organ.

MCs play a pivotal role in medical sciences through the diagnosis and treatment of diseases. Modeling biological processes at the nanoscale has made it possible to generate predictions of disease progression. Compellingly, the design of nanosensors and nanoactuators has made feasible the detection and appropriate treatment of cardiovascular, tumors, gastric, and sclerotic diseases, among others. These nanoelements have the potential to activate the immune system or transport smart drugs to specific areas of the body where other medicinal elements, due to their micro or macroscale, would not be able to transport them.

According to the preceding paragraphs, it is intuitive to think of MCs systems as a particular form of traditional communication systems so that the components and information theories of these conventional systems can be applied to MCs. For example, most MCs systems designs consider the Transmission Control Protocol/Internet Protocol (TCP/IP) architecture as an information transfer platform. Therefore, many researchers perceive MCs as a novel communication paradigm, viewing them as a distinct discipline within the realm of communication networks. This understanding shapes the models, architectures, and tools that information theories support in MCs research. Similar to computer network elements, biological cells can transmit, receive, and process information among themselves, forming interconnected communication networks at the nanoscale.

The above conceptualizes MCs as a scientific discipline whose quantification requires mathematical, statistical, information, communication, and informatic tools. Therefore, MCs research is carried out together with the use of information and communication technology (ICT), which implies the use of markup languages and simulators for design and analysis in MCs, whose results would allow experimental validation.

This is the reason why a large number of computer tools have been created worldwide in the field of MCs, most of them incompatible with each other, which is why the IEEE (Institute of Electrical and Electronics Engineers) has generated the nanocommunications standards 1906.1 and 1906.1.1 with the purpose of enabling compatibility and the reuse of information from one research project in another. Similarly, the CIRCLE project (Coordinating European Research on MCs), supported by the European Commission through the Future and Emerging Technologies (FET) Open program, has coordinated research activities in MCs across the European Union, with additional participation from organizations in the United States, Asia, and Australia. CIRCLE has attempted to establish a common language for programming, simulation, and experimentation in MCs through the markup language MolComML (Molecular Communication Markup Language). This book will extensively discuss the most widely used markup languages and simulators in MCs, including the aforementioned "standardization attempts."

The remainder of the book is structured as follows:

- Chapter 1 introduces the reader to the context necessary to understand the basic elements of MCs systems, their main languages, and simulators, as well as briefly discusses the IEEE nanocommunications standards and the CIRCLE project efforts in MCs.
- Chapter 2 presents the most important markup languages and simulators for MCs.
- Chapter 3 discusses MolComML and the IEEE Standards for Nanocommunications. Additionally, simulations in MolComML and NS-3 are analyzed.

Las Palmas de Gran Canaria, Spain	Yesenia Cevallos
Las Palmas de Gran Canaria, Spain	Paola G. Vinueza-Naranjo
Riobamba, Ecuador	Deysi Inca
Las Palmas de Gran Canaria, Spain	Elvis Pilligua
Riobamba, Ecuador	Luis Tello-Oquendo
Guayaquil, Ecuador	Marcelo León
Guayaquil, Ecuador	Cristian Vacacela Gómez
Loja, Ecuador	Talía Tene
Riobamba, Ecuador	Jiří Svozilík
Riobamba, Ecuador	Elba Bodero-Poveda
Guayaquil, Ecuador	Albert Espinal

Acknowledgment

We appreciate the contributions of the following technical supervisors:

- **Tadashi Nakano**
  Osaka Metropolitan University, Japan
- **Giuseppe Piro**
  Politecnico di Bari, Italy
- **Lin Lin**
  Tongji University, China
- **Jesús B. Alonso, Norberto M. Ramos-Calero**
  Department of Signaling and Communications, University of Las Palmas de Gran Canaria, Spain
- **Jennyfer Granizo**
  Hospital Metropolitano, Ecuador
- **Danilo Chávez**
  Escuela Politécnica Nacional, Ecuador
- **Silvana Guitarra, Luis Miguel Procel**
  IMNE, Universidad San Francisco de Quito, Ecuador
- **Alex Breedy**
  Bellevue University, USA
- **María Isabel Uvidia-Fassler, Andrés Santiago Cisneros-Barahona, Alejandra Pozo, Daniel Haro**
  Universidad Nacional Chimborazo, Ecuador

Competing Interests The authors have no competing interests to declare that are relevant to the content of this manuscript.

Contents

About the Authors

Yesenia Cevallos received her Electronic and Telecommunication Engineering Degree from Escuela Politécnica Nacional in Ecuador in 2001. She received her Master of Information Technology degree from the Universidad Técnica de Ambato in Ecuador in 2010. She received her Ph.D. degree in Science and Technologies of Complex Systems from the University of Calabria in Italy in 2018. She is an Associate Professor at the National University of Chimborazo, Ecuador, where she has worked since 2003. Her research area involves molecular-level nanocommunications. She is a researcher for IDETIC at the Universidad de Las Palmas de Gran Canaria, Las Palmas, Spain. She is an IEEE Senior and HKN member of the Computational Life Sciences, Young Professionals, ComSoc, and WIE societies. She is the vice chair of the IEEE Nanotechnology Council and EMBS Ecuador Chapter. She was a member of the IEEE committee for the approval of the nanocommunications standard 1096.1.1. She is also a member of the IEEE Molecular, Biological and Multi-Scale Committee. She is a reviewer for several journals published by Springer and IEEE. She was a reviewer/editor for the journal *Frontiers in Bioengineering and Biotechnology.*

Paola G. Vinueza-Naranjo received her B.Sc. and M.Sc. degrees in Computer Science from the Universidad Nacional de Chimborazo, Riobamba, Ecuador, in 2005 and 2010, respectively, and her Ph.D. degree in information, communication, and telecommunication (ICT) engineering from the Sapienza University of Rome, Italy, in 2018. She is a Senior Researcher and a Marie Curie Fellow in the ENSURE-6G research group at the Universidad Nacional de Chimborazo, Ecuador. Her research interests include cloud computing, energy-saving technologies, fog-based wireless systems, and Industry 4.0.

Deysi Inca received her degree in Electronics and Computing Engineering from the Polytechnic School of Chimborazo in 2010. She was the winner of first place in the call for papers organized by the National Polytechnic School and the Private Technical University of Loja (2010). She obtained her Master's degree in Network Communication from the Pontifical Catholic University of Ecuador in 2016.

Currently, she is an occasional professor in the Faculty of Engineering at the National University of Chimborazo in Ecuador. Her main research interests are in the fields of antennas, communication networks, and nanocommunications.

Elvis Pilligua received his Electronic and Telecommunication Engineering degree from the Universidad Nacional de Chimborazo in Riobamba, Ecuador, in 2024. He is a Ph.D. student at the IDETIC Institute of the Universidad de Las Palmas de Gran Canaria, Las Palmas, Spain.

Luis Tello-Oquendo received the Electronic and Computer Engineering degree (Hons.) from the Escuela Superior Politécnica de Chimborazo (ESPOCH), Ecuador, in 2010; the M.Sc. degree in Telecommunication Technologies, Systems, and Networks, and the Ph.D. degree (Cum Laude) in Telecommunications from the Universitat Politecnica de Valencia (UPV), Spain, in 2013 and 2018, respectively. From 2013 to 2018, he was a Graduate Research Assistant with the Broadband Internetworking Research Group, UPV. From 2016 to 2017, he was a Research Scholar with the Broadband Wireless Networking Laboratory, Georgia Institute of Technology, Atlanta, GA, USA. He is currently an Associate Professor in the College of Engineering, Universidad Nacional de Chimborazo, Ecuador. His research interests include wireless communications, software-defined networks, 5G and beyond cellular systems, the Internet of Things, and machine learning. He is a member of the IEEE and ACM. He received the Best Academic Record Award from the Escuela Técnica Superior de Ingenieros de Telecomunicación, UPV, in 2013; the IEEE ComSoc Award for attending the IEEE ComSoc Summer School at the University of New Mexico, Albuquerque, NM, USA, in 2017; the Extraordinary Doctoral Thesis Award from the Escuela de Doctorado, UPV, in 2019; and the Best Researcher Award from the Ecuadorian Corporation for the Development of Research and the Academy (CEDIA) in 2021.

Marcelo León holds a Ph.D. in "The Hispanic World: Roots, Development, and Projection" from the University of León; a Master's degree in Economics and Territorial Development from the University of Huelva; a Master's degree in Education and Social Development from the Equinoctial Technological University; and a degree in Business Administration from the National University of Loja. He is a university professor, consultant, and researcher. He has participated as a speaker and presenter at numerous international conferences and academic events and has published scientific articles in leading journals, as well as books and conference proceedings. He currently teaches Environmental Management and Sustainable Development in the Master's program in Environmental Management at Ecotec University. He is also a professor at the Universidad Estatal Península de Santa Elena in Ecuador.

Cristian Vacacela Gómez earned his Ph.D. in Physics and Quantum Technologies from the Università della Calabria (Italy) in 2017 and a Bachelor's degree in Biophysics from the Escuela Superior Politécnica de Chimborazo (Ecuador) in

2013. He is an Associate Professor at the Escuela Superior Politécnica de Chimborazo (ESPOCH, Ecuador) and an Associate Researcher at the National Institute for Nuclear Physics (INFN, Italy) with over 12 years of experience in research and teaching. He has published more than 40 scientific works, including journal articles and chapter books, in high-impact journals covering different areas of interest, such as materials science, photonics, nanosystem simulation, radiation physics, and mathematical models for medicine and the environment. Throughout his career, Cristian has received various awards and recognitions. He was a SENESCYT (Ecuador) scholarship holder from 2013 to 2017 to pursue his doctoral studies and received a scholarship from the Laboratorio Nacional de Frascati (LNF-INFN, Italy) in 2022 for postdoctoral studies. He serves as a reviewer for the journal *Nature Communications* (Springer), an academic editor for the *Journal of Nanomaterials* (Hindawi), and an associate editor for *Coatings* (MDPI).

Talía Tene holds a Bachelor's degree in Biosciences from the Escuela Superior Politécnica de Chimborazo (Ecuador), which was awarded in 2013. Additionally, she obtained her Ph.D in Physics and Quantum Technologies from the Università della Calabria (Italy) in 2017. Currently, she is an Associate Professor at the Universidad Técnica Particular de Loja (UTPL, Ecuador), with over 10 years of experience in research and university teaching. She has more than 20 publications in high-impact journals covering both experimental and theoretical works. Her areas of interest include nanotechnology, quantum biology, and radiation physics. Throughout her career, Talia has received various awards and recognitions. She was a SENESCYT (Ecuador) scholarship holder from 2013 to 2017 to pursue her doctoral studies, received a scholarship from the University of Rome Tor Vergata (Italy) in 2022 for postdoctoral studies, was awarded the "Best Researcher" merit award for 2021–2022 by the province of Loja (Ecuador), and has recently been awarded a new postdoctoral scholarship by the University of Calabria as part of the Horizon 2020–2030 program.

Jiří Svozilík is originally from the Czech Republic. In 2013, he obtained his first Ph.D. in Physics with a specialization in Optics and Optoelectronics from Palacký University (Czech Republic), and in 2014, his second Ph.D. in Photonics was obtained at the Polytechnic University of Catalonia in Barcelona—ICFO-Institute of Photonic Sciences (Spain). His teaching career began in 2009 at the University of Palacký, but later he also worked at the UNAM (Mexico) and the Universidad de los Andes (Colombia). In 2018, he started as a professor at Yachay Tech University (Ecuador) and was later appointed as the Dean of the School of Physical Sciences and Nanotechnology. Since May 2022, he has been working at the Higher Polytechnic School at Chimborazo (Ecuador), where he is currently working as a tenured professor and as the coordinator of the Master's degree in Physics. His research interests cover the areas of quantum computing and information, quantum optics, classical optics, and quantum field theory. He has also directed several research projects.

Elba Bodero-Poveda is a member of the Board of Directors of the Ibero-American Science and Technology Education Consortium (ISTEG), a researcher at the Universidad Nacional de Chimborazo, and a faculty member teaching subjects in the area of computer science. She is also the Deputy Editor of the scientific journal *Novasinergia* at the same university and the Editor-in-Chief of the journal *Esprint Investigación*. She studied Computer Systems Engineering at the Escuela Superior Politécnica de Chimborazo, earned a Master's in Information Technology and Educational Multimedia from the Universidad Técnica de Ambato, a Master's Degree in Distance Education (E-learning) from the Caribbean International University, a Master in Scientiae in Management for Intellectual Creation from the Universidad Politécnica Territorial de Mérida, and a Ph.D. in Computer Science from the Universidad Nacional de La Plata. She holds more than ten international professional certifications. Her research focuses on digital preservation, computer engineering, information and communication technologies, and engineering education. She is also the coordinator of Evaluation and Accreditation for the Faculty of Engineering at UNACH, with experience in institutional and career evaluation according to Ecuadorian accreditation models.

Albert Espinal is a computer engineer who graduated from the Escuela Superior Politécnica del Litoral (ESPOL), Ecuador, in 1996. He obtained a Master's degree in Information Systems in 2000 from ESPOL and a Master's degree in Business Administration from the University of Quebec in Montreal (UQAM), Canada, in 2007. In 2021, he obtained his Ph.D. in Engineering from the National University of Cuyo (UNCUYO), Argentina.

Albert is a Professor in the Faculty of Electrical and Computer Engineering (FIEC) at ESPOL since 1996. He has been an instructor in the CISCO Networking Academy program since 2001, where he currently coordinates the Academy Support Center and the Instructor Training Center in Ecuador. He currently works for the Telematics Engineering program as a faculty member in the data networks area, and as an international accreditation coordinator for programs, such as ABET and EURACE. He also works for other universities in Ecuador and Bolivia as a visiting professor, mainly in Master's programs. His research areas are linked to data networks, traffic patterns, predictive algorithms, mathematical modeling, quality of service, molecular communications, learning ontologies, and engineering education.

Chapter 1
Introduction

1.1 Background Information

MCs have currently become a promising scientific field, mainly due to the wide range of applications it can support; the application fields of MCs are mainly oriented to agricultural, environmental, military, industrial, and materials electronics, and the applications that have primarily allowed great advances and why technologies have focused on MCs are medical applications [1–9].

The underlying principles of signal transmission/reception theories in MCs, which are based on information sensing and control techniques, have enhanced in vivo medical treatments for human patients in the latter field. For instance, abnormalities in the propagation of molecular information during cellular differentiation and proliferation processes can detect the development and progression of cancer. Medical applications supported by MCs are defined by natural or synthetic systems, as visualized in Fig. 1.1 [1–9].

In the natural systems (top-down paradigm) of Fig. 1.1, the direction of informational theories in MCs is established as [1]:

- Body-scale systems, where organs and tissues interconnect to perform a specific physiological function.
- Cellular scale, where cells coordinately perform informational processes.
- Molecular scale, where information is encoded in substances and exchanged through chemical reactions and molecular transport.

In the synthetic systems (bottom-up paradigm) of Fig. 1.1, the information theories behind MCs are instantiated as [1]:

- Components of MCs that use genetic programming tools from synthetic biology and employ technologies that serve as an interface to classical electronics.

Y. Cevallos et al., *Markup Languages and Simulators for Molecular Communications*, Signals and Communication Technology,
https://doi.org/10.1007/978-3-032-33040-6_1

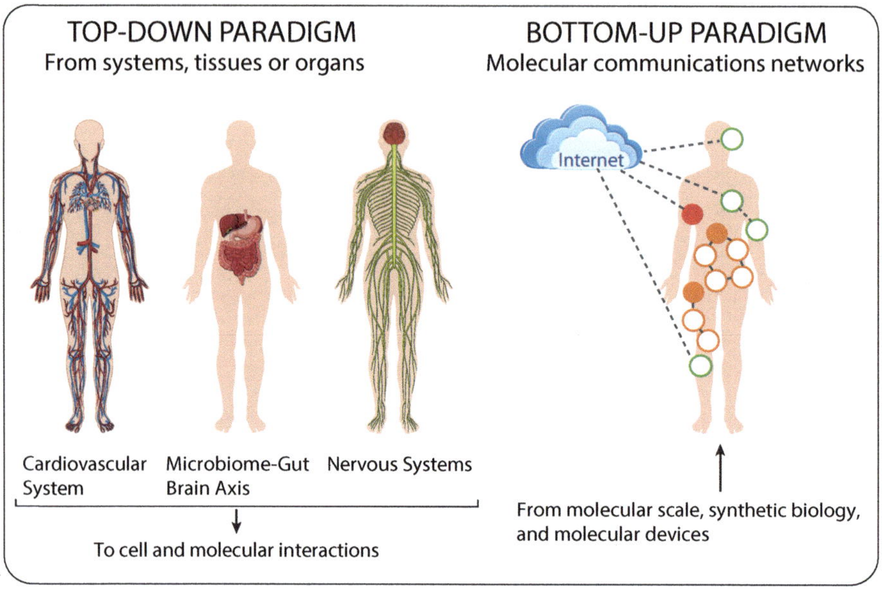

Fig. 1.1 Guidelines for the application of MCs theories to the health area

- MCs elements, which basically consist of biologically inspired engineered systems, are addressed for information transport on a scale ranging from microbes to human cells, which are programmed with different molecular behaviors.
- MCs networks where biologically inspired engineered systems are interconnected with each other to constitute permanent systems for monitoring and controlling medical parameters in human patients; such parameters are transmitted over the Internet [1, 2].

Since MCs were first investigated (approximately 15 years ago), they essentially study the propagation of information at the nanolevel; at that scale, cells are the primordial and intelligent elements that constitute our bodies, and they communicate primarily through transport and molecular bonds. Nature has orchestrated the harmonic behavior of such biomolecules to act efficiently and consistently, despite their noisy environment (which limits MCs) and their mainly diffusion-based transport processes (which also hinder the transmission of molecular information). Despite such limitations, the evolution of species has made it possible for molecular systems, or properly MCs systems, to constitute robust propagation systems, in which the information transferred from one generation to another has allowed survival, which has been carried out through processes of energy efficiency and compatibility with the environment [10–16].

The field of MCs has traditionally been focused on defining fundamental principles and theoretical modeling. Recently, there has been a paradigm shift toward the empirical validation of these theoretical frameworks through experimental methodologies to augment the practical utility of MCs systems. In this regard, synthetic biology and microfluidics have emerged as two pivotal technological domains. Synthetic biology constitutes an interdisciplinary arena that integrates biology, engineering, genetics, chemistry, and computer science. This convergence facilitates the engineering of organisms at the molecular scale to impart novel functionalities and properties, thereby providing the foundational tools to fabricate MC transceivers. Microfluidics, however, offers microscale platforms for experiments [17], allowing precise manipulation and control of molecular transport, which makes it an ideal platform for experimenting with and implementing MCs concepts. This shift from theoretical to practical applications is crucial in demonstrating the real-world feasibility of MCs systems, bridging the gap between abstract models and realistic outcomes in the field. Microfluidics not only substantiates but also propels the evolution of MCs, heralding substantial advancements toward the practical implementation of these systems [17].

Similar to conventional communication studies, scholars have investigated numerous dimensions of MCs theoretically, employing heuristic assumptions as a foundational phase. Theoretical investigations involving channel modeling, channel capacity, secrecy capacity, detection complexities, and localization of sources or receivers have been undertaken to acquire a comprehensive understanding of MCs systems. The identification of conveyed symbols constitutes a fundamental component of all communication systems, encompassing those at the molecular level. A variety of theoretical approaches have been articulated within the literature of MCs to address detection challenges, which include linear memoryless detection, adaptive detection, anti-Inter-Symbol Interference (ISI) demodulation, and matched filtering techniques. These coherent analytical approaches necessitate an understanding of the channel impulse response (CIR). Conversely, non-coherent methods are model free and do not rely on the CIR. In practical MCs systems, deriving a closed-form solution to accurately represent the phenomena and identify the transmitted symbol can prove to be a formidable or even unattainable task due to complications, such as ISI, noise, nonlinearity, and time-variation in channel characteristics [18].

Henceforth, MCs applications have been mainly focused on the health sector due to the nature of information in biological beings, where it has shown potential by bridging the gap between theory and reality, and MCs hold promise for a variety of practical applications [18].

In this context, simulations hold an important role in MCs, due to their unique ability to bridge the gap between theoretical models and practical experiments. Although theoretical models furnish critical understandings, they frequently depend on simplifications and presumptions that may inadequately encapsulate the intricacies of actual scenarios. Conversely, the execution of experiments can be both labor-intensive and financially burdensome, thereby constraining the spectrum of conditions and variables that can be examined. Thus, simulations offer a compelling solution as a compromise between theoretical analysis and experimental validation [17].

Then, in an inertial way MCs systems must be analyzed from communicational theories of traditional communications systems, which suppose the use of markup language and simulators for MCs to validate results for future eventual experimentation. That is the reason for the existence of this book. Thus, we discuss these topics widely through Chap. 2.

1.2 Use of Information and Communications Technologies to Analyze Molecular Communications

Unlike the paradigm used in current telecommunications systems, MCs use molecules as information carrier signals, being precisely such carriers the information itself, unlike what happens in conventional communication systems where the carrier signals are those that allow another signal called information to be transported. At this point, it should be clarified that MCs occur basically in biological entities, so that they can be transported by the carrier signals in biological entities, so it is common in the literature that MCs systems are also known as biological communication systems, or biological systems; therefore, in this book any of these equivalent denominations will be used.

Commonly referred to as bio-nanomachines, biological transmitters encode information into molecules that they release into the surrounding environment, facilitating their propagation to the biological receptor, also known as a bio-nanomachine. Such a receptor reacts biochemically with the molecules arriving at it in order to decode the information signal [19, 20]. Transmitter-receiver pairs in biological communication systems are frequent in nature, where the sending of information (which takes many different forms) is nowadays analyzed from the perspective of ICT, which provides a quantitative basis for the study of the aforementioned biological processes, both for natural biology and synthetic biology. We can consider transmitter-receiver systems, cells, and biomolecules as information-processing agents [21–25]. This perspective of analysis has allowed scientists to quantify molecular signals and has even enabled the prediction of informational behavior in biological communication systems through Shannon's information theories (Fig. 1.2).

According to what is visualized in the upper part of Fig. 1.2, one can notice the different forms of communication in biological systems, in which elements such as proteins, cells, and organs are part of the molecular propagation Scheme [10, 11, 24–28].

In the lower part of Fig. 1.2, the channel capacity theorem (maximum transmission rate) of an analog communication system can be used to describe MCs processes; thus, an incoming signal is defined as S in terms of dosage/concentration (e.g., AHL (acyl-homoserine lactone), dopamine, or even deoxyribonucleic acid (DNA) fragments) and a transmission medium is used that establishes changes in the biological signal. S through "modulation or modifications to the signal itself S." The capacity of such a channel would be: $C = B\log_2\left(1+\frac{S}{N}\right)(bps)$, where B is the bandwidth, which is the operating frequency range of the channel (change in molecule concentration or "Hz"), N is the signal corresponding to the noise, and the

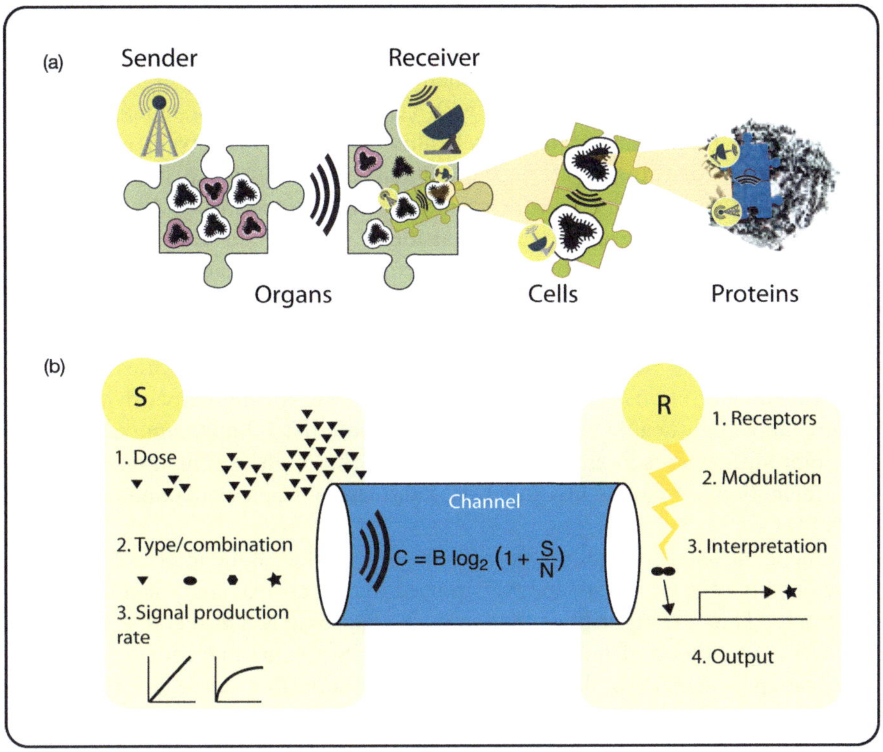

Fig. 1.2 Informational paradigm underpinning the analysis of molecular transmission in biological systems

channel capacity is set in bits per second. Signal processing is determined at the molecular target by cognate receptors, such as cell surface receptors. A modulation mechanism, such as a cell signaling pathway, associates the incoming signal with a "translator" (e.g., a responsive promoter for gene expression), allowing the "meaning" of the received signal to be extracted [26].

We can describe MCs systems by using molecules as messengers between the transmitter-receiver biological entities. This mode of communication proves to be the most promising for general applications, and in particular, in medical treatments. Thus, an example of the model for the communication and information method analyzed so far is precisely the transmission of signals between neighboring cells in the human body, which takes place in a diffusion environment in which different types of molecules propagate, encoding different kinds of messages. This MC allows natural or artificial biological nanomachines to communicate over short distances (intracellularly) or over long distances (intercellularly) at the nanoscale. In this communicational process, the information released by a molecular emitter is transmitted through a propagation medium until it reaches a receptor; the latter, through biochemical reactions, executes a task that allows a physiological action to be carried out to perform a function of a cell, tissue, or target organ in the human body [29, 30].

As ICT has advanced at an unprecedented pace, it is rational to assert that this technological domain is intertwined with nearly every facet of human existence. Consequently, particularly within the healthcare sector, the advent of innovative technologies presents novel prospects for enhanced monitoring, accurate diagnosis, and predictive analytics, all of which possess critical importance within healthcare systems. The increasing integration of artificial intelligence (AI) serves a pivotal function in propelling the healthcare domain into an era wherein decisions informed by empirical data can enhance the caliber of care provided to patients. These technological advancements significantly enhance the development of more efficacious and knowledgeable healthcare methodologies across various species. As health platforms undergo transformation, they pave the way for more cohesive strategies toward innovative health frameworks through interdisciplinary innovations. This paradigm shift underscores the necessity for a thorough examination of existing implementations, ongoing investigations, and the obstacles encountered by both fields [31].

In the mentioned context, the future of healthcare lies at the intersection of technology and human expertise, and AI is poised to be a critical component in shaping that future. Thus, AI algorithms that analyze electronic health records (EHRs) can predict the likelihood of hospital readmission with remarkable accuracy, allowing healthcare providers to implement preventive measures that can reduce readmission rates by up to 20%. This not only enhances patient outcomes but also alleviates the financial burden on healthcare systems, which spend billions annually on avoidable hospital readmissions. Moreover, AI's role in medical imaging is revolutionizing diagnostic practices [32].

Advanced AI algorithms can now analyze medical images with a level of precision that rivals, and in some cases surpasses, that of human radiologists. For example, studies have demonstrated that AI-powered imaging tools can detect certain types of cancers, such as breast cancer, at an earlier stage than traditional methods, leading to earlier treatment and improved survival rates. The integration of AI in imaging diagnostics also facilitates faster and more accurate interpretation of scans, which is crucial in emergency situations where time is of the essence. Additionally, AI's potential to integrate and analyze multimodal data—combining imaging data with genetic, clinical, and demographic information—holds the promise of truly personalized medicine, where treatment plans are tailored to the unique genetic and phenotypic profile of each patient. This approach not only improves the efficacy of treatments but also minimizes the risk of adverse reactions, thereby enhancing overall patient safety [32].

Besides, the consensus in the literature underscores the transformative impact of AI on various aspects of healthcare, from diagnostic accuracy to personalized medicine. For instance, the ability of machine learning algorithms to predict patient outcomes with greater accuracy than traditional statistical models. Currently, AI models, trained on large datasets, could predict the onset of conditions such as sepsis and heart failure with higher precision, reducing mortality rates by up to 15%.

This fact also establishes the importance of the application of deep learning in dermatology. The researchers trained a convolutional neural network (CNN) on over 129,000 clinical images of skin lesions and found that the AI model matched the diagnostic accuracy of board-certified dermatologists in identifying malignant and benign skin conditions. AI holds significant promise in enhancing diagnostic capabilities, thereby improving patient outcomes. Similarly, AI-powered imaging tools have shown exceptional performance in detecting diseases such as lung cancer, diabetic retinopathy, and Alzheimer's disease at earlier stages than conventional methods. AI model outperformed radiologists in detecting breast cancer in mammograms, reducing false positives by 5.7% and false negatives by 9.4% [32].

Subsequently, the coherent amalgamation of digital technologies with medical services has culminated in the emergence of *eHealth*, a transformative paradigm that markedly alters the provision of healthcare as well as the management and engagement of patients within the care continuum. eHealth refers to the integration of digital technologies such as computers, the Internet, and mobile devices to enhance healthcare delivery and improve the distribution of medical information. eHealth is built on a patient-centric vision, with the objective of delivering effective and individualized care trajectories via the implementation of digital technological innovations. The principles underlying eHealth offer a variety of opportunities, encompassing the transcendence of geographical limitations, enhancement of patient engagement and awareness, customization of treatment modalities, early identification of existing pathologies, and the effective management of chronic diseases. Central to this paradigm shift are *digital health platforms*, which are ecosystems that utilize digital technologies (such as electronic data, devices, and software solutions) to improve the efficiency, accessibility, and quality of healthcare services [31].

Along with human population development, eHealth has become an essential domain to improve humanity's life quality. The community is also motivated and has progressed gradually, aiming to satisfy human demands, including the most advanced eHealth applications: Telehealth, Telemedicine, Telesurgery, etc. To allow these applications feasibility, the most challenging network characteristics are mandatory to ensure the medical practitioner's efficiency and the patient's safety. Ultimately, the strict communication requirements in eHealth are resolved, thanks to the deployment of 5G (fifth-generation wireless). Furthermore, the revolutionary technology also introduces a novel network landscape, network slicing, which renovated the community's network perspective. Prior to the advent of 5G, researchers in the eHealth domain largely overlooked the potential for managing backbone network infrastructure. This limited engagement was primarily due to high operational costs and the fact that infrastructure control was typically restricted to Internet service providers (ISPs) and mobile network operators (MNOs).

With the emergence of 5G, these constraints can be addressed through technologies such as network function virtualization (NFV) and network slicing. These capabilities enable third parties to configure and manage virtualized network environments deployed over existing 5G infrastructures. This shift toward programmable and customizable network overlays represents a significant conceptual transformation, empowering the research community with greater flexibility and autonomy in network operation [33].

Consequently, this evolution encourages the research community particularly in the eHealth domain to reassess existing approaches in light of this emerging paradigm. A closer examination reveals that the fundamental enabler of network function virtualization (NFV) and network slicing is the well-established concept of software-defined networking (SDN).

SDN is architecturally based on the separation of control and data planes, where a centralized control entity manages network intelligence, while forwarding devices execute traffic-handling functions. To fully realize the capabilities of NFV and network slicing within 5G infrastructures, SDN functionality must be supported across network components. However, implementing such capabilities through a large-scale hardware replacement of existing Internet infrastructure is impractical given the current global deployment landscape. Hence, the hybrid approach, which takes advantage of both SDN-enabled and traditional network devices, is exploited to provide the SDN-enabled capability to existing infrastructure. The traditional network devices that maintain Internet services usually are switches, routers, etc. The common among all categories is their static and inflexible deployment method. At the same time, substantial research efforts have been directed toward the development of more flexible and programmable network infrastructures, attracting considerable attention from the scientific community [33].

Thanks to wireless technology's advancements (5G and 6G (sixth-generation wireless)) and the shifting trend to embed the communication component in mobile devices, such as mobile phones, vehicles, and unmanned aerial vehicles (UAV), the network architecture similarly evolves itself. The deployment of access-point coverage has become increasingly practical, enabling any device equipped with appropriate communication technologies to function as a network node capable of supporting advanced networking services. Specifically, these entities could deploy network coverage and forwarding capabilities as well as consume network coverage from other nodes as an actual network node. To achieve convergence with the evolution of modern network architecture, the SDN paradigm requires at least equivalent levels of exertion and advancement [33].

As a result, software-defined unmanned aerial vehicles (SDUAV) and software-defined vehicular networks (SDVN) have emerged as an advanced SDN-based architecture designed to extend the software-defined paradigm to highly dynamic and elastic networking environments. These architectures are particularly valuable in mission-critical scenarios requiring rapid and large-scale coverage deployment, such as disaster response, crisis management, and medical emergencies. However, current research trends reveal inherent limitations. By applying the software-defined paradigm to specific operational contexts, the community has primarily focused on defining technical specifications for predetermined environments. While this targeted approach helps narrow research scope and optimize performance within particular domains, it simultaneously restricts the generalizability and broader applicability of SDN-based management frameworks.

Despite the diversity of networking environments, including wireless body area networks (WBANs) and underwater networks, research efforts on SDN integration within these domains remain comparatively limited when contrasted with vehicular and UAV networks. Consequently, the literature still lacks a unified and adaptable framework capable of addressing heterogeneity and ensuring seamless SDN deployment across diverse hardware and software environments [33].

Also in this context, the Internet of Things (IoT) has garnered significant attention within the healthcare sector in recent years, attributable to the myriad benefits it offers. In this regard, wearable technology such as smartwatches is capable of incessantly gathering vital physiological parameters from the human body, including blood pressure, body temperature, and heart rate. This data can be seamlessly transmitted to specialized software, referred to as healthcare services, which systematically analyzes the information and activates alerts whenever the data exhibits anomalous variations. Among the diverse healthcare services that can be operationalized, one noteworthy example involves the automatic notification of emergency medical services when the healthcare system forecasts the imminent risk of cardiac failure, based on historical data correlated with vital signs recorded in the preceding moments. Another illustrative example is the dispatch of alerts to designated family members regarding conditions such as fever or hypothermia. The integration of healthcare services with IoT technologies in smart urban environments has the potential to substantially enhance the quality of life for residents, and in critical scenarios, may serve as a pivotal factor in preserving human lives. Healthcare services possess the capability to anticipate the emergence of significant health issues prior to their onset, thereby facilitating real-time surveillance of individual health conditions [34].

Furthermore, cloud computing and fog computing serve a pivotal function in the processing of ubiquitous IoT data produced by wearable technologies. The cloud offers nearly limitless computing power, which is essential for handling the varying volumes of IoT data generated throughout the day. In contrast, the fog acts as an intermediary layer between end-user devices and clouds. By providing computing resources closer to the data source, fog computing significantly reduces latency compared to sending data to distant cloud servers or executing tasks solely within cloud environments. Ultimately, fog and cloud computing synergistically enhance one another, collaboratively offering solutions for intelligent urban ecosystems that take into account both rapid response times and scalability techniques [34].

This viewpoint presents computational difficulties that necessitate resolution to attain the intended quality of experience for end-users. On the one hand, essential healthcare services may impose stringent latency requirements and ought to be executed without delay, potentially within the fog layer due to its proximity to the individual. Conversely, the machines performing these services possess constrained computational resources within the fog layer, thereby enabling the concurrent processing of only a limited subset of vital signs. Since people move between neighborhoods within a smart city and may gravitate toward particular locations throughout the day, the fog computing infrastructure must possess the capability to accommodate fluctuations and surges in demand. Effectively managing computational resources by discerning which vital signs are imperative and warrant prioritization within the fog environment is crucial for facilitating superior healthcare services. Furthermore, the attributes of scalability, fault tolerance, and the alleviation of network congestion are paramount when designing healthcare systems, as these factors can be instrumental in preserving human life [34].

As preceding paragraphs indicate, the use of digital technologies and the adoption of computer networking paradigms are key factors in monitoring health aspects and have become an essential aspect for enhancing human life. Promoting active and healthy aging has emerged as a critical societal challenge, particularly in light of the growing burden of noncommunicable diseases (NCDs), encompassing cardiovascular diseases, hypertension, chronic respiratory conditions, cancers, renal disorders, and diabetes, which represent the primary cause of mortality worldwide, accounting for approximately 41 million deaths in 2020, according to the World Health Organization. Epidemiological evidence indicates that the adoption of healthy lifestyle practices can mitigate up to 75% of the risk associated with heart disease, stroke, heart failure, and type II diabetes and approximately 40% of cancer incidence. Effective prevention and management strategies for NCDs necessitate the reduction of modifiable behavioral risk factors, including poor dietary habits, tobacco use, and excessive alcohol consumption, while simultaneously promoting regular physical activity and the continuous monitoring of critical health indicators, such as blood glucose, blood pressure, and cholesterol levels [35].

In this context, the integration of IoT-enabled wearable healthcare technologies has gained substantial traction. These devices offer real-time, non-invasive health monitoring capabilities through form factors such as fitness bands, smartwatches, remote patient monitoring systems, specialized diagnostic wearables (e.g., electrocardiogram [ECG] and electroencephalography [EEG] monitors), and smart garments embedded with physiological sensors. Such technologies enable proactive health management, facilitate early diagnosis, and support data-driven interventions, particularly in aging populations and chronic disease management frameworks [35].

Wearable technologies facilitate real-time, on-body acquisition and processing of physiological data, enabling continuous health assessment. Within the realm of IoT-based healthcare systems, remote monitoring solutions play a vital role in supporting individuals facing physical and psychological challenges by measuring and analyzing key physiological parameters, such as blood pressure, oxygen saturation, heart rate, and body temperature. A diverse range of IoT-driven healthcare applications spanning vital sign monitoring, diet and activity tracking, sleep analysis, fall detection, anomaly recognition, geolocation, emergency response, and remote medical consultations has been developed to address the specific needs of the elderly. These technologies aim to compensate for age-related functional decline and reduce the occurrence of preventable health complications [35].

By enabling the seamless exchange of real-time health data between patients and healthcare providers, IoT-enabled devices enhance continuity of care and foster proactive clinical interventions. For older adults, such systems offer access to high-quality, cost-effective monitoring and treatment options. The growing aging population has significantly accelerated the demand for digital healthcare infrastructure capable of delivering sustained, personalized monitoring to support healthy aging and independent living [35].

Recent research has identified IoT as a transformative yet potentially challenging technology, particularly for older populations. A significant barrier for middle-aged and elderly individuals is their limited familiarity with digital technologies, which often demands additional effort to adapt to innovations, such as IoT-based systems. This lack of digital literacy may lead to skepticism regarding the usefulness of IoT healthcare solutions, with perceived low product value deterring adoption [35].

In addition to user-related obstacles, developers of IoT-enabled wearable health devices (WHDs) face numerous technical challenges. These include achieving seamless device integration, ensuring accurate and continuous data acquisition, and maintaining system reliability, an essential requirement in medical contexts. One of the most critical concerns in deploying WHDs is safeguarding data privacy and anonymity. Secure transmission and handling of sensitive health information remain a primary concern among users, especially within vulnerable age groups. To foster broader acceptance, developers must not only address technical limitations but also investigate the behavioral and cognitive factors influencing user acceptance and willingness to adopt such technologies [35].

Also, [35] identifies a demographic disparity in adults' behavioral intentions regarding the adoption of IoT-enabled WHDs. The findings indicate that younger individuals and those with higher levels of health awareness are more inclined to adopt such technologies. In contrast, concerns related to privacy and personal values vary significantly across demographic groups, influencing adoption patterns in different ways. The motivations for and against the adoption of IoT-based wearables in the healthcare of older adults elucidate demographic disparities related to age and health status. Moreover, individual values exert a significant influence on attitudes toward innovation, which differ across various demographic segments. Furthermore, there exists a potential for increased mortality rates among the elderly population in the coming years. Within the extant literature, both the theory of reasoned action and the theory of planned behavior have been rigorously employed to comprehend and forecast a diverse range of health-related behaviors, encompassing physical activities, glucose monitoring, blood pressure monitoring, and individual health practices during the pandemic. Several studies have employed the health belief model owing to its efficacy in clarifying healthcare-related behaviors. Many studies have integrated and augmented various theoretical frameworks, including the unified theory of acceptance and use of technology, the technology acceptance model, health monitoring behaviors, and technology acceptance theory, in an attempt to elucidate the rationale underlying the utilization of information technology within healthcare settings and with health monitoring devices [35].

Currently, empirical evidence has illustrated that individuals who exhibit greater concern for their health exhibit an increased propensity to adopt novel health-related technologies; furthermore, an elevated perception of health value corresponds to a comparatively heightened level of conviction and assurance in the effective management of health status in the face of various health threats. Consequently, there exists a coherent correlation between individual health values and perceptions pertaining to nutritious dietary practices. ICT-based healthcare services indicate that patients who experience significant enhancements in their health through wearable health technologies demonstrate a greater propensity to endorse these innovations to their acquaintances [35].

Modernly, science has developed and transformed societies due to incredible technological advances, and for sure this process involves the use of computing and artificial techniques utilized to realize and turn effective this process. In this sense, for instance, the joining of technologies such as AI and the 6G networks are anticipated to serve as the key to the future paradigm change in the healthcare industry. This forefront segment provides an overview of how the advent of these modern technologies is preparing the healthcare system for tailoring the provision of health services, diagnosis, therapy, and monitoring of patients. The healthcare sector, one of the largest industries in the developing world, is undergoing a profound transformation driven by rapid technological advancements. AI used to be somewhat alien to the medical profession. Still, its role is suddenly becoming decisive as AI opens new horizons in clinical data analysis, diagnostics, and individual treatment personalization [36].

AI systems, while they deal with a huge amount of therapeutic data, can give us their thoughts and decisions that are beyond human capability. Machine learning (ML) algorithms excel at performing image analysis, often surpassing human experts in efficiency, which drives early disease diagnosis and improves patient outcomes. To support the vast data demands of these AI-driven applications, next-generation 6G networks offer ultra-fast Internet connections with minimal delay and high data handling capacity. These features are undoubtedly just some of the most critical ones for AI in healthcare to succeed. The key features of 6G, like low latency and high reliability, are even more critical for applications such as remote surgery and real-time monitoring of patients because, in such applications, only milliseconds matter. The increasing Internet connectivity supported by 6G also creates conditions for the broad spread of IoMT (Internet of Medical Things) utilization whereby patients can be monitored continuously and data collected [36].

AI, along with 6G in the health sector, is anticipated to foster precision, efficacy, and access to healthcare at a rapidly progressing rate. Such integration allows telemedicine, distant diagnostics, and deploying AI-powered predictive models for preventive healthcare. Furthermore, the system is expected to enhance health provisioning services in remote and under-serviced areas where access to professional medical staff is limited primarily. The introduction of AI with 6G technology is one of the key elements of *Healthcare 4.0*. This integration will mark a remarkable transition in the healthcare delivery system as the input capacity of smart healthcare systems is increased. The capability of AI to quickly process data reliably and efficiently combined with high-speed and low-latency data transmission 6G will provide a robust platform for sophisticated healthcare applications. This fusion will ensure that advances in patient care, such as real-time diagnosis, precision medicine, and early intervention, will be realized. It is through the use of AI algorithms that it is possible to analyze complex medical data, guiding the development of personalized treatment plans. Data aggregation and real-time data transmission of 6G will enable doctors and health professionals to make the right decisions in a short time, an essential feature in emergency medicine [36].

Conversely, 6G enables the spread of wide-area telemedicine applications, guaranteeing sophisticated treatment is possible wherever healthcare is rendered. While 6G technology has the potential to significantly enhance digital health solutions, its current deployment in underprivileged areas is limited. As infrastructure evolves, 6G could help address healthcare needs in these regions by providing the necessary bandwidth and robustness to support the IoMT. This includes wearable healthcare monitors and connected medical devices, which can operate continuously to collect health data in real time, offering healthcare providers deeper insights into individual patients. This network of devices can work non-stop and collect health data on the go, which gives healthcare providers a better understanding of individual patients. Through such comprehensive monitoring of health status, there appears the probability of early diagnosis of health problems, preventive medical care, and accurate treatment programs [36].

The mentioned advances in medicine deal with the management of medical information, and this type of information is the result of the analysis of biological entities in human body, and in the last instance, molecules are the "format" of biological information. These molecules are used to transmit, travel, and process information for accomplishing physiological tasks in target cells, tissues, or organs. Hence, MCs are a communication paradigm inspired by the mechanisms and materials found in biological systems. In natural environments, information exchange is predominantly mediated through the propagation of chemical substances, motivating researchers to investigate communication strategies that rely not on conventional carriers such as electromagnetic (EM) waves or photons, but instead on the transmission of particles or molecules. Examples of microscale communication encompass interactions between cells or bacterial entities, whereas macroscale instances involve pheromonal exchanges among various animal species and the functioning of the olfactory system. In contrast to traditional communication modalities, MCs are biocompatible and energy-efficient, thereby positioning themselves as a potentially advantageous paradigm for information transmission within sub-millimeter dimensions [18].

As the IoMT acquires the physiological metrics of a patient, including blood pressure, oxygen saturation, body temperature, heart rate, and glucose concentration, it facilitates the management of therapeutic interventions pertinent to a widespread and global health crisis exemplified by the COVID-19 (Coronavirus Disease 2019) pandemic. IoT typically denotes a constellation of interconnected tangible entities, including medical sensors, surveillance instruments, household appliances, and vehicles equipped with autonomous driving capabilities. These interconnected devices are extensively endowed with the capacity to perceive, process, and disseminate information. This phenomenon engenders data-driven intelligence that facilitates the automation of human activities and interactions. Presently, there is a notable escalation in data traffic, which is projected to attain 4394 exabytes by the year 2030. To meet the requirement of bandwidth, a promising technique known as

CR-IoMT (Cognitive Radio-Internet of Medical Things) is highly preferred. This approach employs the limited spectrum in a dynamic manner to facilitate the interconnection of physical entities within a densely populated environment. Consequently, this technological advancement is particularly well-suited for the current pandemic context, wherein individuals are continuously connected and monitored via an extensive public network [37].

As a result of nationwide lockdowns and limitations on public mobility, many daily activities shifted to digital environments, impacting various sectors including e-commerce, online education, telehealth services, smart energy management, and security monitoring systems. The facilitation of these real-time activities is reliant on wireless networks, which necessitate an increased utilization of bandwidth resources [37].

CR-IoMT facilitates dynamic communication networks capable of acquiring sensory data and autonomously processing information. This functionality is particularly valuable for addressing large-scale and emergent healthcare crises, such as the COVID-19 pandemic. The medical emergency associates IoMT extensively with the provision of online medical services in a dependable manner. Furthermore, it facilitates the creation of a medical infrastructure that links the telecare database to monitor and track patients situated at home or in containment zones. Figure 1.3 illustrates the medi-logic ramifications of CR-IoMT. This framework integrates cutting-edge technologies, encompassing intelligent devices and networks, to address a forthcoming and global health crisis exemplified by the SARS-CoV2 (severe acute respiratory syndrome coronavirus 2) pandemic [37].

Conversely, within the healthcare context, it is crucial to note that many individuals worldwide died, in part, due to the lack of timely medical care following the declaration of the COVID-19 pandemic. This was largely attributed to healthcare systems being inadequately equipped to manage the sudden surge in patient demand.

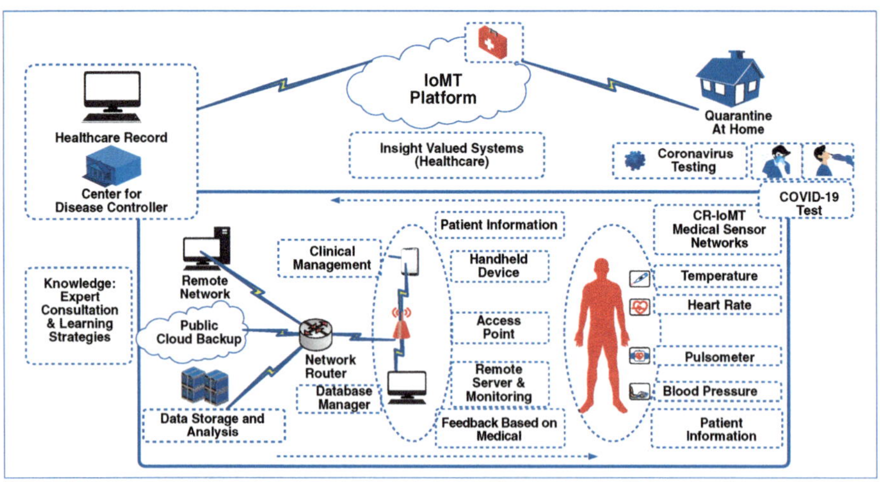

Fig. 1.3 Medi-logic implication of CR-IoMT

COVID-19 has evidenced that an increase in patient caseloads, attributed to a deficiency in hospital infrastructure, inadequate infection control protocols, and delayed responses, culminated in the disintegration of the healthcare system. Even among the most developed nations worldwide, sophisticated healthcare systems designed to deliver timely and secure health services for both personnel and patients encountered significant challenges in alleviating the repercussions of the crisis. The traditional intelligent healthcare system is constrained in its service provision due to insufficient real-time execution and responsiveness to evolving circumstances [38].

For instance, the process of transporting patients via ambulance remains largely similar to that of an ordinary vehicle, with the key distinctions being the administration of oxygen to the patient and the vehicle's prioritized movement through traffic. Patients are unable to receive adequate medical intervention during transit in ambulances en route to healthcare facilities, thereby jeopardizing their well-being due to delays in essential procedures that necessitate immediate medical attention. Moreover, the prevailing healthcare frameworks lack an integrated real-time accident detection mechanism. It is imperative that the accident detection system functions in real time to facilitate the provision of medical aid. In addition, the management of pandemic and epidemic occurrences, such as COVID-19, proves to be formidable due to insufficient technological infrastructure. Therefore, designing an intelligent healthcare system is critical to avoid similar scenarios such in the COVID-19 pandemic with the sceneries shown in Fig. 1.4 [38].

These medical paradigms have evolved into components of an intelligent healthcare system. It is essential to enhance healthcare delivery from the stages of diagnosis to treatment. The advancement of intelligent healthcare services facilitates the provision of superior quality medical care [38]:

- Smart hospital service: Hospital is equipped with advanced medical technology and personnel who have received comprehensive training to provide patients with the highest quality of care achievable. To guarantee the optimal operation of the hospital, proficient communication among management, medical practitioners, hospital personnel, and patients is imperative. Within the hospital environment, interdisciplinary communication is often enhanced by technological tools, such as electronic mail, mobile devices, and two-way radio systems [38].

  In recent times, novel methods aimed at enhancing healthcare services within hospital settings have emerged: (i) the acquisition and preservation of health records in a digital format to mitigate human error; (ii) the augmentation of communication between patients and physicians facilitated by artificial intelligence for improved disease diagnosis; (iii) the implementation of robotic technologies to achieve exceptional precision in surgical procedures, thereby assisting medical personnel. Nonetheless, hospitals rapidly become overwhelmed in crisis situations due to the influx of patients requiring hospitalization. Managing this predicament is exacerbated by constraints in patient capacity, medical apparatus, and healthcare workforce. Given that infectious diseases have the potential to propagate from afflicted patients to healthcare workers during pandemic scenarios, the paramount objective becomes ensuring the safety of staff while

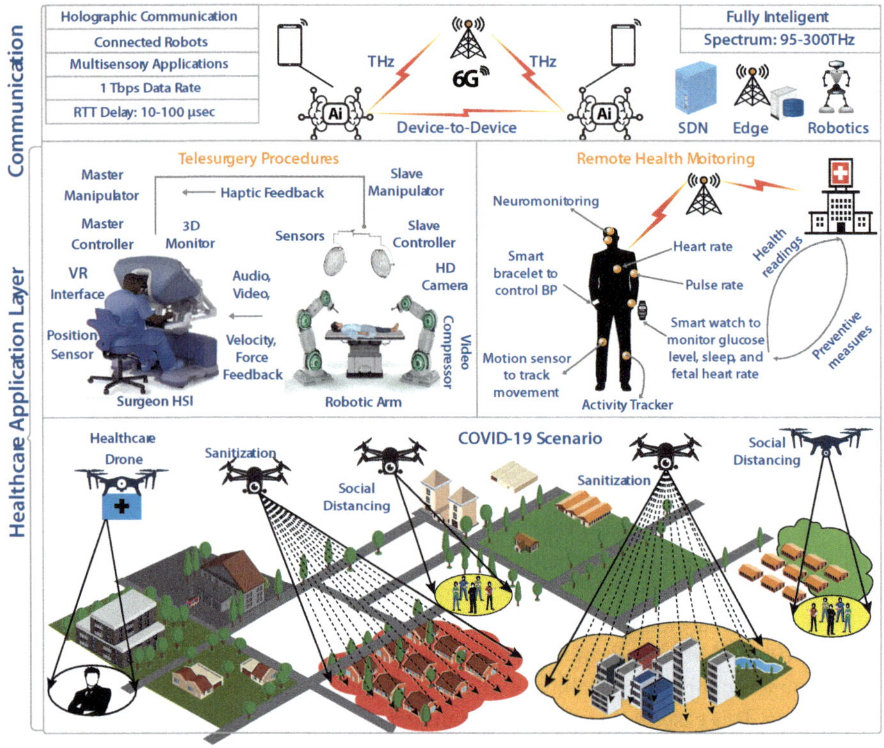

Fig. 1.4 Architecture of healthcare network based on 6G

delivering medical care to those who are ill. One viable approach to addressing this issue is to minimize physical interactions between medical practitioners and patients. This can be effectively realized through the integration of the IoMT and robotics within the healthcare facility [38].

Wearable devices and sensors within the IoT have the capacity to monitor and relay patient health data to healthcare providers via the Internet, thereby facilitating a reduction in in-person interactions. Alternatively, robotic systems may be employed in a multitude of functions, such as conducting medical assessments, delivering essential supplies, and disinfecting patient environments. It is projected that, in the near future, these machines and robotic technologies will significantly augment and safeguard the roles of hospital personnel and medical practitioners. The efficacy of these technological innovations is contingent upon the hospital's communication infrastructure and necessitates a secure, dependable, and rapid data transmission capability [38].

- Remote monitoring service: The utilization of remote healthcare services is becoming increasingly prevalent on a global scale, particularly for the provision of medical care to individuals residing in isolated regions or facing mobility limitations. This service is facilitated through two primary modalities: telemedicine and remote health monitoring. In the context of telemedicine, healthcare professionals assess, diagnose, and manage patient conditions without the necessity for in-person encounters. Within the framework of remote healthcare, patients can be continuously monitored at any location and at any time through the use of healthcare applications and Internet connectivity [38].

At present, telemedicine is facilitated through high-fidelity video teleconferencing; nonetheless, due to the swift progression of holographic technology, holographic tele-presentation is anticipated to be implemented in the imminent future. IoMT technologies, encompassing wearable devices and WBAN, are employed for the purpose of remote healthcare surveillance. In a similar vein, the implementation of systems designed for real-time monitoring significantly enhances the quality of life for geriatric populations by detecting emergency situations through sensor technology and enabling a prompt response. Nonetheless, widespread adoption of innovative technologies, particularly among older adults, constitutes a significant obstacle to achieving this objective. Moreover, dosage adjustments facilitated by digital medicine streamline the daily lives of individuals requiring consistent healthcare, such as those diagnosed with diabetes. Additionally, to safeguard the safety and confidentiality of patients, these technologies necessitate a reliable and secure connectivity infrastructure [38].

Wireless communication links, particularly in the context of remote healthcare, are susceptible to various security threats, including spoofing, jamming, and eavesdropping. All sensors and devices transmit data to medical personnel or transfer information to an Internet gateway. In the absence of secure communication links, an eavesdropper may effortlessly gain access to confidential patient data. Consequently, strategies founded on cryptographic principles are imperative to safeguard the physical communication channels. The promotion of remote healthcare has been catalyzed by the emergence of COVID-19, aiming to mitigate the transmission of infections and alleviate the burden on hospitals. The efficacy of remote healthcare services is entirely contingent upon the robustness of the communication infrastructure. Consequently, many potential applications have yet to be fully realized due to communication challenges, including interference and the high connectivity requirements of the IoMT; the critical need for reliability and ultra-low latency in digital healthcare and tactile-Internet-enabled remote surgeries; energy-efficient communication constraints for biosensors; and the significant bandwidth demands posed by holographic telemedicine [38].

- Disaster response unit service: The establishment of a disaster response unit is imperative for the provision of healthcare services in the context of any emergency or critical scenario. It incorporates disaster management centers, mobile emergency service (MES), medical personnel, and public space surveillance, among other components. Typically, a disaster management center is instituted within the confines of the same hospital or within a separate institution to facilitate rapid response, healthcare services, communication between departments, and observation of the consequences [38].

Whenever a natural disaster occurs, particularly a pandemic, it is imperative to consider the establishment of field hospitals. In such situations, these temporary medical facilities may prove essential as they facilitate the isolation of individuals afflicted by the contagion, thereby preserving capacity within existing healthcare institutions, while also enabling a more effective allocation of limited hospital resources. Currently, these facilities are utilized to isolate and treat infected individuals. The advent of drones and robotic systems is revolutionizing the capacity of healthcare professionals to reduce direct patient contact, thereby enhancing the efficacy of disaster response operations. These advanced technologies are employed to execute a variety of tasks, such as the delivery of food and medical supplies, the collection of diagnostic samples, and the disinfection of patient environments. Nevertheless, it is often observed that field hospitals are impeded by a lack of network connectivity, which arises from inadequate or deficient communication infrastructures [38].

Regarding security threats, signals used for specialized short-range communication in autonomous ambulances may be susceptible to manipulation or control, potentially resulting in the jamming of radar systems or inducing collisions. Moreover, patients in constrained environments can be effectively monitored and tracked through the use of advanced intelligent wristbands. Nevertheless, spoofing and jamming attacks pose significant risks to the integrity of the signals transmitted by these wristbands. From a communication perspective, a multitude of communication challenges have been identified within disaster response units. The advancement of high-mobility aerial drones and automotive ambulances contributes to an increased frequency of handovers, thereby diminishing the stability of communication links and exacerbating latency issues. To guarantee the secure operation of these vehicles, it is imperative to establish resilient vehicle-to-everything (V2X) communication frameworks. Furthermore, the transfer of real-time data derived from medical imaging necessitates a transmission capacity of high data rates [38].

- Intelligent Internet of Medical Things (IIoMT): The IIoMT is poised to undergo significant evolution and fulfill a multitude of roles aimed at enhancing human welfare within the framework of the 6G communication paradigm. A comprehensive (IoE) will emerge, encompassing the IIoMT and the Internet of Bodies (IoB), among others, thereby facilitating the integration of medical devices with the Internet. All medical devices will possess Internet connectivity, which will facilitate expedited decision-making processes. Computed tomography (CT) and magnetic resonance imaging (MRI) scans, when interconnected with the Internet, will be capable of scanning and transmitting data utilizing 6G technology to remote locations. A physician can assess this data in real time. Moreover, another illustration is the convenience with which remote healthcare professionals can administer treatment to patients afflicted with cancer [38].

Currently, the process of ascertaining whether tumors in cancer patients are malignant or benign necessitates a considerable amount of time. Nevertheless, in the foreseeable future, the advent of 6G technology will facilitate the real-time identification of cancer. Moreover, it will no longer be imperative for both physicians and patients to physically visit specialized medical institutions. Consequently, remote healthcare professionals will engage in collaborative efforts with local practitioners to provide treatment for individuals afflicted by cancer. The prompt identification of cancer cases has the potential to significantly diminish mortality rates. This phenomenon is relevant to a multitude of medical conditions, extending beyond cancer to include cardiovascular diseases and numerous other health issues [38].

- Transport from home to hospital service: Traditionally, ambulance services are designed to facilitate the transportation of patients who necessitate oxygen support and preferential treatment in road traffic scenarios. It is inept in providing emergency services due to the absence of essential tools and information. The equipment with emergency signaling devices and oxygen in a standard autonomous vehicle can fulfill a comparable function. Consequently, there exists a pressing need for an innovative ambulance service to enhance response times and deliver prompt medical attention to patients en route to medical institutions. In this context, hospital-to-home (H2H) services are poised to assume the responsibilities traditionally held by ambulance services [38].

  In emergency situations, hospitals can immediately communicate with remote mobile ambulances and direct them to the closest site requiring urgent medical intervention. For this, automobiles will be AI-driven in the near future, yielding sophisticated automobiles that demand exceptionally reliable and low-latency communication networks. As a result, H2H systems will be integrated into a mobile healthcare framework supported by an intelligent vehicle, thereby reducing reliance on traditional hospitals and their medical personnel. The 6G-enabled ambulance will initiate patient treatment prior to arrival at the medical facility. In order to preserve life, a mobile hospital can further detect any calamity in real time and respond promptly to the incident location. Moreover, it will enhance contemporary living standards. Therefore, the implementation of 6G connectivity has the potential to elevate modern lifestyles through H2H services [38].
- Telesurgery service: Telesurgery represents an emerging domain of inquiry, also referred to as physician-assisted remote surgical interventions. The efficacy of telesurgery is dependent upon the establishment of real-time communication channels. Telesurgery is fundamentally reliant on ultra-reliable low-latency communications (URLLC) and networks characterized by high-bandwidth efficiency to mitigate any latency or data loss while performing surgery as it requires precise movements (e.g., for making incisions or stitching of internals) of specialized medical equipment [38].

  Because 6G will make the network reliable and reduce latency further, it will thereby improve the experience in telesurgery. The medical practitioner can provide verbal instructions, utilize telepresence, or employ tele-assist methodologies. With the implementation of 6G, the provision of more proactive oral guidance through holographic communication becomes feasible. The surgeon may be present to offer guidance throughout the procedure and may even reposition themselves to attain a clearer view of the surgical site via holographic communication. Telestration refers to the methodology of demonstrating surgical procedures remotely, for instance, through video transmission. Furthermore, the physicians have the capability to assist the surgical process utilizing tactile or haptic technologies [38].

- Response in pandemic and epidemic service: Communication technology will be critical in the event of a pandemic or epidemic as evidenced by the recent COVID-19 crisis. Generally, pandemics and epidemics are characterized by the transmission of diseases between individuals, thereby posing significant risks to healthcare professionals. The COVID-19 pandemic has resulted in the loss of millions of lives, including those of medical practitioners. The effective management of COVID-19 outbreaks can be achieved through the implementation of the IIoMT [38].
- Tactile/haptic Internet service: Haptic technology enables users to perceive a virtual tactile sensation through the application of force, movement, or vibration. The Tactile Internet needs URLLC and a high-speed connection to facilitate the capture of real-time tactile feedback. Furthermore, it possesses the potential to assist medical practitioners in diagnosing patients through tactile means without necessitating physical presence. The three classifications of haptic human-computer interaction (HCI) encompass computer, interface, and wearable modalities. In the context of desktop HCI, a remote physician can perform surgical procedures or conduct diagnoses utilizing a virtual instrument. The device that conveys commands is typically a flat-screen interface, such as a tablet or smartphone. In this scenario, the remote physician employs wearable HCI technology, such as a haptic glove. The implementation of tactile/haptic technology can significantly benefit the healthcare sector by enabling expert surgeons to execute complex surgical procedures remotely with the assistance of robotic systems. Through the utilization of tactile/haptic technologies, robotic entities may interact with and provide care for patients while medical personnel remain accessible from a remote location [38].
- Precision medicine service: Precision medicine constitutes the systematic endeavor of formulating pharmacological interventions specifically tailored to the unique requirements and circumstances of individual patients, thereby enhancing therapeutic outcomes. In order to facilitate the advancement of precision medicine, healthcare professionals engage in comprehensive research initiatives that involve the stratification of patients based on common attributes. The advent of 6G technology possesses the capacity to significantly augment the efficacy of precision medicine. Furthermore, it necessitates the integration of AI to provide personalized healthcare solutions. The refinement of therapeutic development is contingent upon the incorporation of health data derived from participants in clinical trials [38].

  For instance, cellular therapy is being implemented to address critical health conditions. The iNet Wireless Daemon (IWD) facilitates the acquisition of real-time data by physicians and researchers, thereby providing accurate health information. Moreover, the research may be conducted on a global scale, as geographical factors influence the immune systems of individuals. A singular site for examination can significantly affect both the research outcomes and the subjects under observation. Consequently, medical practitioners and researchers can monitor individuals across the globe by utilizing the IoMT [38].

The investigation in [39] denotes the concept of the Internet of Health Things (IoHT). The applications of IoHT have undergone significant scrutiny due to the recent advancements in IoT. This phenomenon spans various domains within IoHT, including telemedicine, electronic health record management, coordinated care strategies, cost reduction, chronic disease management, disability support, home healthcare administration, personalized health assistance, health monitoring, pharmaceutical supply chain management, healthcare oversight, and remote surgical procedures. The progression within the IoHT ecosystem has manifested a substantial enhancement in quality of life. Nevertheless, amid the myriad applications of the IoHT framework, the equilibrium between security and privacy presents itself as one of the foremost challenges in the context of life-critical applications. In Fig. 1.5, the system architecture of IoHT network [39] is established.

Various methods were followed to incorporate intelligence in IoHT. The general adaptive and intelligent framework within IoHT encompasses the strategies employed to delineate the operational dynamics. The procedure within IoHT is characterized by a tripartite methodology. The primary layer consists of the device

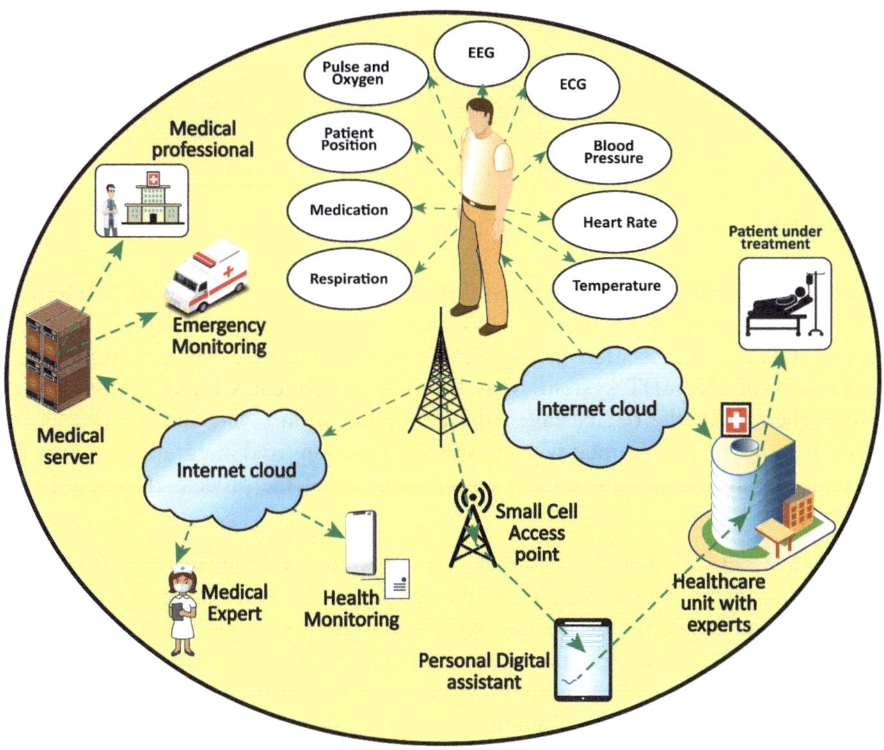

Fig. 1.5 Paradigm adapted by the IoHT

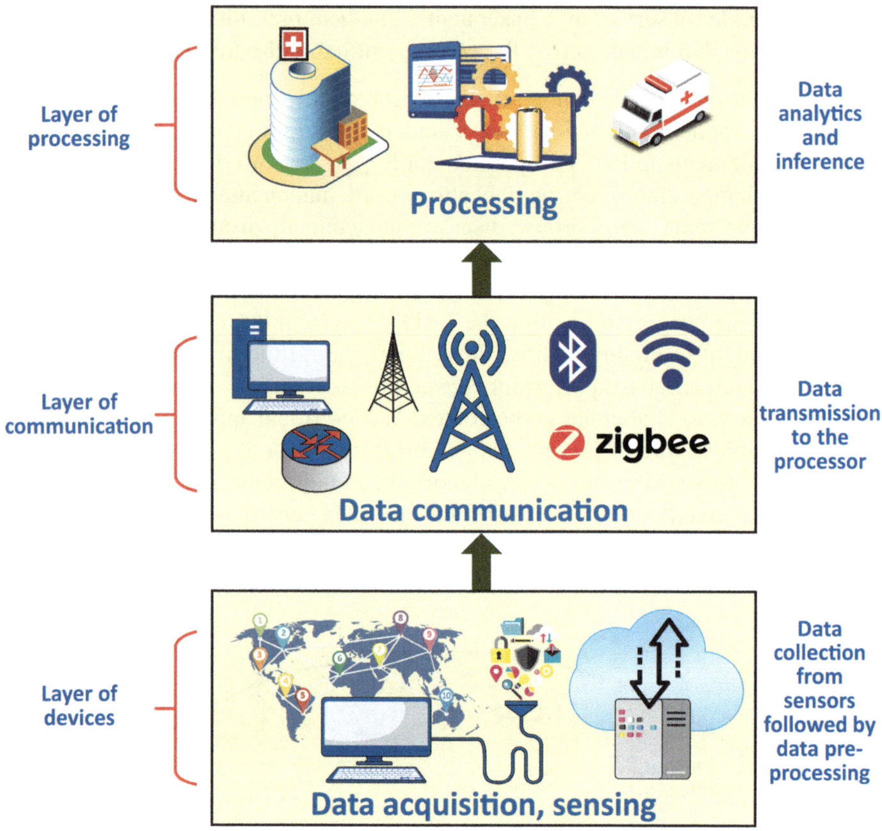

Fig. 1.6 IoHT system represented within a layered architectural framework

layer. The secondary layer encompasses the communication layer, while the tertiary layer pertains to the application process layer. The comprehensive layered architecture of IoHT is represented in Fig. 1.6. It depicts the layer-centric architectural framework of the IoHT system. This framework delineates three distinct layers, which encompass the device layer, the communication layer, and the processing layer. Each layer is explained with its specific functions and integrations within the system, thereby facilitating a lucid comprehension of the process flow within the IoHT network [39]:

- Layer of devices: The device layer is responsible for acquiring data from user-end equipment, which includes a diverse range of sensors, such as temperature sensors, heart rate monitors, and oxygen saturation sensors, among others. The raw sensor data undergoes pre-processing to convert it into a standardized format that facilitates efficient handling and computation. Subsequently, the processed data is transmitted to the communication layer for further processing [39].

- Layer of communication: The communication layer incorporates the procedure of data transmission into the application segment. The conveyance of information may be facilitated through a base station or via alternative wireless communication technologies. These wireless communication technologies encompass Bluetooth, Wireless Fidelity (Wi-Fi), Zonal Intercommunication Global-standard (Zigbee), Light Fidelity (Li-Fi), and radiofrequency mobile communication. The primary function of this layer is to ensure the accurate transmission of relevant data to the application processing layer [39].
- Layer of application processing: The layer of application processing involves the analysis of the acquired sensed data. The interpretation of the incoming data delineates the distribution of appropriate applications, thereby ensuring that patients are systematically linked to medical assistance. This layer facilitates the visualization of the patient, thereby establishing connections with end service providers, which may include ambulances, healthcare facilities, pharmaceutical supply chains, and the attention of medical practitioners [39].

Besides, 6G communicates with the principle of 6G which integrates cognition, communication, caching, computing, capture, and control as its fundamental pillars. The implementation of holographic communication is of paramount importance within the healthcare sector. The acquired information is transformed into a digital format, stored in a local cache, and subsequently transmitted in real time to remote locations. Signals are generated from digital data and conveyed from one device to another for further processing. Nevertheless, prior to computational analysis, cognitive processes aid in formulating viable conclusions derived from the digital input data. These represent astute decisions that contribute to the simplification of computational tasks. In order to enhance the management of smart device functionalities within the healthcare sector, the processed data is relayed to intelligent devices [38].

Additionally, the IoE will need to have a large capacity in order for 6G networks to integrate billions of smart devices, facilitating the capture of tactile sensations and their subsequent digitization for transmission and processing. This infrastructure must be adept at connecting a multitude of actuators and sensors within the realm of healthcare communication. Furthermore, it is imperative that low-latency connectivity is achieved to ensure seamless communication. Upon the commercial launch of 6G technology, a transformative era characterized by big data will emerge. The processing and analysis of vast quantities of minor data produced by healthcare devices will demand significantly enhanced computational capabilities. Figure 1.7 depicts an intelligent healthcare system predicated on the IoE framework [38].

As discussed in previous paragraphs, a decentralized, patient-centered framework is progressively supplanting the conventional healthcare paradigm that is predominantly centered around hospitals and specialists. The advent of communication technologies has facilitated the delivery of personalized and remote healthcare services. With the expansion of the healthcare sector, the proliferation of applications linked to the network will generate data in diverse sizes and formats. The future network is anticipated to encounter intricate demands pertaining to data transmission rates, bandwidth capacity, and latency requirements. Current communication

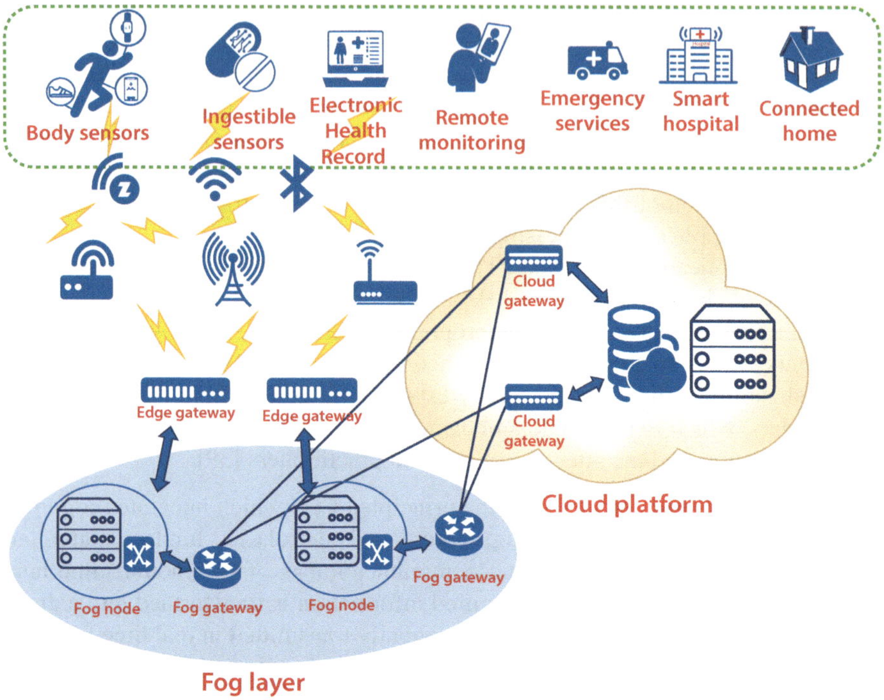

Fig. 1.7 The architecture of a smart healthcare network based on the Internet of Everything (IoE). It will utilize cloud computing for the storage, processing, and analysis of extensive datasets, specifically within 6G-enabled intelligent healthcare systems that leverage fog computing—a decentralized computing paradigm where data processing, storage, and application services are located between the data source and centralized cloud infrastructures

technologies are inadequate in addressing the multifaceted and heterogeneous demands imposed on communication networks by a diverse array of healthcare applications. Consequently, the forthcoming generation of communication technology, referred to as 6G, is projected to furnish essential infrastructure for healthcare by the year 2030. The healthcare sector will increasingly rely on artificial intelligence and 6G connectivity technology, thereby enhancing the quality of life and healthcare services. Moreover, prospective intelligent healthcare networks are expected to integrate a synthesis of 6G and IoT elements, which will augment network performance and expand cellular coverage [38].

Moreover, the acquisition of data from both in situ and wearable sensors, coupled with its transmission to a cloud-based medical facility for analytical processing, necessitates the implementation of novel security protocols. Given the limitations in energy consumption and computational capacity inherent to body sensors, the facilitation of encrypted communication both within the body and on its surface represents a significant avenue of ongoing research that remains to be fully elucidated. The following are the key sources of complexity [38]:

- The constrained availability of resources, including computational capabilities, power supply, and memory capacity, within nanodevices is a significant limitation. Consequently, the implementation of on-demand distribution methodologies becomes unfeasible, while the application of pre-deployed techniques presents considerable challenges [38].
- Requirement for a scalable networking framework within the nanodomain. Consequently, there is an imperative for the development of novel authentication methodologies that are characterized by enhanced energy efficiency and the utilization of cryptographic frameworks with abbreviated key lengths. Several researchers investigated the field of MCs to evaluate the confidentiality properties of diffusion-based MC systems and to develop innovative encryption approaches leveraging biological molecules, such as ribonucleic acid (RNA) and DNA. Nevertheless, additional investigation is imperative to address issues related to the spontaneous reactivity of molecules and unforeseen occurrences, particularly at the nanoscale [38].

Thus, as it seems obvious that MC is an inevitable field of science underlying almost every aspect of human life, with the support of ICT. Consequently, MCs stand at the front of advancing healthcare technologies [40], especially in the domain of nanomedicine, intrabody networks, specifically the Internet of Bio-Nano Things (IoBNT), and targeted drug delivery (TDD), as shown in Fig. 1.8. The corresponding information for the mentioned platform is analyzed in [17].

IoBNT represents a significant prospective application domain within the forthcoming 6G mobile network [18, 41], which anticipates the facilitation of interactions between biological cells or nanoscale devices and the Internet. MCs may provide a suitable modality for communication among these nanodevices,

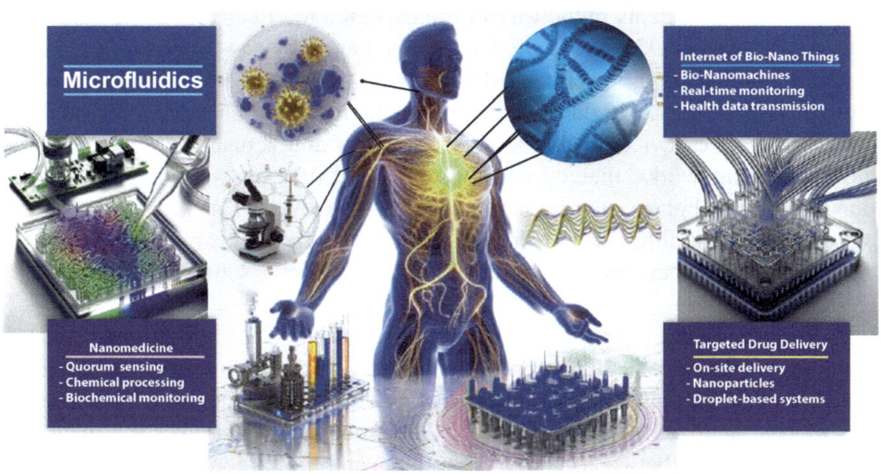

Fig. 1.8 Illustrating the most promising in-body applications of molecular communications, including nanomedicine, the IoBNT, and targeted drug delivery. Microfluidics serves as the key enabling platform for experimentally validating and practically implementing these applications

employing chemical molecules as carriers of information. Nevertheless, the intricate nature of experimental procedures and the lack of adequate interdisciplinary collaboration have resulted in the predominance of theoretical research in the field of MC, thereby substantially impeding the progression and prospective applications of MCs. Consequently, it is imperative to investigate methodologies for the development of transmitter and receiver nanomachines to actualize operational MCs systems. This investigation is predicated on the advancements observed in micro/nanomachines (MNMs) over recent years [41].

In the field of biotechnology, digital twins (DTs) signify a revolutionary advancement that precisely represents human beings, organs, microorganisms, and medical systems, alongside procedures within cyber-physical healthcare environments. DTs incorporate a diverse array of components, encompassing the IoT in healthcare, pharmaceutical industries, clinics, hospitals, and biomedical devices. They orchestrate the comprehensive lifecycle management of biological entities or equipment assets, thereby enabling monitoring, behavioral assessment, regulation, and adaptation. A fully realized DT surpasses a mere static representation or simulation of patients, ailments, organs, or biological assets. It transforms into a dynamic, intricately detailed, and sophisticated digital equivalent of physical and living entities within these sectors, encapsulating their interactions with the surrounding environment [42].

Consequently, DTs signify an integrated amalgamation of the tangible and digital domains, transcending the traditional IoT paradigm characterized by restricted interconnectivity and unidirectional data transmission from the tangible to the digital spheres. In forthcoming developments, industrial DTs are anticipated to assume a crucial function across a myriad of applications, ranging from pharmaceutical innovation to digital healthcare, while simultaneously promoting the expansion of the Metaverse [42].

Recent developments in the IoBNT are expected to effectively tackle and possibly overcome the existing challenges encountered by the digital twins (DTs)-enabled biotechnology sector. IoBNT refers to a network of interconnected biological and nanoscale devices capable of facilitating communication and data exchange. These devices may encompass sensors and actuators integrated within biological frameworks, thereby permitting the real-time observation and modification of biological phenomena at micro or nanoscale levels. This innovative technology markedly amplifies the proficiency of DTs in the biotechnology domain by delivering high-precision, real-time information directly sourced from biological entities [42].

IoBNT devices possess the capability to observe and react to alterations at the molecular or cellular scale, thereby providing a degree of precision and detail that significantly enhances the modeling and simulation proficiency of DTs. Moreover, IoBNT can facilitate the amalgamation of heterogeneous data types by functioning as an intermediary between biological systems and digital representations, thereby guaranteeing an uninterrupted flow of information. Additionally, IoBNT can integrate sophisticated hardware-level security mechanisms, providing strengthened data protection and privacy, an essential requirement when handling sensitive biological information. In summary, the advancement of IoBNT not only enhances but also substantially augments the capabilities of DTs within the biotechnology sector, thereby tackling critical obstacles and facilitating the progression toward more precise, instantaneous, and secure modeling and analysis of biological systems [42].

Very outstanding research with prospective applications is presented in [43]; it encompasses advanced health-monitoring systems predicated on bio-nanomachines that operate and interact with one another within the biological milieu. These systems necessitate considerable considerations regarding biocompatibility and energy efficiency. Due to the communication function, traditional robotic systems utilized in eHealth predominantly depend on EM wave transmission; this approach presents numerous challenges, including the fabrication of nanoscale actuators, antennas, and the physiological absorption of terahertz (THz) frequency bands. Thus, [43] introduces MCs as an alternative technology for *eHealth* applications. The study further examines the IoBNT, which leverages MCs, particularly within blood vessels. Given their extensive distribution throughout the body, blood vessels provide effective communication pathways for IoBNT deployment, enabling the collection of physiological data from multiple anatomical locations. Figure 1.9 conceptually illustrates the IoBNT-based health-monitoring system in four main steps [43], and the corresponding explanation is defined subsequently.

- Various sensors measure human body signals from different regions of the human body. These sensors must ensure safety and function effectively within stringent energy limitations. Consequently, nanotechnology and biotechnology constitute the foundational elements for an array of bio-nano-integrated sensing systems [43].
- Bio-nanotechnology entities are strategically introduced within the human body to establish a network via microcontrollers situated within the vascular system and transmit the signals detected by the sensors [43].
- The platform developed in Fig. 1.10 can be employed for the integration of the IoBNT network with the subcutaneous relay microchip [43].
- The system operates through the principles of galvanic coupling and facilitates communication across biological tissues [43].
- Finally, the relay chip establishes a connection with external electronic devices via radiofrequency technology, including Bluetooth. In this way, the acquired physiological signals can be effectively utilized in various medical applications [43].

For instance, individuals suffering from cardiac pathologies might require the implementation of a continuous monitoring apparatus after cardiac surgical procedures. In this context, a sensor positioned adjacent to the heart quantifies cardiac signals, including the rhythm of the heartbeat. The MC network transmits these cardiac signals, thereby enabling users to effectively monitor their cardiac function through a smartphone application. In a similar vein, individuals diagnosed with diabetes may utilize the system to routinely assess their glycemic levels. The continuous surveillance of biological signals emanating from the body can facilitate

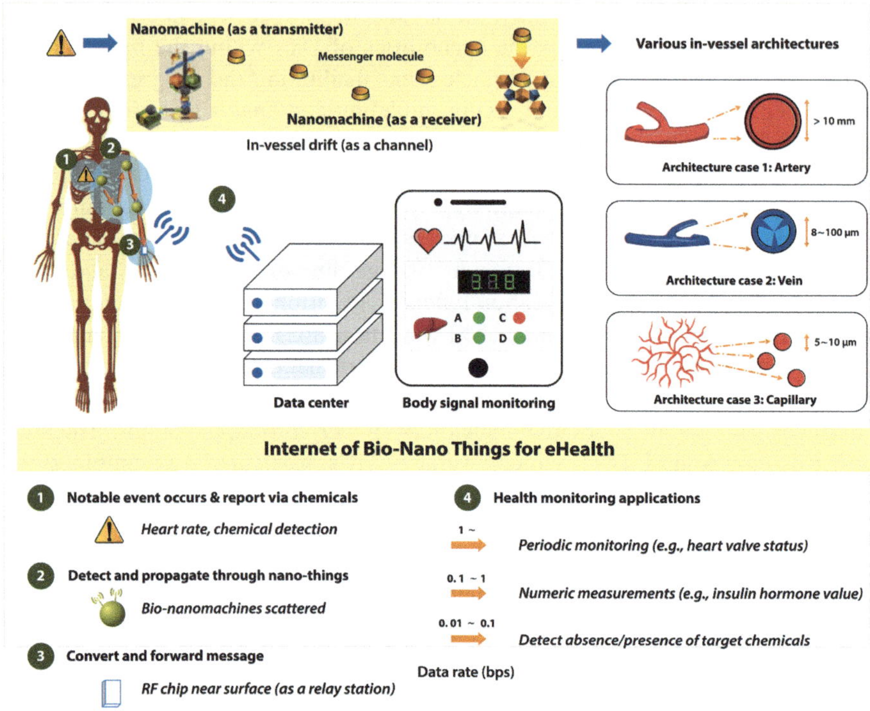

Fig. 1.9 A conceptual illustration of IoBNT pertaining to eHealth. Within the context of health monitoring, nanomachines that are capable of intercommunication via molecular interactions are utilized to enhance the biocompatibility of the system when compared to a system that incorporates EM components (These devices possess the ability to adaptively transmit and receive specific target molecules within in-vessel environments through the application of cutting-edge bio-nano machine technologies.)

physicians in establishing diagnoses and result in improved prescriptions and/or treatments for previously elusive symptoms. From the perspective of system requirements, it is imperative that such a communication system is characterized by a high degree of reliability and poses no harm to the patient [43].

These systems do not necessitate elevated transmission velocities, as the pertinent data are often uncomplicated in most application contexts. Hence, authors in [43] mainly address the communication method in the blood vessels. The human blood vessel network consists of three distinct categories of vessels, specifically arteries, veins, and capillaries, each exhibiting unique characteristics pertaining to their respective channels. Therefore, authors in [43] have built an IoBNT network by considering the channel characteristics of blood vessels. Furthermore, they confirmed the feasibility of the proposed IoBNT systems by developing and implementing a nanoscale MC prototype as Fig. 1.10 indicates [43].

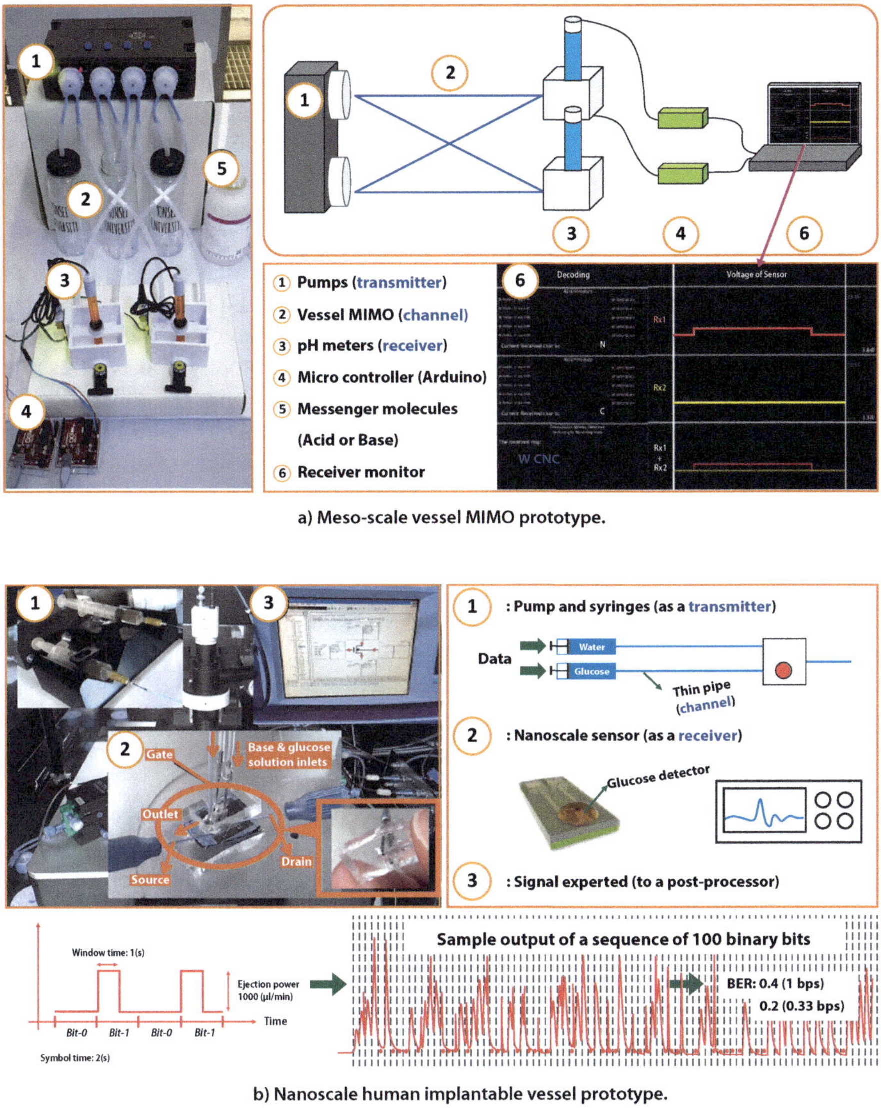

Fig. 1.10 Prototypes of blood vessel networks with an MIMO (multiple-input multiple-output) system

The preceding paragraphs have allowed us to establish the fundamental framework of analysis for MC systems, that is, such systems are investigated precisely under informational and communicational theory approaches, both for natural biology and synthetic biology. In this perspective, for example, it is necessary to mention the challenges that involve key technical aspects in embedded computing, where devices such as microcontrollers cannot be used to process information; thus, in [44] the open image system Window CRN++ (discrete and real-valued computing) is introduced. This is a new deterministic programming language based on chemical kinetics to execute computing processes, although this type of scientific advances is certainly impressive, it is no less true the great challenge that means producing embedded components based on a set of chemical reactions that occur simultaneously. Incredibly, new commands can be incorporated into this language, adding new types of chemical reactions; this will originate the operational framework for robust molecular programming [44].

Another important example of the relationship between MCs and computational resources for healthcare application is found in [45]. Consequently, contemporary investigations in MCs have incorporated more sophisticated computational frameworks to accurately represent the dynamics of real-world biophysical systems. [45] concentrates on designing realistic models of MC transceivers inspired by spheroids-three-dimensional cellular aggregates widely employed in organ-on-chip experimental platforms. Potential applications that may be utilized or simulated with spheroids encompass nutrient transportation within organ-on-chip systems, the liberation of biomarkers or the reception of pharmaceutical compounds at neoplastic tumor sites, or transceiver nanomachines engaging in the exchange of information [45].

Accordingly, a simplified diffusion-based MC system is examined, consisting of a spheroidal transmitter and a spheroidal receiver immersed in an unbounded fluid medium. These spheroidal structures are modeled as porous media that facilitate the transport of diffusing signaling molecules, enabling the analysis of boundary conditions and the estimation of effective diffusion coefficients. Furthermore, with respect to either a point source or a spheroidal transmitter, Green's function for concentration (GFC) has been analytically deduced and expressed in the form of an infinite series, corroborated through a particle-based simulator (PBS). The established GFCs facilitate the calculation of both the transmitted and received signals within the proposed spheroidal communication framework [45].

Authors in [45] show that the porous architecture of the receiving spheroid serves to magnify diffusion signals while simultaneously causing their dispersion, thereby indicating a compromise between porosity and the rate of information transmission. Furthermore, the findings indicate that the porous configuration of the transmitting spheroid not only results in the dispersion of the received signal but also causes a reduction in its intensity when compared to a point source transmitter. The performance of the system is further assessed through the lens of the bit error rate (BER). It has been demonstrated that a reduction in the porosity of the receiving spheroid contributes positively to the enhancement of system performance. In contrast, a decrease in the porosity of the transmitting spheroid may negatively impact the overall performance of the system.

Currently, the world around us generates an impressive amount of information, in online and offline processes, in human activities that include commerce, academia, medical care, entertainment and social networks. The processing of information regardless of its source obeys the techniques known as "Big Data," which store and manage an amount of information of approximately 2.5 quintillion bits that are generated on the Internet daily [46], and which according to the International Data Corporation will grow to approximately 175 zettabytes by 2025 [47]. This large amount of information is currently stored in magnetic or optical media; however, the explosive speed of increase in the production of information is expected to be insufficient to store it. Thus, science has turned its gaze to biological storage alternatives, among which DNA storage is found [48, 49]. Studies in [50] apply computational parameters such as value, variety, speed, variability, veracity, and volume, to evaluate and manage information in sustainable big data systems, generated with nanosensors (with molecular components), that share information through nanocommunication networks, in order to be used for future applications in medical care and biomedicine.

A notable prospective trend in IoT-based healthcare frameworks is the emergence of Biological Internet of Things (Bio-IoT), which encompasses the application of the body's intrinsic mechanisms for MCs. This methodology facilitates the identification of atypical conditions within the organism by capitalizing on its preexisting biological systems. MCs, in this context, depend on signaling molecules categorized into primary and secondary reporters. Primary messengers, including hormones and growth factors, promote intercellular communication, whereas secondary messengers regulate intracellular communication [51].

These systems operate in a manner akin to biological processes, with the velocity of communication being contingent upon the spatial alignment of cellular entities. For instance, in the case of neuronal cells, the conduction of signals can transpire at significantly accelerated rates owing to the presence of axonal extensions. Notwithstanding its nascent development, Bio-IoT possesses immense potential. Analogous to the seemingly improbable miniaturization and advancement of mobile devices during the 1990s, the prospective evolution of Bio-IoT may bring about a transformative impact on the healthcare sector. Nevertheless, a multitude of substantial obstacles remain, particularly concerning issues of security. Continuous monitoring of the human body may pose novel risks, including threats related to biological warfare and terrorism, underscoring the necessity for robust security protocols and the revision of Internet security standards. Additionally, an emerging area of interest is the development of Nano-IoT, also known as the Internet of Nano-Things (IoNT). This principle, utilizing materials such as graphene, facilitates the interlinking of nanoscale entities, thereby establishing nanonetworks with diverse applications, particularly within the realm of healthcare. Nano-Bio IoT signifies the amalgamation of biological and electronic communication systems, permitting intrabody sensing as well as the activation of nanodevices for therapeutic applications. This communicative framework integrates biological systems with networking technologies, thereby enhancing interaction and data transmission within a biochemical context [51].

Consequently, within the domain of healthcare, the advent of the IoNT has precipitated the advancement of the Internet of Nano-Medical Things (IoNMT). IoNMT harnesses nanodevices and sensors that possess the capability to collect and relay data at the molecular scale. For instance, nanodevices may be deployed for the enhanced targeting of cancer cells or for the early identification and surveillance of diseases. In IoNT, two key communication techniques are nanoelectromagnetic communication and MCs [52].

Nanoelectromagnetic communication employs EM waves, such as radiofrequency waves, at the nanoscale, presenting significant challenges attributable to the diminutive size of devices and constraints related to power consumption. Radiofrequency identification (RFID) and RedTacton exemplify this domain, with RedTacton providing enhanced data throughput and more reliable connections. THz communication, an emergent technological paradigm, utilizes EM waves situated within the terahertz frequency spectrum, facilitating high levels of penetration and data transmission rates, which are advantageous for facilitating communication between nanoscale devices and their respective gateways. MCs utilize molecular entities as vehicles for the transmission of information, thereby facilitating elevated concentration and precise localization. This characteristic is particularly advantageous for the IoNT, wherein the devices are diminutive, situated in inaccessible locations, and possess significantly constrained processing, storage, and energy capacities. The integration of both methodologies serves to augment communication efficacy and bolster network dependability. Nanoelectromagnetic communication guarantees rapid transmission across extensive distances, whereas MCs are characterized by their energy efficiency for nanodevices, delivering redundancy and ensuring continuous communication within IoNT networks [52].

Although applications of the IoNT are being developed across a multitude of sectors including military, agriculture, robotics, energy, food production, biology, and chemistry, the primary domains of IoNT applications lie within the medical and healthcare fields, and that can proficiently identify harmful tumors, viruses, and bacteria, as well as monitor physiological parameters such as body temperature, glucose concentration, blood insulin levels, heart rate, oxygen saturation, blood pressure, and hormone levels [53].

The nanosensors employed in IoNT-based healthcare applications are termed bio-nanosensors, which incorporate either antibody or aptamer-based modules for the in vivo detection of chemical and biological parameters. These miniature devices have emerged as a crucial component of IoNT-based healthcare applications, offering real-time chemical and biological information for swift analyses, thereby surpassing conventional methodologies. Bio-nanosensors can be directly administered into the human organism, enabling the collection of medical information through interactions with biological mechanisms. The early detection of illnesses, particularly malignancies, is crucial for patient management, especially given the significant rise in incidence observed in recent years. Healthcare applications leveraging the IoNT present enhanced opportunities for both patients and healthcare professionals to initiate treatment prior to the dissemination of disease throughout the organism. Furthermore, IoNT technologies are also employed in pharmacological delivery within medical applications. These systems aim to ensure the precise delivery of therapeutics to targeted sites for the remediation of pathological conditions or the restoration of compromised tissues, including osseous structures, musculature, and neural pathways, through the utilization of nanomachines [53].

Consequently, an innovative paradigm has emerged known as the Medical Internet of Nano-Things (MIoNT), wherein nanosensors gather health-related data via EM means in vivo utilizing the THz frequency band, subsequently transmitting this information to centralized data repositories and, thereafter, to medical server applications for analytical processing. The MIoNT-based framework is pivotal in a multitude of healthcare applications and is anticipated to gain even greater significance in the imminent future. With the utilization of bio-nanosensors, the detection of tumors, cancerous cells, asthma exacerbations, as well as various viruses and bacteria, can be conducted with enhanced efficacy. Furthermore, these sensors are adept at facilitating the study of interactions among DNA and RNA, antibodies, enzymes, nucleic acids, proteins, and lipids in vivo [53].

The advancements in nanotechnology have facilitated the development of nanoscale devices with unique and advantageous properties, playing a crucial role in numerous advanced smart city applications. The previously mentioned applications encompass a wide range of domains; however, they are primarily confined to intelligent drug delivery systems, body area networks (BANs), and environmental monitoring frameworks. Nevertheless, while nanotechnology exhibits significant potential for revolutionizing future developments, its practical implementation faces substantial challenges that necessitate critical examination. For instance, nanosensors represent an essential element in the realization of nanoscale networks aimed at healthcare applications. It is imperative that these sensors leverage the unique properties of nanomaterials to execute their functions, all while complying with stringent limitations concerning dimensions and energy consumption. The constrained sensing, processing, storage, and networking capabilities of nanosensors arise from the restrictions they encounter, despite their relatively simple design. To mitigate the aforementioned limitations, a viable strategy entails the deployment of nanosensors in densely packed configurations within interconnected environments, commonly referred to as nanosensor networks or nanonetworks [54].

Hence, the application of molecular and EM communications has emerged into prominence as key wireless technologies enabling communication among nanomachines. Currently, scholarly pursuits are primarily focused on the improvement of MC systems, which encounter a considerable constraint in attaining elevated data transmission rates. This limitation potentially hinders the efficacy of nanosensor networks. Plasmonic nanoantennas remain relatively underdeveloped, notwithstanding their capacity to enable high-frequency wireless communication among nanodevices from an EM perspective. This situation stands in contrast to optical-based methodologies, which continue to be predominantly employed in most nanosensing applications. Although nanosensors have significantly improved traditional healthcare methodologies, the capacity to execute precise measurements at the nanoscale remains inadequate during earlier developmental stages. The deployment of nanosensor networks offers a potentially effective remedy to the limitations imposed by microscale devices, as it enables measurements at the cellular level, consequently providing superior accuracy. Furthermore, the adoption of nanosensor networks introduces innovative possibilities for healthcare institutions. For example, it is conceivable to enhance the effectiveness of targeted drug delivery systems that minimally affect adjacent healthy cells [54].

Furthermore, the deployment of implantable nanosensors possesses the potential to significantly elevate the quality of life for individuals afflicted with chronic health conditions through the persistent observation and documentation of their vital physiological metrics. The distinctive attributes inherent to nanoscale devices necessitate a comprehensive reassessment of the communication methodologies employed among them, as these methodologies diverge from those associated with conventional nanodevices. The application of EM wireless communication represents a prominent strategy for information exchange among sensors. Nevertheless, due to the diminutive scale of these devices, substantial alterations are required in the channel models, communication frameworks, and architectures. MC is a new type of paradigm that employs molecules to encode and receive data within nanosensor networks. Moreover, the application of wireless nanocommunication is employed within intrabody nanonetworks as indicated in the cited reference [54].

Another very important concept in the new era of computing is quantum computing (QC), which is fundamentally grounded in the principles of quantum mechanics, which is frequently elucidated through the paradigms of superposition, interference, and entanglement. In the realm of quantum physics, a singular bit possesses the capacity to exist in multiple states concurrently (i.e., 1 and 0) at any given moment, and a QC system capitalizes on this phenomenon, conceptualizing it as a qubit (quantum bit). With its foundations rooted in quantum physics, QC harbors the potential to constitute the underpinning of future high-performance computing architectures, thereby facilitating the real-time processing of vast quantities of data. Recently, quantum computing has experienced a significant increase in scholarly interest from researchers aiming to elevate computational capabilities beyond the limitations imposed by Moore's law; nonetheless, there exists a pressing necessity for a comprehensive systematic investigation to delineate the opportunities, challenges, and obstacles associated with this field [55].

QC is exceptionally appropriate for a multitude of computation-intensive applications within the healthcare sector, particularly in the contemporary highly interconnected IoT digital healthcare framework, which includes a variety of interconnected medical devices (such as medical sensors) that may be linked to the Internet or cloud-based systems. The substantial augmentation in computational capacity not only serves to enhance healthcare IoT but also enables quantum computers to facilitate significant advancements in this particular field [55].

When transitioning from classical bits to qubits, there exists the potential to enhance pharmaceutical research within the healthcare domain, which encompasses the examination of protein folding, elucidating the compatibility of molecular structures, such as drugs and enzymes, as well as ascertaining the affinities of binding interactions between individual biomolecules, for instance, proteins or DNA, and their respective ligands or binding partners, such as pharmaceuticals or inhibitors, while simultaneously expediting the clinical trial process. A limited number of prospective applications are succinctly delineated below for the purpose of illustration. A quantum computer can perform DNA sequencing at unprecedented speeds, thereby facilitating the advent of personalized medicine. It can enable the development of new therapies and pharmaceuticals through intricate modeling processes. Furthermore, quantum computers have the potential to establish highly efficient imaging modalities that can furnish clinicians with enhanced resolution and clarity in real-time contexts [55].

Furthermore, it possesses the capability to address intricate optimization challenges associated with the formulation of an ideal radiation strategy aimed at eradicating malignant cells while preserving the integrity of adjacent healthy tissues. QC is poised to facilitate the examination of molecular interactions at the most fundamental level, thereby creating avenues for pharmaceutical innovation and biomedical investigation. The process of whole-genome sequencing is inherently time-consuming; however, with the utilization of qubits, the execution of whole-genome sequencing and associated analytics could be conducted within a constrained timeframe. QC possesses the potential to transform the healthcare paradigm by facilitating advanced modalities of on-demand computing, redefining security for medical data, predicting chronic diseases, and enabling accurate drug discoveries [55].

Figure 1.11 establishes a differentiation of QC paradigms with classical computing approaches in terms of their strengths, weaknesses, and applicability. In contrast to traditional computers that function based on bits, the fundamental operational units within a quantum computer are designated as quantum bits, or "qubits," which

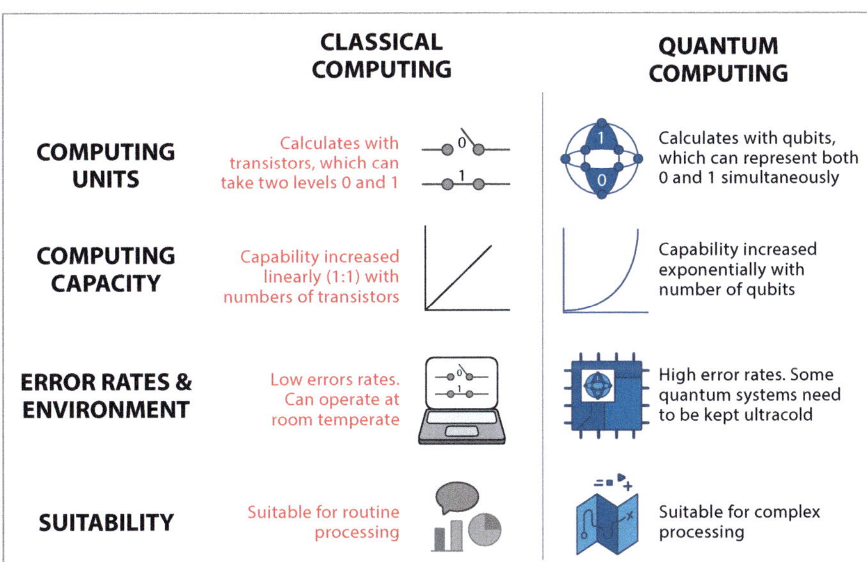

Fig. 1.11 A comparative analysis of classical computing and QC can be conducted through the examination of four fundamental attributes: (a) computational units; (b) computational capacity; (c) appropriateness; and (d) error rates

exhibit two states or levels; specifically, they possess the capability to concurrently represent a single bit in both "1" and "0" [55].

Recent research shows that QC possesses a distinct superiority over traditional computing frameworks. QC facilitates a significant acceleration in the processes of disease diagnosis and therapeutic intervention, and in select instances, it can substantially diminish computational durations from years to mere minutes. It stimulates novel methodologies for achieving enhanced proficiency in specific tasks, as well as fosters the development of innovative architectures and strategic approaches. Consequently, QC possesses substantial potential to be utilized across a diverse array of applications within the healthcare domain broadly and for healthcare providers specifically, particularly in the realms of expedited diagnostics, tailored medical therapies, and cost optimization strategies. A discernible surge in the adoption of both classical modeling techniques and quantum-based methodologies is evident, largely attributable to enhanced access to global health-relevant data repositories and their availability. The subsequent paragraphs delineate several potential applications of QC within the healthcare sector; a visual representation of these applications is illustrated in Fig. 1.12 [55].

- Molecular simulations: QC ends data processing by utilizing quantum bits, in contrast to classical computing systems, where the speed of processing is dictated by integrated circuits. Unlike traditional methods that encode information solely in binary digits (0s and 1s), QCs leverage the principles of quantum entanglement, thereby facilitating the development of quantum algorithms that challenge classical computing paradigms, which are inherently ill-equipped to exploit this unique phenomenon. In the domain of healthcare, quantum computing has the potential to leverage ML, optimization techniques, and AI to execute intricate simulations. Healthcare processes frequently involve intricate correlations and highly interconnected molecular structures characterized by interacting electrons. The computational demands for simulations and other operations in this field inherently escalate exponentially with the increase in problem size, with temporal constraints perpetually serving as the limiting factor. Consequently, systems predicated on QC inherently align with the specific requirements of this use case [55].
- Precision medicine: Aims to deliver tailored prevention and treatment strategies addressing the specific healthcare requirements of individuals. Given the complexity of the human biological system, future medical approaches will necessitate personalized therapies that extend beyond conventional standardized treatments. Classical ML methodologies have demonstrated considerable efficacy in forecasting the likelihood of future medical conditions through the utilization of electronic health records (EHRs). Nonetheless, persistent challenges exist in the application of traditional ML techniques attributable to issues related to data quality and noise, the dimensionality of features, and the intricate interrelations among various features [55].

This incites the idea of using quantum-enhanced ML, which has the potential to improve the accuracy and resolution of early disease detection. Healthcare

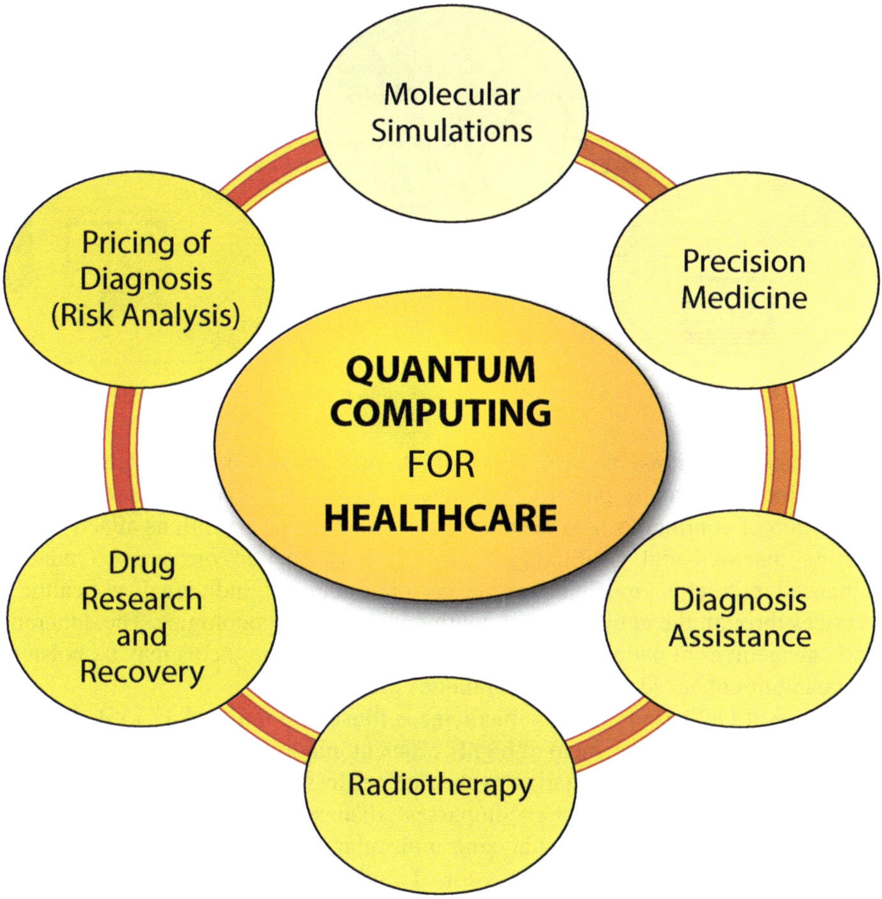

Fig. 1.12 Taxonomic classification of various prospective utilizations of QC within the domain of healthcare

professionals may subsequently utilize instruments to ascertain the ramifications of risk factors on individuals amid specific alterations in their conditions through ongoing virtual diagnostics predicated on incessant data streams. The sensitivity of pharmacological agents remains a pertinent area of inquiry at the cellular level, particularly with respect to the genomic characteristics of neoplastic cells. Current investigations are elucidating the chemical characteristics of pharmacological models that may be employed to forecast the efficacy of cancer treatments at a detailed level. The application of quantum-enhanced ML could accelerate advancements within the healthcare sector, primarily by facilitating drug inference models [55].

Precision medicine seeks to elucidate the relationship between causes and treatments, enabling the prediction of individualized therapeutic interventions and the optimal subsequent steps in patient care. Conventional diagnostic methodologies that rely on the subjective reporting of symptoms by patients often culminate in

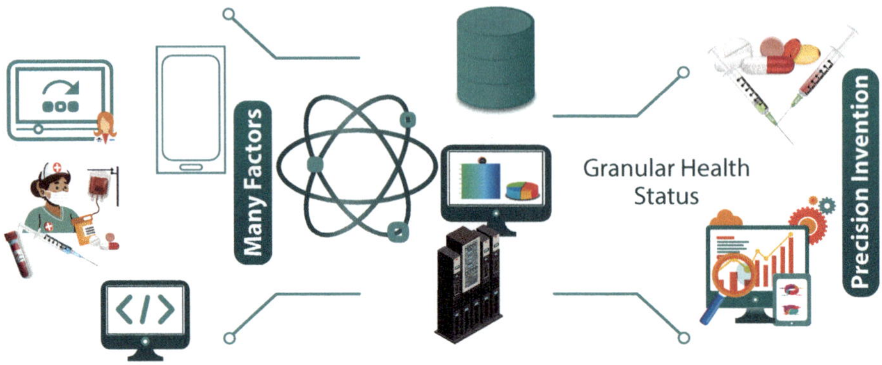

Fig. 1.13 An illustration of QC that harnesses vast multimodal data to enable precision medicine

broad categorizations of diagnoses, which may result in associated treatment modalities occasionally proving ineffective. QC has the potential to facilitate the utilization of continuous data streams through tailored interventions aimed at forecasting diseases and facilitating appropriate therapeutic measures. Quantum-enhanced predictive medicine serves to optimize and individualize healthcare services through the application of continuous care methodologies. The adherence and engagement of patients in individualized treatment protocols may be bolstered by quantum-enhanced modeling techniques [55].

A case of QC-based precision medicine is illustrated in Fig. 1.13 [55].

Moreover, the advancement of highly efficient imaging systems characterized by enhanced resolution for utilization by medical professionals in real time is promised by quantum computers. This encompasses advancements such as understanding protein folding mechanisms, analyzing molecular interactions between drugs and enzymes, and expediting the progression of clinical trials. The rapid DNA sequencing capabilities afforded by quantum computers enable the emergence of personalized therapeutic strategies, particularly in combating genetic disorders. Their precision and efficacy facilitate the exploration of novel treatment modalities. QC holds significant potential for the domains of pharmaceutical research and development, as it possesses the capability to analyze and replicate intricate chemical and biological processes with unprecedented accuracy. By advancing the boundaries of scientific innovation and discovery, the application of QC within the healthcare sector has the capacity to enhance patient care and accelerate the progression of medical advancements [56, 57].

Concomitantly, with the outstanding importance of QC in health area, in [58] the QC application to IoMT is highlighted especially to vulnerable people. Hence, IoMT is becoming increasingly significant in remotely monitoring the health of older individuals or those with specific health issues. Within IoMT networks, sensor nodes collect health data, such as the patient's blood glucose levels, heart rate, and other signs that may suggest the presence of a disease [58].

To facilitate the monitoring, diagnosis, or treatment process, this information is broadcast to remote servers that may be accessed by healthcare staff via communication protocols. Since the inception of the IoMT concept, considerable focus has been placed on fortifying data transmission security (as happens in all areas of computers and digital equipment networks) while concurrently reducing the expenses linked to IoMT communication and computation. The IoMT utilizes sensing devices to transmit up-to-the-minute health data from any location. Hence, ensuring the dependable transmission of sensory data from the individual is crucial. The sensors must securely transmit health-related data through a wireless channel to monitor health-related data remotely. Nevertheless, in the practical application of IoMT technology, sensitive patient data is susceptible to unauthorized access by hackers, presenting a significant security concern. To address these security concerns, [58] presents a new blockchain-enabled quantum trusted organization. Figure 1.14 illustrates an example of an IoMT system [58].

Therefore, the system model under examination shown in Fig. 1.14 of [58] comprises three principal entities:

- The trusted entity (TE)
- The mobile control unit (MCU)
- The users

The architecture of this scrutinized system model for IoMT according to [58] is illustrated in Fig. 1.15. The TE functions as an entirely reliable and secure authority,

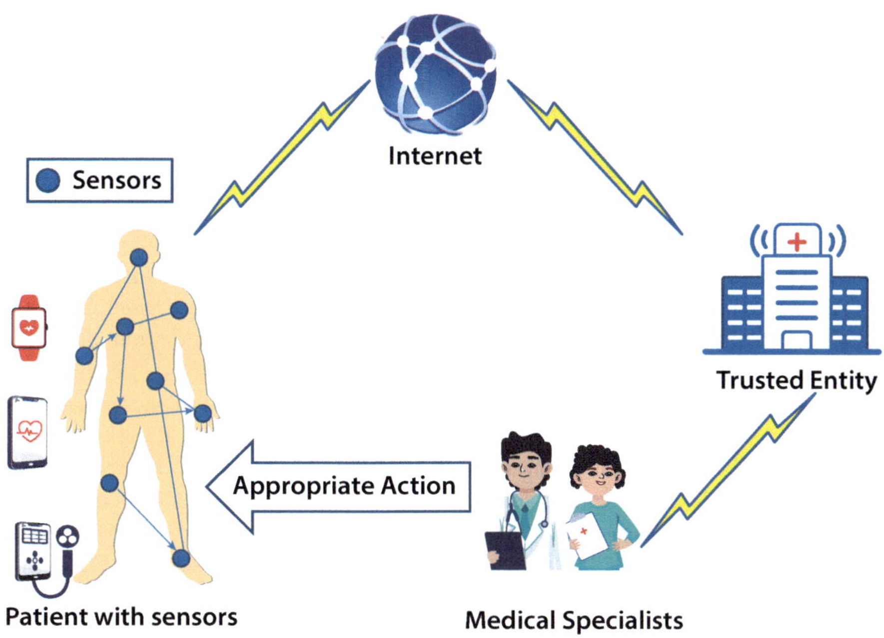

Fig. 1.14 An example of an IoMT system that must be supported by a proposed blockchain-enabled quantum trusted organization in [58]

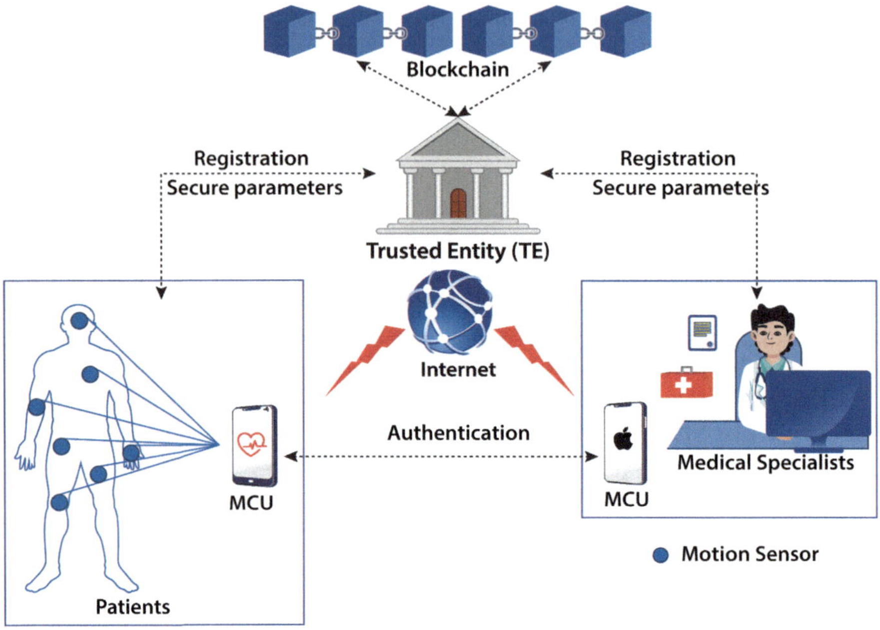

Fig. 1.15 Framework model for the blockchain-enabled quantum trusted organization proposed in [58]

impervious to compromise by any individual. This entity generates network parameters and furnishes authentication parameters to authorized users within the IoMT network (Fig. 1.14). Both patients and medical specialists (MS) must initially register with the TE by providing their requisite authentic credentials. Upon successful registration, authenticated users are supplied with the necessary parameters by the TE. Additionally, the TE assumes the responsibility of maintaining a blacklist of malicious doctors. Consequently, these malicious actors are expelled from the IoMT, and their subsequent communication within the network is thwarted [58].

The authenticated end users within the IoMT network are provided with the mobile control unit (MCU). The MCU boasts considerable computation, data storage, and parameter generation capabilities. Both patients and MS, i.e., doctors, are equipped with distinct types of MCUs. The MCU is empowered to encrypt medical prescriptions and transmit them to the patient's device. Moreover, it collects sensitive biological data from the patient's MCU. Within the patient's MCU resides a controller capable of gathering sensitive data from internally or externally connected sensors on the patient's body. The acquired sensitive data is encrypted and relayed back to MS MCU. The patient's MCU incorporates an integrated analog-to-digital converter to convert analog sensor data from sensors into digital signals, subsequently processed within the patient's MCU. The individuals inside the IoMT network include MS and experts, as well as patients [58].

In order to gain authorization as a user on the network, one must first complete the registration process. Users are authenticated only when their registration is successfully completed. The TE only provides a unique MCU to authenticated users. The MCU facilitates communication and information exchange between end users in the network [58].

Consequently, [58] presents a highly effective certificateless blockchain-based quantum authentication protocol that ensures anonymity and privacy. Then the proposal in [58] aims to ensure the secure and efficient authentication data transmission between doctor and patient user devices within the IoMT environment. It offers a detailed examination of an algorithm employed for authentication within the IoMT network according to Figs. 1.14 and 1.15. The authentication algorithm is founded on generating private keys, which are essential to verifying encrypted data during the authentication process. Moreover, these keys are generated using the quantum key distribution (QKD) method. Blockchain technology is leveraged to achieve efficient patient authentication without necessitating a TE. A streamlined QKD mechanism is devised within the proposed authentication scheme framework. This approach facilitates the secure transfer of patient data between healthcare providers. Thus, [58] proposes an efficient revocation method to remove any doctor engaging in improper behavior from the IoMT network.

The relationship between digital electronic systems and biological components appears to follow an evolutionary trajectory, as discussed in [59]. As conventional silicon-based computing nears its physical and energy-efficiency limitations, biocomputing is increasingly viewed as a viable alternative. This paradigm leverages biomaterials such as DNA and neuronal structures for data processing and storage applications. It explores the potential of neuronal biocomputing to rival silicon-based systems [59]. The authors in [59] investigate the feasibility of neuronal biocomputing as a competitor to silicon-based architectures. The study proposes neuronal logic gates and sequential circuits designed to replicate traditional computer architectures. Through mathematical modeling, optimization techniques, and computer-based simulations, the authors validate the functional performance of neuronal sequential circuits, including implementations of a neuronal NAND gate, an SR latch flip-flop, and a D flip-flop memory unit.

The approach in [59] involves manipulating neuron communication, synaptic conductance, spike buffers, neuron types, and specific neuronal network topology designs. The experimental results demonstrate the feasibility of encoding binary information through specific patterns of neuronal activity, while addressing synchronization challenges using neuronal buffering mechanisms and inhibitory control strategies. Results in [59] further validate the effectiveness and scalability of neuronal logic circuits, indicating that they sustain a stable metabolic load even when implemented in complex data storage configurations. Moreover, the study not only illustrates the concept of embodied biocomputing by harnessing neuronal properties for digital signal processing but also lays the groundwork for next-generation biocomputing technologies. These architectures highlight the potential of neurons as energy-efficient computational substrates, offering a biologically viable and potentially carbon-neutral alternative to conventional silicon-based systems.

Furthermore, within the conceptual framework delineated in [60], there is an exposition of research that draws upon recent innovations in organs-on-chip technology, specifically examining the cancer-on-chip model, wherein neoplastic cells are subjected to chemotherapeutic agents and subsequently release biochemical signals into the microenvironment, thereby eliciting an immune response. Thus, [60] exposes a scenario-based investigation conducted through a simulation algorithm founded on a hybrid model that integrates a coupled PDE-ODE (partial differential equations-ordinary differential equations) framework, which elucidates the concentration dynamics of chemical agents and the migration of immune cells within a cancer-on-chip environment. The main goal of [60] is to investigate numerically the qualitative impacts on the overall dynamics of immune cells instigated by coefficients that govern cell alignment, growth rates, and the influx of chemical substances. To achieve this objective, various simulated scenarios derived from manipulating the values of the relevant model parameters are showcased in [60].

Another highly significant investigation pertaining to biological communications and technology is documented in [61] which explores organophosphorus compounds (OPP), defined as organic chemicals characterized by the presence of carbon-phosphorus bonds. These substances, known for their high toxicity, are extensively employed as insecticides within the agricultural sector. It is estimated that nearly 70% of the pesticides utilized in everyday practices fall within the OPP pesticide classification. Widespread OPPs can cause severe pollution to water, air, soil, and agricultural products, thereby endangering the ecological equilibrium. OPPs predominantly infiltrate the human organism through mechanisms such as inhalation, dermal exposure, or oral intake, inhibiting the enzymatic activity of acetylcholinesterase, which results in the pathological accumulation of acetylcholine, thereby disrupting neural function and potentially culminating in paralysis or fatality [61].

OPP abuse can pose a severe threat to both environmental and food safety. Research indicates that fatalities associated with OPP toxicity in rural regions of developing nations surpass 200,000. Among the class of OPPs, malathion has emerged as one of the most prevalent commercial broad-spectrum insecticides for the management of agricultural pests and diseases, attributable to its comparatively low level of toxicity. The International Agency for Research on Cancer (IARC) has designated malathion as a probable carcinogen in humans (group 2 A). Consequently, numerous nations have instituted maximum residue limits for malathion. In order to avert public health crises attributable to malathion and to safeguard ecological well-being, there exists an imperative to develop methods for malathion detection that are convenient, precise, highly selective, and sensitive [61].

At present, conventional techniques for detecting organophosphorus pesticides (OPPs) primarily involve chromatography and mass spectrometry. While these methods exhibit elevated sensitivity, they prove impractical for expedited on-site detection of OPPs due to their substantial detection expenses, protracted time requirements, dependence on extensive instrumentation equipment, and professional operators. To overcome this challenge, researchers have developed enzyme inhibition techniques and immunological assays. Enzyme inhibition methods detect the presence of OPPs by exploiting their capacity to inhibit the activity of specific enzymes. In the context of immunological detection methodologies, signals are produced through specific interactions that occur between antibodies and their respective targets, facilitating the identification of OPP molecules. These methods are characterized by their simplicity and rapidity, necessitating neither costly large-scale instrumentation nor cumbersome sample pretreatment processes, thereby rendering them exceptionally suitable for on-site screening essays involving multiple samples. Nevertheless, it is noteworthy that these methodologies typically exhibit limitations regarding accuracy when employed in practical applications [61].

These methods are fundamentally predicated upon singular sensing modalities, rendering them vulnerable to a myriad of testing environments, instruments, and procedural variations. Consequently, detection methods that rely on a single mode are deficient in the reliability and precision requisite for quantitative assessments. In addition, dual-mode sensing approaches encompass the intrinsic attributes of each sensing modality and facilitate the reciprocal validation of measurement outcomes derived from disparate modes. Furthermore, dual-mode detection methods demonstrate extensive applicability due to their dual selectable signal output modalities, which permit adaptable responses to diverse detection scenarios and enhance the reliability of the resultant detection outcomes. For example, a colorimetric and fluorescent dual-mode visual imaging sensor based on H_2O_2-free 2D nanoenzymes has been proposed, which can reliably and accurately detect OPPs in situ while mitigating the incidence of false positives. Nonetheless, the detection sensitivity, stability, and specificity of the dual-mode sensor necessitate further enhancements. Thus, there exists a pressing requirement to develop quantitative OPP detection technologies characterized by elevated reliability and sensitivity capable of satisfying the exigencies of rapid on-site detection [61].

In recent years, DNA hydrogels have attracted considerable scholarly interest as innovative biomaterials due to their remarkable biocompatibility, substantial loading capacity, elevated diffusion rates of small molecules, and notable mechanical stability. Through the incorporation of nucleic acid aptamers as molecular probes, DNA hydrogel biosensors exhibit the capability for target-specific detection. Nucleic acid aptamers are generated via the systematic evolution of ligands by exponential enrichment (SELEX) technology, which provides benefits including heightened sensitivity, extensive in vitro synthesis potential, excellent reproducibility, and significant stability. The superior biocompatibility and loading capacity inherent to DNA hydrogels facilitate the encapsulation of diverse signaling molecules within the sensor, thereby enabling the achievement of multimodal signal output [61].

As a new type of analytical technology, microfluidic chips are characterized by microchannels that exhibit diverse structures and functionalities, facilitating the integration of reaction, separation, and detection processes for samples within a singular chip, thus contributing to the miniaturization and automation of analytical apparatus. This technology presents several advantages, including the utilization of minimal sample and reagent volumes, elevated sample handling capacity, enhanced portability, reduced costs, and accelerated detection rates. Consequently, a microfluidic chip characterized by tailored microchannels and specialized functional regions can serve as a platform for the amalgamation of dual-mode detection methodologies, enabling rapid on-site detection [61].

The digital technology is being so extended that the parallelism between biological communications systems and traditional communication systems seems to be inertial. Thereby, implantable devices have been increasingly used inside human bodies to deal with chronic conditions, such as diabetes or coronary heart diseases. These devices can form an Internet of Implanted Things (IoIT) to perform therapeutic functions or to monitor critical physiological parameters. IoIT is a subset of IoT focused on implantable medical devices that can communicate wirelessly for various health monitoring and therapeutic purposes. These devices include pacemakers, glucose monitors, cochlear implants, and neurostimulators. The role of IoIT is to enhance patient care by providing continuous monitoring, real-time data, and improved medical intervention capabilities [62].

Then, with the continuous development of technologies including human-computer interfaces, artificial intelligence, and microelectromechanical systems (MEMS), the concept of *Digital Humans* has been recognized as one of the emerging technology fields for the next decade, and implantable devices are certainly one of the key enabling factors in achieving this concept. The advancements of medical implants have been transforming healthcare delivery. These devices are designed to replace, support, or enhance a biological function in the human body. They are used to treat a wide range of medical conditions such as cardiovascular diseases, neurological disorders, hearing loss, and diabetes. Medical implants play a crucial role in improving the quality of life for patients, enabling them to perform daily activities with ease, and reducing the healthcare costs associated with prolonged hospitalization [62].

The use of ICT and communication theories of conventional transmission systems and computer networks is essential to understanding the behavior of the different elements of an MCs system [63–73]. However, the practical quantification of communication processes in molecular systems also necessitates the use of computer tools, such as programming languages and simulators, to create relevant abstraction models. While these models may simplify MCs systems, they serve as a foundation for experimentation. Based on numerous investigations in the world of MCs, there are a series of elements for their theoretical analysis, programming tools, and simulation (analyzed in Chaps. 2 and 3), which include [74, 75]:

- MolComML (Molecular Communications Markup Language)
- SBML (Systems Biology Markup Language)
- SBOL (Synthetic Biology Open Language)
- CellML (Cellular Modeling Markup Language)
- NeuroML (Neuroscience Markup Language)
- BiNS (Biological and Nano-Scale Communication Simulator)

- N3Sim (Simulation framework for diffusion-based Molecular Communication nanonetworks)
- CalComSim (Calcium Signaling Molecular Communication Sim)
- COMSOL Multiphysics
- NS-3 (Network Simulator-3)
- BNSim (Bacteria Network Simulator)
- HLA (High-level Architecture)
- MUCIN (MolecUlar CommunicatIoN)
- AcCoRD simulator (Actor-based Communication via Reaction-Diffusion)
- N^4Sim (Nervous NaNoNetwork Simulator)
- Simulator nanoNS3 (A Network Simulator for Bacterial Nanonetworks)

1.3 Protocol Standards for Molecular Communications

Numerous simulation platforms exist for MCs as it was indicated in the preceding paragraph, and biological scenarios emulate different layers of the TCP/IP architecture's communication protocols. Given this context, the IEEE Institute of Electrical and Electronics Engineers has established, via its two standards 1906.1 and 1906.1.1, a theoretical model and simulation platform for nanocommunication systems supported by Network Simulator-3 (NS-3). This simulator will be analyzed in Chap. 2, Sect. 2.3.5. Such a model, inspired by biology, extends the theoretical framework of ICT to the world of communication networks at the nanolevel [72].

In its first version (approved in 2015 [46]), IEEE 1906.1 recommends generic communicational aspects for guided, ad hoc, and molecular networks at the nanoscale and defines a conceptual model that includes [64, 72]:

- Definition of communication nanonetworks
- Common terminology for communication nanonetworks
- Nanoscale definition of communication channels, differentiating them from their macro counterparts
- Abstraction of communication channel interfaces in nanocommunication systems
- Performance metrics for nanonetworks
- Mapping between conventional communication networks and those at the nanolevel, considering, explicitly, encodings, information structures, addressing, routing, localization, interlayer functions of the International Organization of Standardization/Open System Interconnection (ISO/OSI) model, and reliability parameters.

According to [64, 72, 76], the IEEE 1906.1 standard establishes a common abstract model that allows for the theoretical processing of information using a common language, regardless of the scientific discipline analyzing it. As the industry increasingly seeks commercial integration of technology, the standard's abstract structure establishes a recommendable general foundation for the practicality of nanoscale communications and other standards. For instance, the biomedical industry needs standards at the nanolevel to generate diagnostics and convenient treatment methodologies, and the standard's definition of abstraction can adequately address these processes.

According to [64, 72, 77], at the end of 2020, a new IEEE version for a nanocommunications standard is approved and thus appears the 1906.1.1 version, which establishes a YANG (Yet Another Next Generation) information model for the implementation of the conceptual model defined in the 1906.1 standard. In line with its predecessor, 1906.1.1 defines a series of modules that describe nanocommunication systems and the parameters that enable their quantification. The YANG model punctuates the physics behind the nanoscale transmission of its predecessor. For remotely executed nanoprocesses, YANG specifies configuration and management formats and describes the structure of the information, allowing for shared interpretation from a wide range of media and associated technologies [64, 72, 77]:

- Support of the requirements that make up the IEEE 1906.1 standard
- Description of nanoscale systems
- Representation of the fundamental physics generated in IEEE 1906.1 standard processes
- Definition, configuration, and management of simulations in NS-3 and their relevant analysis
- Definition of a self-description of the data structure used in experimental nanocommunications repositories

Such IEEE standards will be explicitly discussed in Chap. 3, Sect. 3.3 [64, 76].

Since there is a large number of programming languages and simulators for nanocommunications systems that are generally not interoperable with each other, as previously indicated, then, as with IEEE standards supported by NS-3, the MolComML emerges with the purpose of presenting an efficiency in the repeatability of results because its operation provides compatibility, reproducibility and exchange of information between different platforms to contrast and validate results obtained in experiments in biological research laboratories [74, 75].

MolComML (which could be considered as the attempt of IEEE standards 1906.1 and 1906.1.1) normalizes MCs schemes, programming languages, simulators, and experimentation tools. In order to facilitate the possibility of researchers to configure experiments, define the same information input/output interfaces, collect, share, and validate results, and rewrite models, ensuring the repeatability of experiments using different simulation tools, the CIRCLE project was created with the participation of organizations from Europe, Asia, and Australia. CIRCLE was organized primarily to promote MolComML worldwide and the benefits derived from its use [74, 75].

MolComML is deployed as an XML-based language that brings together both numerical analysis and experimental synthesis ensuring flexibility. Its design is oriented toward the convergence of all research fields through the integration of platforms and offering interoperability to represent elements, interactions, configuration, results of experiments, and simulations in the field of MCs. MolComML will be analyzed in Chap. 3, Sect. 3.2 [74, 75].

References

1. I.F. Akyildiz, M. Pierobon, S. Balasubramaniam, Moving forward with molecular communication: from theory to human health applications [point of view]. Proc. IEEE **107**(5), 858–865 (2019)
2. A. Zadeh Shirazi et al., The application of deep convolutional neural networks to brain cancer images: a survey. J. Pers. Med. **10**(4), 224 (2020)
3. D. Romanchenko, M. Tartie, B.Q. Le, J.T. Gómez, F. Dressler, Molecular communication channel modelling in FPGA technology. Electron. Commun. EASST **80** (2021)
4. I. Isik, M.B. Er, E. Isik, Analysis and classification of the mobile molecular communication systems with deep learning. J. Ambient. Intell. Humaniz. Comput. **13**(5), 2903–2919 (2022)
5. Z. Cheng, J. Sun, J. Yan, Y. Tu, Optimizations for mobile MIMO relay molecular communication via diffusion with network coding. KSII Trans. Internet Inf. Syst. **16**(4), 1373–1391 (2022)
6. I. Isik, How mobility of transmitter and receiver affects the communication quality. AIP Adv. **12**(2) (2022)
7. Y. Bayraktar et al., Analyzing of Alzheimer's disease based on biomedical and socio-economic approach using molecular communication, artificial neural network, and random forest models. Sustainability **14**(13), 7901 (2022)
8. C. Fonseca, M. T. Barros, A. Odysseos, S. Kidambi, S. Balasubramaniam, Quasi-Spherical Absorbing-Reactive Receiver Model of Glioblastoma Cells for Exosome-based Molecular Communications
9. D. Vimalajeewa, S. Balasubramaniam, D.P. Berry, G. Barry, Virus particle propagation and infectivity along the respiratory tract and a case study for SARS-CoV-2. Sci. Rep. **12**(1), 7666 (2022)
10. K.R. Pilkiewicz, P. Rana, M.L. Mayo, P. Ghosh, Molecular communication and cellular signaling from an information-theory perspective, in *Nanoscale Netw. Commun. Handb.*, (2019), p. 235
11. D. Bi, A. Almpanis, A. Noel, Y. Deng, R. Schober, A survey of molecular communication in cell biology: establishing a new hierarchy for interdisciplinary applications. IEEE Commun. Surv. Tutor. **23**(3), 1494–1545 (2021)
12. N. Dalchau et al., Computing with biological switches and clocks. Nat. Comput. **17**(4), 761–779 (2018)
13. L. Khaloopour, M. Mirmohseni, M. Nasiri-Kenari, Joint sensing, communication, and localization of a silent abnormality using molecular diffusion. IEEE Internet Things J. **10**(6), 4871–4882 (2022)
14. S. Figueiredo, N. Souto, F. Cercas, Low-complexity channel codes for reliable molecular communication via diffusion. Sensors **22**(1), 41 (2021)
15. F. Vakilipoor, F. Ratti, H. Awan, M. Magarini, Low complexity receiver design for time-varying Poisson molecular communication channels with memory. Digit. Signal Process. **124**, 103187 (2022)
16. L. Chouhan, M.-S. Alouini, Rescaled Brownian motion of molecules and devices in three-dimensional multiuser mobile molecular communication systems. IEEE Trans. Wirel. Commun. **21**(12), 10472–10488 (2022)
17. M. Hamidović, S. Angerbauer, D. Bi, Y. Deng, T. Tugcu, W. Haselmayr, Microfluidic systems for molecular communications: a review from theory to practice. IEEE Trans. Mol. Biol. Multi-Scale Commun. **10**, 147 (2024)
18. D. Scazzoli, F. Vakilipoor, M. Magarini, Molecular communication data augmentation and deep learning based detection. Nano Commun. Netw. **40**, 100510 (2024)
19. D.B. Menendez, V.R. Senthivel, M. Isalan, Sender--receiver systems and applying information theory for quantitative synthetic biology. Curr. Opin. Biotechnol. **31**, 101–107 (2015)
20. M. Kuscu, O.B. Akan, Detection in molecular communications with ligand receptors under molecular interference. Digit. Signal Process. **124**, 103186 (2022)

21. J.T. Gómez, K. Pitke, L. Stratmann, F. Dressler, Age of information in molecular communication channels. Digit. Signal Process. **124**, 103108 (2022)
22. S. Sarkar, M.Z. Ali, S. Choubey, Efficacy of information transmission in cellular communication. Phys. Rev. Res. **5**(1), 13092 (2023)
23. S.S. Somathilaka, D.P. Martins, S. Balasubramaniam, Information flow of cascading bacterial molecular communication systems with cooperative amplification, in *ICC 2022-IEEE International Conference on Communications*, (2022), pp. 1728–1733
24. Z. Cheng, J. Yan, J. Sun, Y. Tu, K. Chi, Joint optimizations of relays locations and decision threshold for multi-hop diffusive mobile molecular communication with drift. IEEE Trans. Nanobiosci. **21**(3), 454–465 (2022)
25. S.S. Almutairi, R. Ullah, Q.Z. Ullah, H. Shah, Inhalation configuration detection for COVID-19 patient secluded observing using wearable IoTs platform. KSII Trans. Internet Inf. Sys. (TIIS) **18**(6), 1478–1499 (2024)
26. T. Nakano, M. Moore, A. Enomoto, T. Suda, Molecular communication technology as a biological ICT, in *Biological Functions for Information and Communication Technologies*, (Springer, 2011), pp. 49–86
27. M. Javad Salariseddigh, U. Pereg, H. Boche, C. Deppe, V. Jamali, R. Schober, Deterministic identification for molecular communications over the Poisson channel (2022), Preprint at https://arxiv.org/abs/2203.02784
28. V. Jamali, H.M. Loos, A. Buettner, R. Schober, H.V. Poor, Olfaction-inspired MCs: molecule mixture shift keying and cross-reactive receptor arrays. IEEE Trans. Commun. **71**(4), 1894–1911 (2023)
29. D.C. Ferreira, L.P. Reis, N.V. Lopes, A nanocommunication system for endocrine diseases. Clust. Comput. **20**(1), 689–706 (2017)
30. S. Ahmed, J. Hu, S.M.Z.A. Naqvi, Y. Zhang, L. Linze, A.M. Iderawumi, Molecular communication network and its applications in crop sciences. Planta **255**(6), 128 (2022)
31. P.-B. Bök, D. Micucci, The future of human and animal digital health platforms. J. Reliable Intell. Environ. **10**, 245–256 (2024)
32. P. Kakker, Revolutionizing patient care: the impact of AI on healthcare outcomes. Int. J. Mach. Learn. Res. Cybersecurity Artif. Intell. **15**(1), 238–271 (2024)
33. T.-T. Chu, M.A. Labiod, B. Augustin, A. Mellouk, A self-manageable network infrastructure towards supporting healthcare transport communication. IEEE Trans. Veh. Technol. **74**(3), 4639–4655 (2024)
34. G.A.S. Cassel, R. da Rosa Righi, C.A. da Costa, M.R. Bez, M. Pasin, Toward providing a priority-based vital sign offloading in healthcare with serverless computing and a fog-cloud architecture. Futur. Gener. Comput. Syst. **157**, 51–66 (2024)
35. Q. Yang, A. Al Mamun, M. Wu, F. Naznen, Strengthening health monitoring: Intention and adoption of Internet of Things-enabled wearable healthcare devices. Digit. Heal. **10**, 20552076241279200 (2024)
36. I.A. Khan et al., Big data analytics model using artificial intelligence (AI) and 6G technologies for healthcare. IEEE Access **12**, 97924 (2024)
37. F. Al-Turjman, B.D. Deebak, Privacy-aware energy-efficient framework using the internet of medical things for COVID-19. IEEE Internet Things Mag. **3**(3), 64–68 (2020)
38. A. Ahad, Z. Jiangbina, M. Tahir, I. Shayea, M.A. Sheikh, F. Rasheed, 6G and intelligent healthcare: taxonomy, technologies, open issues and future research directions. Internet of Things **25**, 101068 (2024)
39. M. Shafi, R.K. Jha, S. Jain, M. Gupta, Z. Zahra, Safeguarding the internet of health things: Advancements, challenges, and trust-based solution. Telecommun. Syst. **87**(3), 779–808 (2024)
40. H. Elayan, S. Elmaadawy, A.W. Eckford, R. Adve, J. Jornet, A thermal study of terahertz induced protein interactions. IEEE Trans. Nanobiosci. (2024)
41. L. Kong et al., A survey for possible technologies of micro/nanomachines used for molecular communication within 6G application scenarios. IEEE Internet Things J. **10**, 11240 (2023)

42. M. Jamshidi, D.T. Hoang, D.N. Nguyen, CNN-FL for biotechnology industry empowered by Internet-of-BioNano Things and digital twins. IEEE Internet Things Mag. **7**(5), 54–63 (2024)
43. C. Lee, B.-H. Koo, C.-B. Chae, R. Schober, The Internet of bio-nano things in blood vessels: system design and prototypes. J. Commun. Netw. **25**(2), 222–231 (2023)
44. M. Vasić, D. Soloveichik, S. Khurshid, CRN++: molecular programming language. Nat. Comput. **19**(2), 391–407 (2020)
45. M. Rezaei et al., Spheroidal molecular communication via diffusion: signaling between homogeneous cell aggregates. IEEE Trans. Mol. Biol. Multi-Scale Commun. **10**(1), 197–210 (2024)
46. S. Taluja, J. Bhupal, S.R. Krishnan, A survey paper on DNA-based data storage, in *2020 International Conference on Emerging Trends in Information Technology and Engineering (ic-ETITE)*, (IEEE, 2020), pp. 1–4
47. Y. Hao, Q. Li, C. Fan, F. Wang, Data storage based on DNA. Small Struct. **2**(2), 2000046 (2021)
48. P.Y. De Silva, G.U. Ganegoda, New trends of digital data storage in DNA. Biomed. Res. Int. **2016**, 8072463 (2016)
49. L. Sun, J. He, J. Luo, D.H. Coy, DNA and the digital data storage. Health Sci. J. **13**(3), 1–7 (2019)
50. M. Rezaei, H.K. Rudsari, M. Javan, N. Mokari, E.A. Jorswieck, M. Orooji, Molecular communication transmitter design in limited-capacity storage regime. IEEE Trans. Nanobiosci. **22**(2), 212–222 (2022)
51. S.M. Algfari et al., Real-time health monitoring using IoT devices for patients with chronic conditions. Int. J. Health Sci. (Qassim). **4**(S1), 214–228 (2020)
52. N. Kumar, R. Ali, A smart contract-based 6G-enabled authentication scheme for securing Internet of Nano Medical Things network. Ad Hoc Netw. **163**, 103606 (2024)
53. O. Gulec, Distributed routing and self-balancing algorithm for Medical IoNT. Simul. Model. Pract. Theory **129**, 102833 (2023)
54. A.F.A. Abedi, P. Goh, A. Alkhayyat, Nano-sensors communications and networking for healthcare systems: review and outlooks. J. Comput. Sci. **81**, 102367 (2024)
55. R. Ur Rasool, H.F. Ahmad, W. Rafique, A. Qayyum, J. Qadir, Z. Anwar, Quantum computing for healthcare: a review. Futur. Internet **15**(3), 94 (2023)
56. U. Ullah, B. Garcia-Zapirain, Quantum machine learning revolution in healthcare: a systematic review of emerging perspectives and applications. IEEE Access **12**, 11423 (2024)
57. B. Singh, P.K. Dutta, C. Kaunert, Boosting quantum computing to redefine supply chain in the hospital industry: pitch the vision for medical and healthcare needs, in *Quantum Computing and Supply Chain Management: A New Era of Optimization*, (IGI Global, 2024), pp. 196–219
58. A. Noel, D. Makrakis, A. Hafid, Channel impulse responses in diffusive molecular communication with spherical transmitters (2016), Preprint at https://arxiv.org/abs/1604.04684
59. G. Basso, R. Scherer, M.T. Barros, Embodied biocomputing sequential circuits with data processing and storage for neurons-on-a-chip (2024), Preprint at https://arxiv.org/abs/2408.07628
60. G. Bretti, E. Campanile, M. Menci, R. Natalini, A scenario-based study on hybrid PDE-ODE model for cancer-on-chip experiment, in *Problems in Mathematical Biophysics: A Volume in Memory of Alberto Gandolfi*, (Springer, 2024), pp. 37–57
61. Z. Liu et al., A colorimetric/electrochemical microfluidic biosensor using target-triggered DNA hydrogels for organophosphorus detection. Biosens. Bioelectron. **263**, 116558 (2024)
62. H. Fu et al., Piezoelectric wireless power transfer using Halbach Array for the internet of implanted things. IEEE Internet Things J. (2024)
63. A. Rizwan et al., A review on the role of nano-communication in future healthcare systems: a big data analytics perspective. IEEE Access **6**, 41903–41920 (2018)
64. Y. Cevallos et al., Theoretical basis for gene expression modeling based on the IEEE 1906.1 standard, in *International Conference on Bio-inspired Information and Communication Technologies*, (2021), pp. 145–162
65. Y. Cevallos, L. Molina, A. Santillán, F. De Rango, A. Rushdi, J.B. Alonso, A digital communication analysis of gene expression of proteins in biological systems: a layered network model view. Cogn. Comput. **9**(1), 43–67 (2017)

66. Y. Cevallos et al., *Modelamiento comunicacional de la expresión genética y el transporte de proteínas mediante un sistema de transmisión digital extremo a extremo* (Editorial CEDIA, 2022). https://doi.org/10.48661/CEDIA-TIC02
67. Y. Cevallos, L. Tello-Oquendo, D. Inca, C. Palacios, L. Rentería, Genetic expression in biological systems: a digital communication perspective. Open Bioinform. J. **12**(1) (2019)
68. Y. Cevallos, L. Tello-Oquendo, D. Inca, D. Ghose, A.Z. Shirazi, G.A. Gomez, Health applications based on molecular communications: a brief review, in *2019 IEEE International Conference on E-health Networking, Application & Services (HealthCom)*, (2019), pp. 1–6
69. Y. Cevallos et al., On the efficient digital code representation in DNA-based data storage, in *Proceedings of the 7th ACM International Conference on Nanoscale Computing and Communication*, (2020), pp. 1–7
70. Y. Cevallos et al., A brief review on DNA storage, compression, and digitalization. Nano Commun. Netw. **31**, 100391 (2022)
71. Y. Cevallos et al., Modeling gene expression and protein delivery as an end-to-end digital communication system. Open Bioinform. J. **14**(1), 21 (2021)
72. Y. Cevallos et al., *Comunicaciones Moleculares Un análisis desde el paradigma protocolario de IEEE* (Editorial CEDIA, 2022). https://doi.org/10.48661/CEDIA-TIC03
73. M. Egan et al., Toward interdisciplinary synergies in molecular communications: Perspectives from synthetic biology, nanotechnology, communications engineering and philosophy of science. Life **13**(1), 208 (2023)
74. L. Felicetti, S.S. Assaf, M. Femminella, G. Reali, E. Alarcon, J. Sole-Pareta, The molecular communications markup language (MolComML). Nano Commun. Netw. **16**, 12–25 (2018)
75. E. Alarcon et al., MolComML: The molecular communication markup language, in *Proceedings of the 3rd ACM International Conference on Nanoscale Computing and Communication*, (2016), pp. 1–6
76. IEEE standard data model for nanoscale communication systems, IEEE Std 1906.1.1-2020 (2020), pp. 1–142. https://doi.org/10.1109/IEEESTD.2020.9285373
77. S.F. Bush, Interoperable nanoscale communication [future directions]. IEEE Consum. Electron. Mag. **6**(2), 39–47 (2017)

Chapter 2
XML Languages and Simulators for Analyzing Molecular Communications Systems

2.1 Background Information

Mother nature serves as the foundational milieu for the existence of living organisms, evolving over millions of years through meticulous care and an exquisite balance, such that each component within the natural world appears to be intricately designed to fulfill its essential role in sustaining life. This phenomenon indicates the transmission of information not only to subsequent generations but also to the environment and other organisms for the purpose of survival. Consequently, contemporary scientific inquiry seeks to comprehend these behaviors in order to apply them to human endeavors, thereby enhancing both individual and communal well-being within advanced and technologically sophisticated societies. With this paradigm in mind, we present the underlying relationship between MCs and digital technologies, examining them through significant case studies highlighted in the most recent and relevant academic literature. Following this, we will investigate markup languages and simulators pertinent to the study of MCs.

In this context, throughout history, both philosophers and scientists have endeavored to construct increasingly sophisticated artificial systems by meticulously observing, comprehending, elucidating, and emulating numerous intriguing phenomena present in the natural world. Over hundreds of millions of years, an extensive array of species has emerged within the universe, the inherent complexity of which transcends human understanding. These species frequently engage with one another under various constraints and constitute integral components of natural systems [1].

Consequently, the progression of complexity within the natural world tends to exhibit a distribution that ranges from low to high. To date, both recognized and unrecognized natural phenomena have served as a fertile source of creativity across a multitude of emerging disciplines and technological arenas. Following the advent

Y. Cevallos et al., *Markup Languages and Simulators for Molecular Communications*, Signals and Communication Technology,
https://doi.org/10.1007/978-3-032-33040-6_2

of computational technology, scholars have persistently sought inspiration from nature to design general-purpose machines that either rival or exceed human capabilities. Owing to its inherent parallelism, absence of derivatives, and straightforward scalability, nature-inspired intelligent computing paradigms have demonstrated considerable efficacy in addressing intricate problems and black-box optimization challenges, such as aerodynamic optimization which entails a costly evaluation process. A multitude of investigations has established that these paradigms serve as robust methodologies for cognition and optimization, including the enhancement of complex dynamical systems and the construction of cognitive models. Nature-inspired intelligent computing paradigms emulate various phenomena found in the natural environment to address the diverse practical challenges encountered by humanity, effectively constructing a "bridge" between natural sciences and artificial systems. For instance, the evolved human brain in nature facilitates complex social collaboration among humans. In a similar vein, it is a significant application of the nature-inspired intelligent computing paradigm to employ evolutionary algorithms (EAs) for the optimization of brain-inspired artificial neural networks (ANNs), thereby enabling chaotic neural networks to evolve into generalized neural networks, as illustrated in Fig. 2.1 [1].

Computer simulation represents a highly advantageous approach for the execution of MCs research. Although experimental procedures conducted in wet laboratories are indispensable within this discipline, they often entail substantial financial implications and pose challenges in achieving reproducibility. Furthermore, wet-laboratory experiments possess intrinsic constraints concerning the breadth of

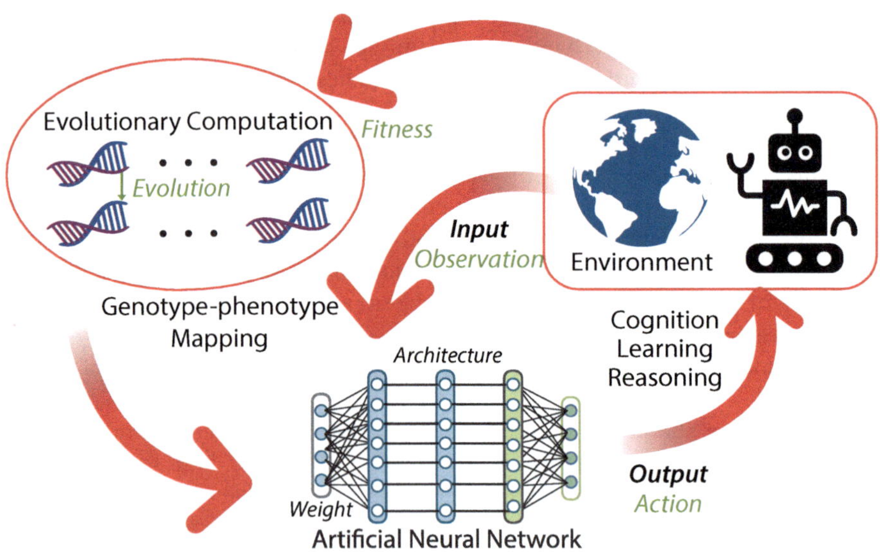

Fig. 2.1 The nature-inspired intelligent computing paradigm provides an approach to evolving the architecture and parameters of ANN, thereby enabling such systems to engage in dynamic interactions with their surroundings for the purposes of cognition, learning, and reasoning

information they can yield. Conversely, simulation experiments, carried out on personal computing devices, significantly diminish financial expenditures, while efficient algorithms can be leveraged to optimize temporal resource allocation. In addition, the settings employed in simulation studies can be reutilized, thereby facilitating the reproducibility of findings. Moreover, simulation experiments provide the distinct benefit of acquiring data with elevated spatiotemporal resolution [2].

Computational modeling assumes an increasingly pivotal function in both biological and medical research domains, in addition to its significance within the medical device sector. Similar to other industrial contexts, the successful advancement of medical devices and implants necessitates not only rigorous empirical testing (including bench tests, animal studies, and human clinical trials) but also comprehensive computational simulations that enable engineers to economically explore system dynamics and refine device designs iteratively. These methodologies are intrinsically multidisciplinary: biomedical engineers must possess a profound comprehension of not only the fundamental physical principles governing their devices but also the intricate interactions these devices have with complex physiological systems across cellular, tissue, and organ levels. Computational models representing such systems are generally characterized by their multiphysics nature. In the foreseeable future, bioengineering modeling is anticipated to be employed routinely within clinical settings to customize an array of individualized therapies and treatment strategies, predicated on models that are specific to each patient [3].

An essential facet concerning ICT pertains to molecular property prediction (MPP), which constitutes a vital undertaking across diverse fields, including pharmaceutical development and material engineering. Conventional methodologies, encompassing first-principle computations and experimental laboratory techniques, necessitate substantial financial investment, thereby hindering the capability for high-throughput evaluation of molecules exhibiting favorable characteristics. Consequently, deep learning methodologies have emerged as a promising avenue for cost reduction and significant enhancement of the screening procedure. A primary challenge encountered by deep learning approaches in MPP is the limited availability of annotated molecular datasets. In order to mitigate this challenge, an array of self-supervised molecular pre-training techniques has been introduced to leverage the inherent information contained within unannotated molecular data. The current pre-training methodologies are predominantly influenced by strategies from computer vision (CV) and natural language processing (NLP) domains, such as contrastive learning and masking techniques. Contrastive approaches aim to amplify the representation concordance of augmentations derived from identical molecules while simultaneously minimizing the representation concordance between disparate molecules; in contrast, masking techniques compel the model to reconstruct the complete molecular entity from a partially obscured molecular string, graph, or structure [4].

Unfortunately, the direct application of existing pre-training tasks commonly utilized in CV and NLP may prove to be inadequate for molecular structures, as they fail to account for the intrinsic chemical properties and physical laws governing molecules. For example, graph-level augmentations employed in contrastive learning, including edge perturbation and subgraph extraction, have the potential to significantly modify molecular attributes, resulting in suboptimal representations for property prediction tasks. Regarding the masking process, the task of recovering a masked atom type can become relatively straightforward when the three-dimensional coordinates are accessible [4].

Consequently, it is imperative to integrate chemical priors and the principles of physics within artificial intelligence (AI) frameworks aimed at scientific discovery, in order to devise appropriate pre-training methodologies for molecular structures and to enhance the generalization and robustness of molecular representations. Recent developments in denoising techniques have unveiled a physically interpretable pre-training task, which corresponds to the acquisition of approximate atomic forces derived from sampled noisy conformations. Within these methodologies, equilibrium conformations are subjected to initial perturbations by noise, after which neural networks are trained to predict the noise based on the perturbed conformation. The type of noise utilized in the prior denoising framework was confined to coordinate Gaussian noise (CGN) characterized by isotropic noise variance, thereby preserving the interpretative validity of force learning [4].

However, the application of isotropic CGN results in a skewed molecular distribution, emphasizing isotropic oscillations about equilibrium states, given that molecules engage in not only minute vibrations but also rotational movements along rotatable single bonds on a comparatively larger scale. The modeling of this skewed molecular distribution precipitates inaccuracies in force targets and constrains the sampling range surrounding equilibriums, thereby ultimately impeding the model's efficacy in subsequent tasks. Consequently, the ensuing challenge resides in the effective modeling of the comprehensive molecular distribution, while concurrently maintaining the fundamental physical interpretation pertinent to force learning [4].

Therefore, molecular informatics, which integrates methodologies from both the discipline of chemistry (chemo-informatics) and the field of biology (bioinformatics), has emerged as a sophisticated instrument and a pivotal element in the contemporary structure-based design of novel chemical entities. Chemo-informatics and bioinformatics synergistically amalgamate technologies to extract and analyze information pertaining to chemical structures, molecular and physicochemical properties, alongside biological activities and their corresponding effects. More specifically, chemo-informatics entails methodologies aimed at constructing databases of chemical structures, complemented by analytical tools designed to interrogate these datasets and formulate predictive modeling frameworks. Consequently, chemo-informatics has become essential in various industrial domains, facilitating the application of chemical knowledge to foster new scientific advancements [5].

Conversely, numerous databases containing chemical structures are interconnected with biological data that have been quantified through a diverse array of assays encompassing omics methodologies, in vitro evaluations, and in vivo reactions. The efficacy of informatics methodologies lies in their capacity to capture, harness, and scrutinize information in order to generate new insights and facilitate decision-making processes informed by that knowledge, subsequently reintegrating the newly acquired insights back into the system beyond the initial cycle. The initial emphasis was on drug development to find and enhance pharmacological efficacy; however, several computational approaches based on molecular informatics have also been utilized in the safety assessment of chemical compounds. These devices capitalize on the benefits of molecular informatics, facilitating the effective use of resources for navigating complex datasets, aiding in inference and conclusion formulation, identifying possible risks, and enhancing safety assessments. Informatics tools are very advantageous in product creation and are also employed in regulatory filings across several sectors [5].

Therefore, the methodology to replicate biological communication systems is through simulation; given that in vivo experiments are often prohibitively expensive and invasive, while in vitro experiments may lack realism, mathematical models and simulations present a cost-efficient in silico alternative to illuminate complexities while minimizing the necessity for experimental procedures [6].

In order to circumvent the necessity of conducting wet experiments within laboratory settings, computational models may be utilized for their inherent predictive capabilities. In this context, it is imperative to acknowledge that, regardless of the potential high costs, extensive time requirements, and ethical implications associated with living organisms, wet experiments can occasionally yield unsatisfactory results due to limitations in *operability* (certain compounds may be unmanageable) and *observability* (the expression of specific compounds may be unobservable). Consequently, models can assist in forecasting the values of certain entities that elude direct observation. Furthermore, they possess the capacity to anticipate the responses of the system under particular conditions, such as when subjected to external influences, including pathogens, pharmaceuticals, and environmental modifications. The anticipated behaviors must be rigorously delineated to facilitate automated reasoning regarding them [7]. Besides, computational methods play a crucial role in synthetic biology, providing powerful tools that significantly improve the design, analysis, and construction of synthetic biological systems, with a particular emphasis on multicellular synthetic systems [8, 9].

In this context, models constitute a fundamental component of the scientific methodology and are employed for various purposes, including research, economic assessment, quality assurance, scaling processes, simulation, as well as design and optimization. Within the realm of systems biology, models can encompass an extensive spectrum of temporal and spatial dimensions, ranging from subcellular molecular frameworks to comprehensive ecosystems. A pivotal function of models is their capacity to forecast forthcoming occurrences. Such models encompass mechanistic frameworks, empirical constructs, and those that incorporate artificial intelligence or machine learning methodologies [10].

Mechanistic models are predicated on well-established scientific laws and theoretical frameworks. When employed appropriately, they can yield profound insights into the functioning of the system. One of the most venerable and arguably most esteemed of these is the Hodgkin and Huxley model of neural conduction, which mathematically introduced multiple conductance terms essential for the establishment of ion channel gradients and the facilitation of action potential propagation. It was only subsequently that these conductance terms were demonstrated to correspond to ion channels, which had not been identified at the time of the model's initial formulation. The parameters of mechanistic models are also derived from the physical characteristics of the system being modeled, which are either approximated from empirical observations or computed utilizing probabilistic methodologies. These models are frequently articulated mathematically, thereby permitting their implementation through computational software to simulate the dynamics of the system. Nevertheless, achieving reproducibility in such studies has proven to be a formidable challenge [10, 11].

To let advances in science as a universal resource and then be accessible for research in the whole world, there must be some very important conditions when digital tools are incorporated, some of these aspects are discussed briefly in the following paragraphs.

Numerous ramifications may emerge when published models lack reproducibility. Most critically, there exists a deterioration of credibility. Should model simulations prove unrepeatable, the dependability of the predictions or insights obtained from such models is likely to be called into question. Discrepancies in results among diverse research groups also necessitate extended periods for resolution, as evidenced by investigations into supercooled water. Moreover, scholarly contributions may be rendered obsolete or inaccessible due to the swift evolution of software, thereby impeding progress for researchers who must, rather than build on prior investigations conducted by others, replicate previous efforts to advance the discipline [10].

The deficiency of reusability becomes progressively untenable as the objectives within the scientific realm escalate in intricacy. It can be posited that the primary purpose of disseminating research findings is to convey one's scholarly contributions to the broader academic community. Consequently, to enhance the accessibility of one's model, it is imperative that the models are systematically archived and disseminated in accordance with prevailing best practices that will render them findable, accessible, interoperable, and reusable (FAIR), i.e., see Fig. 2.2. The FAIR principles endeavor to address three principal obstacles that researchers encounter when striving to comprehend a published biological model [10]:

- Finding
- Understanding the model, its simulation code, and data
- Obtaining the same results as the model's authors when re-simulating the model

In this manner, the establishment of a foundational comprehension of the intrinsic characteristics of elements within biological systems through computational simulation methodologies necessitates expertise from diverse domains, including but not limited to physics, materials science, chemistry, mechanical engineering, mathematics, computer and nanomaterial sciences, as well as telecommunications engineering. The precise modeling of the attributes of a specific system typically encompasses multiple scales and consequently necessitates the integration of methodologies from various disciplines into bespoke simulation workflows. The conventional strategy for formulating ad hoc solutions on an individual case basis demands

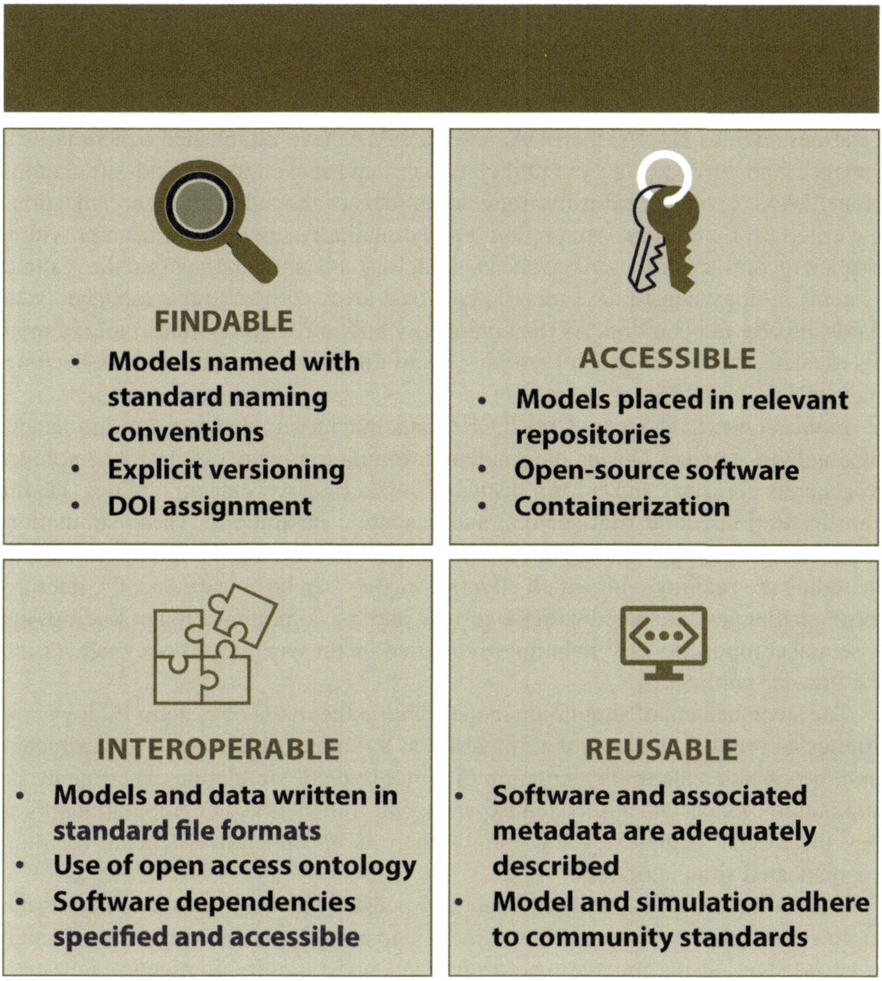

Fig. 2.2 Strategies for making FAIR models

substantial proficiency in scripting, command-line operations, and an extensive understanding of all methodologies and instruments involved in data preparation, inter-module data transfer, module execution, and subsequent analysis. Hence, multiscale simulations utilizing cutting-edge methodologies are impeded by constraints related to scalability, reproducibility, and adaptability [12–14].

Computational models are progressively being employed in high-stakes decision-making processes across the domains of science, engineering, and medicine. The National Aeronautics and Space Administration (NASA) utilizes computational models to conduct intricate experiments that would otherwise be prohibitively costly or necessitate a microgravity setting. In a similar vein, both the Food and Drug Administration (FDA) and the European Medicines Agency (EMA) have initiated the acceptance of models and simulations as a valid form of evidence in the approval processes for pharmaceuticals and medical devices [15].

It is imperative that computational models adhere to a standard of credibility when employed in the context of significant decision-making. Consequently, organizations such as NASA, the FDA, and the EMA have established benchmarks to enhance and evaluate the credibility of computational models and simulations. Nonetheless, given the extensive range of models evaluated by these organizations, the criteria for credibility are predominantly qualitative and refrain from providing explicit recommendations. Conversely, modeling and simulation within the realm of systems biology constitute a more specialized area, with several established standards already in operation. As the complexity and impact of systems biology models escalate, the formulation of a robust credibility assessment framework becomes increasingly vital [15].

In this context, the recent COVID-19 pandemic has emphasized the significance of simulation studies. During the pandemic, simulations were extensively employed to generate forecasts and inform decisions made by governmental entities, healthcare institutions, industrial sectors, and academic institutions. These simulations exposed certain contemporary limitations inherent in the execution of such studies, including the rapidity with which effective models can be constructed, the interpretative challenges associated with the results, and the communication of findings and essential components of simulation studies to domain specialists, policymakers, and the broader public [16].

The advancement of simulation models within the realm of systems biology constitutes a complex endeavor that, in addition to empirical laboratory experimentation, necessitates the implementation of a multitude of simulation experiments. As simulation models and associated experimental frameworks in computational biology increase in complexity and variety, there is an imperative demand for more sophisticated simulation tools [17].

Thus, simulation has emerged as an indispensable element across numerous disciplines. Simulation, empirical assessment, and analytical reasoning are universally acknowledged as the foundational triad of scientific inquiry. Simulation investigations depend on meticulously executing and effectively integrating the phases of system analysis, model development and refinement, execution of varied simulation experiments, and interpretation of the (intermediate) findings. In this framework, beginning with the research inquiry and the system under scrutiny, inputs are selected and altered, assumptions and simplifications are reassessed to determine an appropriate abstraction for the model, requirements concerning outputs are articulated and modified, and potential data sources are identified for use as inputs or to calibrate or validate the simulation model until a satisfactory approximation is attained (Fig. 2.3). Consequently, every phase is both augmented and restricted by the methodologies employed, in conjunction with the knowledge and expertise of the modeler. Besides the modeler who may be assisted by data analysts and programmers, visualization specialists, domain experts, and decision-makers may also engage at various stages of the process [16].

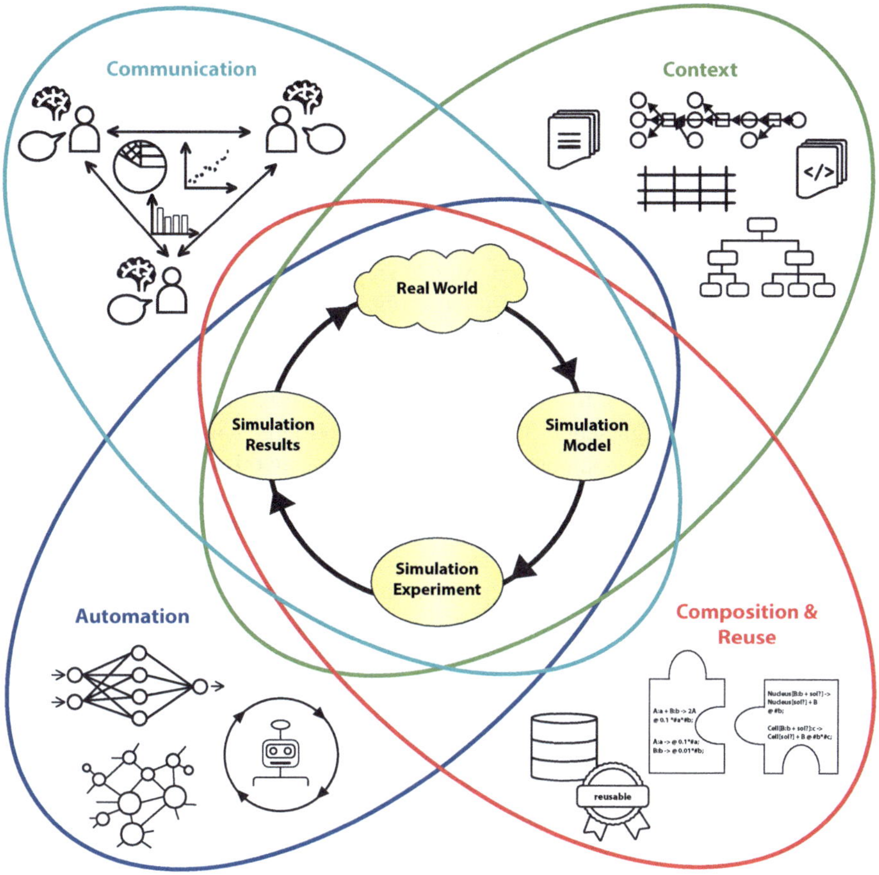

Fig. 2.3 The proposed roadmap advocates for the comprehensive support of the modeling and simulation life cycle through (**a**) the augmentation of contextual information that extends beyond the conceptual framework, which can be utilized for the development of the simulation model, the execution of simulation experiments, and the interpretation of outcomes, (**b**) the provision of mechanisms for the composition and reutilization of various artifacts generated during the simulation study, including but not limited to the simulation model, simulation experiments, and behavioral requirements, alongside the requisite software and methodologies, (**c**) the automation of substantial segments of the modeling and simulation process, leveraging recent advancements in artificial intelligence, and (**d**) the meticulous adjustment of the presentation of results, models, and activities associated with the simulation study to align with the cognitive frameworks and requirements of diverse users and stakeholders [16]

Then, for example, the utilization of modeling and simulation techniques, such as in silico (clinical) trials, significantly facilitates the acceleration of drug research and development, concurrently mitigating associated costs, thus leading to the establishment of the term "model-informed drug development." Approaches that are driven by data and inferential methodologies are increasingly being augmented by the advent of sophisticated physiologically based and knowledge-driven models pertaining to diseases (and pharmaceuticals), albeit with variations in their configuration, limitations, data prerequisites, and applications, which also reflect the distinct scientific communities from which they originated. Concurrently, within the context of model-informed drug development, regulatory bodies and pharmaceutical developers are progressively recognizing in silico trials as a viable instrument for the refinement, reduction, and eventual substitution of traditional clinical trials. Consequently, the barriers that have historically separated various modeling methodologies are beginning to diminish. The broader implementation of in silico trials requires enhanced collaboration among diverse stakeholders as well as the establishment of precedent use cases in critical applications, a process currently hindered by a fragmented array of tools and methodologies [18].

With the substantial advancements and developments in the domain of nanotechnology, the necessity for the establishment of communication protocols among nanodevices, also designated as nanomachines, has emerged organically. A nanodevice is regarded as the fundamental functional unit within this framework. The characterization of a nanodevice encompasses a mechanical or electromechanical apparatus, whose dimensions are quantified in nanometers. Additionally, it may comprise minuscule components that consist of an organized assembly of molecules. Nanodevices are alternatively referred to as nanonodes. The constrained dimensions of nanomachines, along with their limited capabilities, necessitate that the execution of more intricate functions relies on the collaborative efforts of nanomachines within extensive collectives, which is termed nanonetwork. There exist two principal modalities for nanocommunication, namely electromagnetic communications within the terahertz band and MCs [19].

The terahertz band represents a significantly expansive transmission spectrum capable of facilitating exceptionally elevated transmission rates over limited distances. The existence of such an extensive bandwidth engenders substantial alterations in traditional networking protocols. In MCs, cells, along with numerous living organisms, engage in the exchange of information through molecular mechanisms. These entities employ molecules as the means to encode, transmit, and receive information [19].

An instance of the underlying application of digital technologies and molecular analysis is exemplified in [20]. Motivated by the principle of over-the-air computation (AirComp), [20] introduces an innovative computational paradigm for MCs situated within microvascular networks. In this paradigm, the root node functions as the receiver, while the remaining nodes operate as transmitters, utilizing the microvascular network as a conduit for communication to directly acquire the aggregate of the transmitted data, thereby facilitating computational operations in MCs. This framework establishes a significant advancement, providing a foundational structure for the execution of computational activities within MCs. Through simulation experiments, [20] assesses its model in terms of computational error rates, thereby demonstrating its feasibility and effectiveness.

Then, as previously articulated, akin to conventional communication systems, the MCs systems delineated in [20] are characterized as an amalgamated design of source and channel that concurrently enables communication and computational functionalities. In the context of traditional MCs systems, as illustrated in Fig. 2.4a, the receiver is required to sequentially detect the digitally modulated information and subsequently execute the necessary computations. Nevertheless, the constraints imposed by molecular computational capabilities render the integration of computational functions within biological systems considerably arduous [20].

Consequently, [20] suggests an MC system with computational capabilities. This system features several transmitters that release molecules, which can be aggregated within the channel, using the channel as a medium, while the system directly demodulates the aggregated data, thereby integrating computing capabilities into the communication process.

The method in [20] eliminates the necessity for separate reception and computation units, facilitating the execution of computational activities directly within the MCs transmission architecture. Figure 2.4b illustrates the core concepts of the MCs computational model, employing summation as a case study. In this scenario, it is expected that two transmitters and the receiver share mapping databases. At the transmitter end, data is allocated to a certain number of molecules based on a mapping table, with differing quantities signifying distinct bits of information. Molecules released by two transmitters aggregate in the transmission medium and reach the receiver via advection and diffusion. Ultimately, at the receiver's end, post-mapping the aggregate number of received molecules enables the execution of the summing function [20].

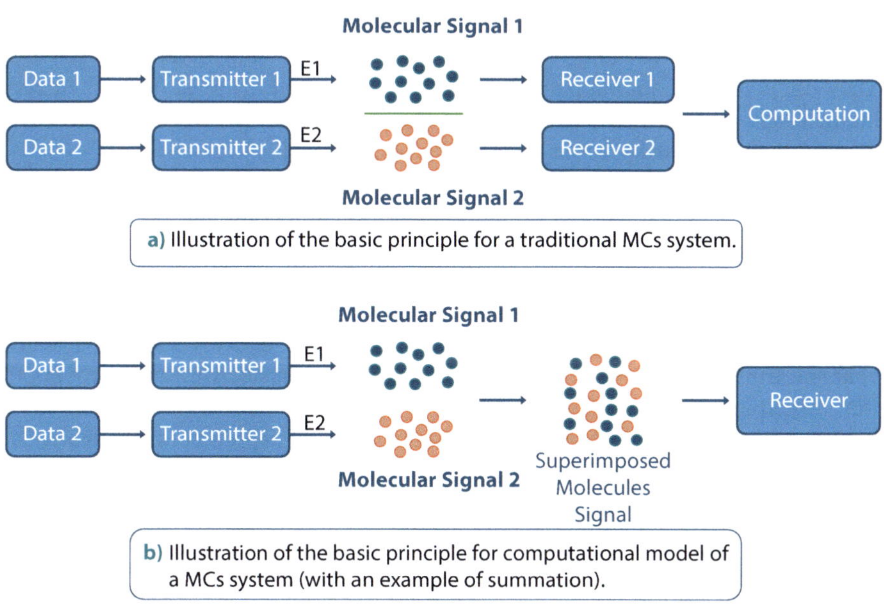

Fig. 2.4 Two paradigms. (**a**) A traditional MCs system. (**b**) An MC systems with computation function

The transmission of information from within the body to the external environment has become a popular and pressing issue that needs to be addressed. Neural communication has been posited as a viable mechanism, employing action potentials (APs), transient fluctuations in membrane potential, as the principal unit of communicative function. Contemporary investigations delineate the membrane potential into two distinct states: an excited state that elicits APs in response to stimulation and a resting state that is devoid of such stimulation. Nonetheless, this premise neglects the intrinsic oscillatory dynamics of membrane potential, which demonstrate a marked sensitivity to particular input frequencies [21].

Hence, the authors in [21] incorporate the oscillations characterized by the Izhikevich model into the channel model. Here, APs are generated when the inter-spike interval closely aligns with, or is a multiple of, the oscillatory period. Following this, an adaptive coding and modulation strategy is introduced. To represent "1," a pair of pulses, separated by one period, is used to stimulate APs. In contrast, to represent "0," two consecutive pulses, distanced by half a period, are employed to inhibit the membrane potential. The transmission efficiency of the "digital MCs systems" is evaluated by bit error rate (BER) and mutual information (MI). Numerical simulation results demonstrate that the proposed neural communication system is biologically plausible and exhibits higher resistance to interference. This research opens up possibilities for connecting Internet of Nano-Things (IoNT) to external networks [21].

Consequently, [21] introduces for the first time the Izhikevich model from the field of computational neuroscience as a communication channel model. This model can accurately simulate the firing pattern of neurons while ensuring high computational efficiency. Based on this channel model, the authors in [21] construct a neural communication system and propose optimized coding and modulation schemes. They use communicational parameters such as BER and MI to evaluate their communications system, demonstrating that this model can exhibit better biological characteristics and show better transmission efficiency [21].

Therefore, in the scenario specified in [22], a framework for a physical system is presented. This system simulates the process of neural communication, consisting of three main components: a transmitter, a neuron channel, and a receiver, as illustrated in Fig. 2.5 The transmitter conveys information by generating electrical stimulus characterized by specific voltage and frequency, thereby initiating nerve stimulation and triggering APs. These APs propagate actively along the neuron channel. Subsequently, the receiver detects the APs on neurons and recovers the transmitted information [22].

Reference [23] proposes a novel ray tracing routing approach for elongated and narrow nanonetworks (such as pipes, blood arteries, and the gastrointestinal tract, where existing routing techniques for nanonetworks are ineffective) that employs a minimal number of forwarders to conserve the limited resources of the nanonodes. This protocol enhances the fundamental version by including four control packets: Request To Forward (RTF), Clear To Forward (CTF), FORGET, and FORGET2. A comprehensive examination of the technique demonstrates that it circumvents the majority of the shortcomings of the basic algorithm: RTF and CTF packets rectify

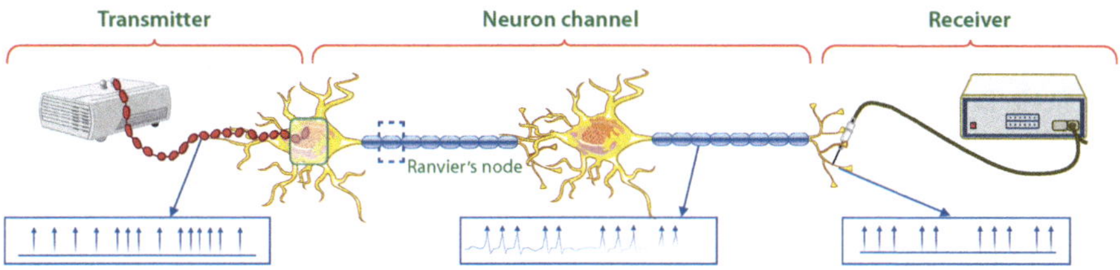

Fig. 2.5 System framework. This system replicates the process of information transmission within a neural communication system, comprising three primary components: a transmitter, a neuronal channel, and a receiver. (The transmitter emits electrical signal carrying information, stimulating nerves to generate APs. These APs actively propagate along the neuronal channel. Subsequently, the receiver identifies the APs present on neurons and reconstructs the transmitted information)

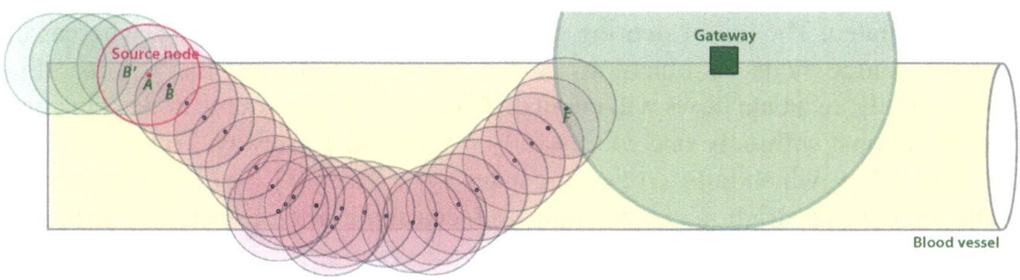

Fig. 2.6 An application of the ray tracing routing in the human cardiovascular system

the double propagation problem, while FORGET and FORGET2 packets facilitate the recovery of the transmission to address die-out scenarios. By restricting the selection of excessively proximate forwarders, the problem of substantial forwarding angles is also resolved. In [23], the authors numerically model the forwarding angle and examine the effects of nanonode distance and transmission range. The enhanced protocol is executed within a nanonetwork in Bit Simulator. The relevance of ray tracing in heterogeneous and three-dimensional nanonetworks is examined and validated through simulations. A practical use of the suggested protocol in the HCS is envisioned as a proof of concept [23].

The proposed ray tracing routing methodology encompasses several attributes that are advantageous in various applications. It is characterized as a coordinate-free protocol and does not necessitate a setup phase, in contrast to coordinate-based protocols, such as sustainable longevity routing (SLR), which demand a preliminary setup phase prior to the commencement of data routing. Coordinate-free protocols are applicable in dynamic environments with mobile nodes, exemplified by the human cardiovascular system, as each node endeavors to transmit its data packet to the intended destination without prior knowledge of the data path. Additionally, ray tracing is classified as a stateless protocol, with the exception of maintaining the identifiers of previously encountered packets within nodes [23].

Consequently, considering the technique illustrated in Fig. 2.6, [23] anticipates an application for ray tracing routing inside the human circulatory system. Node A has detected an event (e.g., a malignant cell) that requires notification to the gateway and subsequently to external entities. Subsequently, node A arbitrarily picks a steering node B and initiates a multi-hop collinearity ray tracing toward one terminus of the blood artery. The diagram illustrates a gateway positioned at one end of the blood artery, equipped with enhanced resources and notably high reception sensitivity, enabling it to intercept packets transmitted by nearby nodes within its strip width [23].

Upon receiving a packet, node F transmits a high-power signal to all nearby nodes, instructing them to cease further propagation, since the gateway has acknowledged receipt of the message. The gateway will transmit the received packet to a macro global wireless protocol, such as Wi-Fi or Bluetooth, to execute the necessary action. Nevertheless, as node A is unaware of the gateway's location and randomly picks steering node B, the propagation may go in the other direction from the gateway. To address this issue, a reverse direction forwarding from node A must also be initiated. The initial steering node B serves as the source node, whereas the primary source node A functions as the steering node for the reverse flow. Consequently, two data flows will emanate from the primary source node A in divergent directions, with only one successfully reaching the gateway. To terminate the alternative flow, which fails to reach the gateway, a Time To Live (TTL)-style maximum hop count restriction can be employed, ceasing it after a sufficient number of retransmissions [23].

As is common, literature regarding MCs establish layered architecture and describe research issues that MCs face at each layer of the architecture with the open research issues that need to be addressed at each layer, adapting communicational theories from typical networking in telecommunications to MCs to expand communications scenarios to technological areas such as nanomedicine [19, 24].

Therefore, nanotechnology offers a novel array of instruments to the engineering discipline for the purpose of designing nanomachines. Nanomachines represent the fundamental operational nanodevices that possess the capability to execute exceedingly simple tasks. An independent nanodevice is significantly limited by constraints related to energy, processing capabilities, and communication range. Furthermore, a nanonetwork constitutes a collective of nanomachines, enhancing the functionalities of an individual nanodevice by facilitating cooperation and the sharing of information. Generally, nanocommunication refers to the transmission of information at the nanoscale, predicated on any form of wired or wireless interconnectivity among nanomachines within a nanonetwork [19].

Furthermore, a hierarchical bottom-up architecture can be employed to elucidate the conventional methodologies by which the performance criteria of nanonetworks have been addressed historically, encompassing the functionalities of the physical layer (including information modulation, coding, and transmission), the data link layer (encompassing medium access control and error management), and the network layer (which involves information forwarding). Nanonetworks exhibit diverse applications across multiple domains, ranging from environmental monitoring, agro-food production, industrial sectors, and military/defense operations to extensive applications in biomedicine and healthcare, such as drug delivery systems, diagnostic procedures, tissue regeneration, and surgical interventions. Within the healthcare arena, nanonetworks can aggregate crucial patient data and relay this information to computational systems, thereby facilitating a more precise and efficient approach to health monitoring. The incorporation of the IoNT within healthcare infrastructures will enhance diagnostic capabilities and assist in patient treatment through precise and localized drug administration, alongside facilitating the tumor detection process [19, 25].

Nanonetworks exhibit enhanced communication and processing capabilities that surpass the constraints of isolated nanomachines through the collaborative efforts of nanodevices. A nanonetwork possesses the ability to transmit data to an external apparatus, such as a smartphone or gateway, thereby facilitating wireless communication between nanodevices and more robust external processing units. This gateway can establish a connection to the Internet. Upon the integration of nanonetworks with an Internet gateway, a novel network paradigm is established, known as the Internet of IoNT. Due to the inherent limitations of computational capabilities, nanodevices are not equipped to manage complex communication protocols, particularly in terms of processing and power management. The operational capabilities of a nanonetwork can be augmented by the resources afforded by supplementary technologies, enabling the organization and optimization of nanodevice behavior through decisions derived from strategies and algorithms executed using high-performance computing (HPC) [19].

Consequently, the implementation of hierarchical architecture is essential for the effective management of nanodevices and the execution of intricate computations externally, thereby facilitating the regulation of an extensive nanonetwork. Furthermore, the adoption of a hierarchical architecture will transfer the processing intricacies from the nanodevices to more robust layers, wherein each layer possesses sophisticated capabilities that alleviate the complexity associated with any protocol, algorithm, or operation deployed throughout the nanonetwork by leveraging cutting-edge networking technologies [19]. Consequently, nanoscale devices possess the capability to engage in interactions with one another, integrate with prevailing communication networks, and, ultimately, connect to the Internet. Consequently, the established Internet of Things (IoT) framework is progressively enhancing its capabilities at the nanoscale, thereby facilitating the emergence of the IoNT, which transcends the already prevalent concept of "smart things" and advances toward the tangible implementation of "smart nano-things" [26].

Thus, contemporary investigations in nanotechnology encompass the identification, modeling, analysis, and organization of communication mechanisms at the nanoscale. Given that the ultimate objective is to enhance the capabilities of nanomachines to facilitate information exchange, it is imperative to examine appropriate communication and networking models at the nanoscale. The field of biomedicine significantly drives the deployment of nanotechnologies to facilitate interactions across different organs and tissues. The proposed uses encompass immune system enhancement against infections, plant biohybrid systems for medication delivery, and health monitoring, among others. To provide a sampling of the simulator's capabilities, for example, [25] offers a simulated model of a lymph node segment.

This simulation illustrates the transmission of information during the humoral immune response mediated by antibody molecules generated by the immune system [25].

In recent years, there has been a marked increase in the scholarly interest surrounding nanotechnologies and nanoscale communication across various application domains. Current advancements in bioresorbable and biocompatible miniature systems present promising opportunities for the creation of innovative nanorobots that function within the human body. It is anticipated that these nanoscale entities will biodegrade in vivo and produce no adverse effects upon the completion of their designated tasks. Furthermore, the integration of biodegradable semiconductor materials and genetically engineered microorganisms, alongside nanorobotic technologies, may yield significant advancements within the medical sector. These inorganic or organic miniature robots have the potential to physically vanish from the human body post-task completion, ensuring that they inflict no harm [27].

Reference [28] details the first modular and programmable device capable of conveying a text message by chemical signaling a technique for MCs. This mode of communication is appealing for situations where traditional wireless systems are ineffective, ranging from nanotechnology to urban health surveillance. This platform is intended for MCs and demonstrates the characteristics of various communication systems through experiments. The aforementioned approach addresses a significant gap in the MCs literature by offering a straightforward and cost-effective method for conducting experiments, contrasting with the prevalent reliance on simulations utilizing simplified system models. A premise for utilizing this platform is that MCs systems are frequently nonlinear, but some simulations and analyses generally presume the system to be linear. Nevertheless, as demonstrated by the authors in [29], reliable communication remains feasible despite the nonlinearity. Moreover, the research in [29] encourages subsequent investigations into more realistic modeling, analysis, and construction of theoretical frameworks and algorithms for complex systems.

This viewpoint presents a study by [27] that has established the analytical framework of Touchable MC (TouchCom) to elucidate the process of direct drug targeting (DDT) utilizing externally controlled nanorobots. The cited research proposed an innovative latticed channel model of TouchCom for an interconnected capillary network adjacent to a targeted tumor region. It delineates a two-dimensional grid for synthesizing the microcirculation environment, utilized to characterize the propagation process of nanorobots. Additionally, by implementing the principles of multiple-input multiple-output (MIMO) systems in wireless communications to the therapeutic window in cancer therapy, an MIMO DDT technique may be utilized in the structured channel to improve targeting efficacy while reducing the detrimental effects of drug toxicity, [27] examines the impact of blood flow direction on the efficacy of DDT and proposes a compensatory technique utilizing an external guiding field to alleviate the mismatch between blood flow direction and tumor site.

The simulator detailed in [28] offers a framework for modeling various forms of nanonetworks. The first phase of its development was identifying the primary characteristics of the communication methods utilized by nanomachines in most settings. The requirements analysis was conducted by considering the MCs posited in some nanonetwork models; however, this methodology does not preclude the simulation of other forms of nanocommunications. The primary characteristics of the package are [28]:

- The possibility for different entities (nanomachines) to send, receive, and process information. Each nanomachine has well-defined physical properties, such as mass, size, shape, motion pattern, and lifetime [28].
- The signal transduction is done by specific structures placed on the outer surface of nanomachines, referred to as receptors. Their size and affinity make them specialized for particular types of received objects [28].
- The information is transferred by carriers, which can be modeled through different attributes, such as shape, size, motion pattern, and lifetime, similarly to nanomachines [28].
- Carriers can be bound to compatible receptors, which may transduce the associated signal and transfer information from/to the interior of a nanomachine [28].

The implementation of the simulator as delineated in [28] necessitates the discernment of environmental characteristics within which the information dissemination occurs, including the attributes that may influence such dissemination. Additionally, this simulator elucidates the overarching principles governing the specific interactions among nanomachines and carriers, carriers with carriers, as well as nanomachines interacting with one another. These interactions facilitate a generalized behavior and do not impede the progression of intricate or customized scenarios [28].

Furthermore, in [28], information is conveyed by carrier absorption by cells. The IL-4 cytokines transfer across a fluid medium from the sender, a basophil, to the recipient, a B lymphocyte cell, traversing many relay nodes, T lymphocyte cells. The surfaces of the receiving and relay nodes contain carrier receptors. Utilizing this model, [28] assessed the efficacy of particular immune system interactions across varying receptor sizes and cell concentrations. The results pertain to the information transmission duration, corresponding channel gain, and probability of successful communication. The simulator provides not only a verification of qualitative expectations but also facilitates a quantitative analysis for design objectives [25, 28]. This simulator evolves to BINS2 that will be analyzed in this chapter, Sect. 2.3.1.

The objective in [29] was to maintain simplicity in the communication protocol and algorithms, enabling researchers from many scientific disciplines to replicate the findings. Furthermore, the sensors and apparatus were selected for their cost-effectiveness and widespread availability. Consequently, their primary contribution was the creation of the platform, which illustrated the viability of MCs at macroscopic sizes. This platform also aimed to inspire future research and connect theory with practice. Following the meticulous selection of requisite materials for this platform, the comprehensive system response was evaluated. Moreover, this platform revealed a linear association between flow velocity and the delay until the system's response peak is attained. The platform also demonstrated a linear association between flow velocity and the ratio of the peak's maximum to the whole width at half maximum [29].

Furthermore, reference [29] demonstrates that laminar flows have a narrower system response, which is advantageous. A significant discovery was the nonlinearity of their platform. This discovery is significant given the majority of contemporary communication theory relies on linear systems [29].

Inspired by the intricacies of the biological realm, novel interdisciplinary fields and technological advancements have emerged across international borders. Pertinent illustrations encompass systems chemistry, which introduces a bottom-up methodology toward the intricacies of chemical complexity, alongside bio/chemical Information and communication technology (bio/chemical ICT), which investigates the requisite conditions for the transmission of signals among discrete microreactors that are delineated by selectively permeable membranes. The fabrication of precise arrangements of microreactors has been evidenced as effectively achieved through the utilization of microfluidics [30].

In particular, droplet-based microfluidics constitutes an advanced methodological framework for the encapsulation of biological entities and chemical reagents within synthetic microenvironments, predominantly characterized by water-in-oil microdroplets. In these configurations, the interfaces are constituted by liquid-liquid interactions, and their physicochemical properties emerge as critical determinants for modulating the coupling between molecular diffusion processes. Simple and double emulsions, wherein aqueous domains achieve equilibrium with oil domains through boundary layers composed of amphiphilic molecules, represent organized assemblies exhibiting elevated interfacial-area-to-volume ratios. These membranes can be meticulously engineered to yield diverse surface charge characteristics, arrangements of single or multilayer stacking, and a variable extent of defects within the molecular packing structure. Emulsions have found extensive applications across various domains, including the food industry, pharmaceuticals, and cosmetics. Moreover, micro- and nanoemulsions serve as effective models for elucidating the propagation of chemical species over considerable distances, which is not only essential for cellular signaling mechanisms but also holds significant implications in the realm of molecular computing [30].

While the theoretical paradigm is suitable for modeling MCs systems under ideal conditions to assess communicational characteristics, simulations provide settings that allow for more realistic circumstances than those addressed by the theoretical formulation [31]. Simulations play a crucial role in MCs, since they effectively connect theoretical models with real investigations. Theoretical models offer valuable insights but can depend on simplifications and assumptions that may inadequately represent the intricacies of real-world situations. Conversely, performing experiments may be laborious and costly, so constraining the variety of situations and parameters that may be investigated. Consequently, simulations present an effective resolution as a synthesis of theory and experimentation [32–38].

In [28], a simulation platform designed for modeling information exchange at nanoscales was presented. This platform exhibits a high degree of adaptability to various forms of nanocarriers, specifically any mechanism employed for the transmission of information, including but not limited to electromagnetic (EM) waves and calcium ions. Furthermore, it encompasses a comprehensive suite of configuration functions designed to accommodate diverse biological environments. The functionalities of the platform are elucidated through the modeling of a segment of a lymph node, alongside the information transfer occurring within it, which transpires between antibody molecules synthesized by the immune system during the humoral immune response [28].

Given that MCs utilize molecules as carriers of information to transmit data between nodes, it is particularly advantageous for applications where conventional wireless communication is nearly infeasible or hazardous. Application domains for MCs encompass medical environments, such as communication within the human body, and industrial settings, such as intricate piping systems in the chemical industry. Therefore, modeling is an essential tool for assessing communication networks at the nanoscale [39]. Established models exist for modeling radio channels; nevertheless, due to the distinct nature of molecular channels, it necessitates a reevaluation of appropriate methodologies. Molecules transmit information within a molecular channel in fluid environments, such as air or water. The mobility of these molecules must be represented, and for accurate modeling of transmitter and receiver activities, they require precise tracking. In this context, both physical and chemical interactions between the molecules and the environment must be considered [39].

It is crucial to establish an optimal balance between accuracy and scalability in the representation of a molecular channel model to develop simulation tools for MCs. The authors in [39] focus on the modeling of molecular channels by these simulators, which are significantly influenced by the geometric configuration of the environment between the transmitter and receiver, where carrier molecules convey information. They compare molecular channels in terms of operation, experiment with their parameters, and analyze their source code [39].

The adaptability of MCs is demonstrated throughout the course of development. It is utilized for both intrabody communications, such as enzymes, hormones, medicine administration, and neurotransmitters, and for out-of-body communications, including pollen, spores, and pheromones. The EM communications inside the neurological system are regulated by ion mobility, which is likewise examined by a subfield of MCs. Consequently, computer simulations have been employed for decades to elucidate the stochastic characteristics of chemical processes [40].

The reaction kinetics of glutamatergic receptors and quantal size are among the several factors that were precisely calculated using simulations with minimal observation. Currently, MCs simulations are employed not only to elucidate intrabody communication mechanisms but also to analyze illness development and evaluate various proposed models of molecular transmitters and receivers. Furthermore, the predominant majority of molecular dispersion models employ the assumption of Brownian motion [40].

Thus, the approach mentioned by [40] is completely justified for heavy molecules with a negligible charge-to-mass ratio and a small polarity. Synaptic MCs are particularly suitable for the use of Brownian motion. Following molecular dispersion, molecular receiver and information-carrying molecules interact. Interactions, specifically the events of binding and unbinding, transpire with a probability that is contingent upon the concentration of the molecules, their availability, and the presence of binding sites on the molecular receptor. When a molecule that conveys information binds to a molecular receptor, it is subsequently extricated from the surrounding medium. Conversely, when an information-bearing molecule unbinds, it ought to be reintroduced into the medium. Nonetheless, this procedure may become excessively laborious if fractional molecules are employed in accordance with binding/unbinding probabilities. Consequently, binding and unbinding events are predominantly represented through Monte Carlo simulations. In other terms, when utilizing Monte Carlo simulations, binding and unbinding events are delineated by a binary characteristic, namely, whether they occur or do not occur at a specific temporal juncture [40].

Also, in [40] inspired by the wave-particle duality of light, it is reconciled the solution of the boundary value problem with the discrete time models for MCs for localized molecular transmitter and molecular receiver. The authors model the molecular dispersion with the wave model and calculates the base concentration in the vicinity of the molecular receiver by solving the boundary value problem. Also, they model the interactions with the particle model of light, which collapses the molecular concentration when a binding/unbinding occurs. To do so, a Markov model is used for the molecular receiver where interaction probabilities depend on the simulation time-step and the kinetic model of the molecular receiver. Then, inspired by the electron-hole propagation in semiconductors, the authors in [40] assume that each binding releases a -1 molecule and each unbinding releases a $+1$ molecule, at the molecular receiver location [40].

2.2 PART I: Extensible Markup Language (XML[1]) for Molecular Communications

Bioinformatics is presently undergoing a significant transformation from a supplementary role for experimentalists to an independent scientific discipline, endowed with extensive applications, innovations, and capabilities, thereby revolutionizing the domain of biological research. The utilization of ontologies in information science for the categorization and characterization of diverse data types through semantic frameworks has provided biologists with an exceptionally powerful instrument to facilitate the representation of the multifaceted knowledge that arises from

[1] XML is a markup language, but without predefined tags to use. It is possible to define own tags designed specifically for programing's need. This is a powerful way to store data in a format that can be stored, searched, and shared. Most importantly, since the fundamental format of XML is standardized, if you share or transmit XML across systems or platforms, either locally or over the Internet, the recipient can still parse the data due to the standardized XML syntax (https://developer.mozilla.org/en-US/docs/Web/XML/Guides/XML_introduction).

their laboratory investigations. Especially XML[2] has demonstrated far greater expressiveness compared to conventional structured database languages, thereby enabling not only the semantic representation of knowledge pertaining to various species but also the elucidation of the relationships and interconnections among them. This advancement has culminated in the development of intricate models that delineate the various levels of biological processes occurring within organisms [41].

In 1997, the International Union of Physiological Sciences initiated the Physiome Project with the objective of addressing pertinent challenges by establishing a comprehensive framework for the modeling of human physiology. As part of this project, the specification of the CellML[3] Markup Language was released in 2001. This language serves the purpose of storing and facilitating the exchange of computational mathematical models. Originating from the domain of cardiac modeling, CellML aspires to encompass a diverse array of biological phenomena, predominantly focusing on cellular functions. Simultaneously, a separate consortium developed Systems Biology Markup Language (SBML[4]) over a series of workshops and was released in 2002. SBML is also an XML-based markup modeling language and was developed to capture biochemical processes at the molecular scale [37]. SBML and CellML will be discussed in Sects. 2.2.1 and 2.2.3, respectively[5].

Alternative notational systems have emerged, each establishing constructs pertinent to the physiological domain they seek to represent; however, the fundamental concept persists: the creation of a notation that is tailored to a specific domain and distinct from general-purpose programming languages or mathematical computation tools, such as MATLAB (Matrix Laboratory); facilitating the capabilities of editing, storing, and simulating these models. For instance, the Pharmacometrics Markup Language[6] (PharmML) has been designed as the exchange medium for pharmacometric models driven in part by the success of SBML; FieldML[7] proposes a standard for modeling the physics of structures and fields in physiology, such as muscle fibers in heart muscle, as well as linking to other scales of model; and NeuroML[8] for modeling biophysical and anatomical properties of the neuron and brain [42]. NeuroML will be discussed in Sect. 2.2.4.

As the quantity and intricacy of models escalate, there arises a heightened necessity for a uniform encoding format to facilitate the searching, comparing, and integrating of these models. Although standards such as SBML and CellML elucidate the mathematical framework of a model, they fail to provide insight into the specific variables and mathematical expressions that they embody. Basic textual

---

[2] http://www.w3.org/XML/

[3] http://www.cellml.org

[4] http://sbml.org

[5] There are many XL for biological systems, we will discuss the most important cited in literature of MCs applications explicitly.

[6] http://www.ddmore.eu

[7] http://physiomeproject.org/software/fieldml

[8] http://www.neuroml.org/

descriptions of these representations are prone to inaccuracies and ambiguities, necessitating text-mining techniques for computational interpretation. Standardized metadata annotations encapsulate the biological significance of a model's components and articulate its simulation, provenance, and layout details for visualization purposes. Such annotations enhance model interoperability, reutilization, comparability, and understanding. In order to circumvent the complexities associated with a multitude of annotation formats and methodologies, the establishment of standard annotation protocols is imperative. The facilitation of standardized protocols is made possible through the implementation of ontology frameworks specific to systems biology [15].

Ontologies delineate a standardized lexicon and a framework of principles that facilitate the unequivocal representation of information by delineating concepts, attributes, and particular instances of those concepts, as well as elucidating the interrelations among these concepts and instances. Beyond mere annotation, ontologies serve a vital role in the formal depiction of simulations, dynamics, and various dimensions of systems biology models. Annotations enhance the interoperability, reusability, comparability, and comprehensibility of models. To mitigate the complexities associated with diverse annotation formats and methodologies, the establishment of standardized annotation protocols is imperative [15].

In this scholarly framework, Minimum Information Requested in the Annotation of Biochemical Models (MIRIAM) was conceived to promote the uniform annotation of computational models by offering comprehensive guidelines for such annotation. The MIRIAM guidelines advocate that metadata associated with models should explicitly cite the documentation of relations (for instance, journal articles), that there exists a significant degree of alignment between the documentation and the encoded model, and that the model should be represented in a format conducive to machine readability (such as SBML or CellML). Furthermore, annotations must encompass the designation of the model, the reference for its corresponding journal publication, the contact details of the creators, the date of creation, in addition to a declaration regarding the terms of distribution [15].

Furthermore, models must possess precise annotations that unequivocally associate model components with the relevant structures found in existing open-access bioinformatics repositories. The cited information should be articulated utilizing a triplet format, comprising data collection, a collection-specific identifier, an optional qualifier, and represented as a Uniform Resource Identifier (URI), which is a distinct sequence of characters that identifies a resource utilized by web technologies. The optional qualifier segment serves to elucidate the relationships among the model constituents and the associated knowledge, employing terminology such as "has a", "is a version of," "is homologous to," and similar expressions [15].

The implementation of efficient physiological markup notations has culminated in MIRIAM, a concerted endeavor aimed at standardizing the curation and annotation methodologies of biological models at the community level. MIRIAM encompasses a comprehensive set of guidelines applicable to any structured format. Consequently, this promotes collaboration among various groups and the exchange of resultant models. Adherence to these guidelines facilitates the dissemination of software and service architectures that are predicated upon modeling activities [42].

In conjunction with community guidelines, it is imperative to incorporate within the model description information that contextualizes the model within a broader framework. This supplementary information is referred to as metadata, which is essentially "data concerning data." Such physiological markup notations necessitate metadata for two principal justifications [42]:

- To enable *reuse*: If a modeler wants to use a model written by someone else, then they need to know about the phenomenon the component describes, such as what biological entity it represents. Where possible these metadata annotations should link with publicly accessible ontologies of such concepts, such as the Gene Ontology,[9] the PROtein ontology,[10] or UNIPROT.[11] The modelers might also need to know when the model was created and from which experimental datasets it was validated with [42].
- To enable *curation*: Metadata provides a means for locating particular models and components. It is also important to document the model, and binding this information with the model itself keeps the metadata from becoming obsolete as the model is refined [42].

Sitting on top of these efforts is the BioModels Database[12]. This database transcends the conventional definition of a mere repository of models. It encompasses a plethora of meticulously curated models, enhanced with semantic metadata and cross-referenced with external data repositories, including scholarly publications. The models, along with their systematically controlled annotations and all pertinent information, are systematically organized within a series of MySQL tables. The database facilitates the capabilities for scientists to search, store, and retrieve mathematical models effectively. It enables the automated processing of CellML and SBML files and incorporates an integrated SBML simulator [42].

Notwithstanding the accomplishments of the markup language initiatives delineated in the preceding section, challenges emerge when endeavors are made to augment the original notations. Extensibility represents a fundamental characteristic intrinsic to systems design. It serves as an indicator of the extent and effort required for a system to incorporate novel functionalities while minimizing disruption to its pre-existing behaviors. Both SBML and CellML encapsulate internal components and models to a certain extent. They employ relatively straightforward methodologies to ensure backward compatibility with earlier models. Intricate models that simulate multifarious processes accomplish this by either distributing the biological concepts across diverse segments of the code or encapsulating multiple concepts within a singular segment of the code. Neither CellML nor SBML furnish a

[9] http://www.geneontology.org/

[10] http://pir.georgetown.edu/pro/

[11] http://www.uniprot.org/

[12] http://www.ebi.ac.uk/biomodels-main/

mechanism that facilitates the direct interconnection of data between modules. Consequently, any concept of cohesion is not directly endorsed. The aggregation of concerns to attain enhanced modularity and encapsulation is ultimately relegated to the discretion of the developer [42].

Standard module features have been added to CellML and these enable the embedding of submodels. Current work is looking at dealing with variation and stochasticity. For solving an ODEs based model, then these features will allow the representation of any model type you like within a CellML framework [42].

To address the need for being able to fuse different tumor models, the authors in [42] developed TumorML; it is an XML-based markup language targeted at the tumor modeling domain. The TumorML XML schema[13] was developed out of the European Commission Transatlantic Tumor Model Repositories project (TUMOR). The schema facilitates the documentation of metadata pertinent to cancer model descriptions in the form of TumorML XML documents. TumorML derives elements from several other XML standards. Models may be categorized as either "simple" or "complex," which is represented in the XML schema by offering a selection of two patterns that may be encapsulated within a model XML element. To summarize, a simple model description permits the delineation of a singular computational cancer model, whereas a complex model articulates a composite model; an integrated entity composed of simple or complex models encapsulated within TumorML [42].

A fundamental model is comprised of two essential descriptors: an input parameter specification and an output parameter specification, articulated through a parameters XML block that includes in and out elements delineating the input and output parameters. The parameters may reference system-level files to facilitate the reading and writing of data by the implemented models. Subsequently, a minimum of one implementation block is employed to elucidate the metadata pertaining to the software implementation of the model [42].

A sophisticated model bears resemblance to a rudimentary model; however, it encompasses numerous model declarations. The Multiscale Modeling Language (MML), along with its XML counterpart, xMML, establishes a normative framework for delineating both multiscale models and the methodology for coupling models that operate at varying simulated scales. The XML syntax utilized for constructing intricate models within TumorML is derived from xMML. This framework enables computational engines to accurately interpret and execute oncological models, encompassing the communication of inputs and outputs to the process, as well as promoting inter-model communication [42].

The quantity of accessible computer models of biological systems is expanding at an accelerating rate. Concurrently, their dimensions and intricacy are also rising. Consequently, the necessity to expand upon current research through the reutilization of models becomes increasingly critical. It is now well acknowledged that one

[13] http://github.com//tumorml/TumorML-schema

must be capable of interchanging the biological and mathematical frameworks of models [43].

The initiatives to standardize the representation of computational models in many biological domains, such as SBML and CellML, have significantly enhanced the sharing and reutilization of models. The description of the model structures is inadequate for reproducing simulation results. It is also necessary to delineate the techniques to which the models are treated, as outlined by the Minimum Information About a Simulation Experiment (MIASE). Therefore, reference [43] discusses the Experiment Description Markup Language (SED-ML[14]), a computer-readable format for encoding simulation experiments. SED-ML files are encoded in XML [43].

At present, contemporary biological investigations are progressively guided by computational simulation experiments, which require the establishment of protocols for the annotation, archiving, dissemination, and replication of the experiments that have been executed [44].

The aforementioned simulations increasingly necessitate comprehensive collaboration among modelers, experimentalists, and engineers. Additionally, the MIASE guidelines delineate the requisite information for the dissemination of simulation experiments. Subsequently, SED-ML, functioning as a computer-readable format for the information specified by MIASE, was developed as a collaborative initiative and enjoys the endorsement of numerous researchers and software tools. Level 1 Version 5 of SED-ML enhances the capacity of modelers to articulate simulations in SED-ML by employing the Kinetic Simulation Algorithm Ontology (KiSAO). Although Version 4 permitted the complete definition of a simulation utilizing KiSAO, Version 5 now empowers users to delineate tasks, alterations to models, parameter ranges, and outputs through the ontology as well. SED-ML is bolstered by an expanding ecosystem of researchers, modeling languages, and software tools, encompassing a variety of languages for constraint-based, kinetic, qualitative, rule-based, and spatial models, along with numerous simulation tools, visual editors, model repositories, and validators [44].

The guidelines are respected by the SED-ML[15], an XML-based format for simulation experiment encoding[16]. A SED-ML document delineates the requisite models for an experimental framework, the requisite transformations that must be applied to the models prior to their utilization, the simulation methodologies mandated for execution on each model, the analytical functions necessary for result interpretation, and the specifications regarding which results should be outputted, culminating in the manner in which the output is to be presented. Each of these delineations is independent of the foundational implementation of the model. The model-agnostic nature implies that one can conduct identical experiments across various implementations that seek to represent the same phenomena [42].

---

[14] https://sed-ml.org/

[15] http://sed-ml.org

[16] We discuss briefly SED-ML because there are no works in literature using this ML for MCs applications explicitly.

As SED-ML constitutes a software-agnostic framework elucidating the methodology for executing simulation experiments, it remains unconfined to any specific simulation environment or toolkit. The endorsement of SED-ML has proliferated within the biological modeling sector, thereby facilitating the interchange of simulation descriptions to enable the replication of identical experiments across various simulation platforms. A salient application of SED-ML is functional curation, predicated on the premise that during the development and refinement of mathematical and computational models, the paramount objective should be the ongoing validation of these models against empirical data. An augmented iteration of SED-ML provides for this close integration, permitting the concurrent curation of both datasets (experimental and simulated); furthermore, as novel competing models of the same biological system are formulated, they can subsequently be assessed directly against pre-existing models by employing identical protocols. So from a behavioral point of view, functional curation enables models to be extended and reused by other members of the community with confidence. This technique has been demonstrated on cardiac electrophysiology cell models [42].

An approach for difference identification in biological system models was reported in [45]. This method pre-processes the model documents, delineates the hierarchical model structures, and post-processes the resultant mappings. The mapping can thereafter be exported in both machine-readable and human-readable forms. The algorithm is executed within a software library. The approach, implemented in the software package BiVeS[17], may be promptly utilized with current model repositories and model management platforms. Furthermore, the author in [45] demonstrates the algorithm's capabilities in their web-based utility BudHat for version management of models encoded in SBML or CellML. The algorithm and BudHat illustrate their contributions to the effective provenance of model-driven research in computational biology. The technique for identifying and conveying discrepancies analyzes two iterations of an XML-encoded model. It delineates the subsequent phases [45]:

- Pre-processing the XML documents.
- Mapping the hierarchical structures and post-processing the resulting mapping.
- Based on the identified mapping, the algorithm computes a delta. This delta can be converted into both machine-readable and human-readable formats.

The algorithm works with top-down and bottom-up propagations and it is available as a web service to facilitate the integration with non-Java applications [45].

In [46], OpenMCSISS[18] was presented, which is a mathematical modeling environment designed to solve field-based equations and link subcellular and tissue-level biophysical processes to organ-level processes. It employs a general-purpose parallel design, in particular distributed memory, for its computations. OpenMCSISS facilitates the assessment and integration of CellML models across diverse spatial

[17] GitHub - binfalse/BiVeS: the BiVeS framework

[18] www.opencmiss.org

and temporal dimensions. In particular, CellML is designed to encode lumped parameter biophysically based systems of ODEs and nonlinear algebraic equations [46].

Given its substantial inherent parallelism, hardware acceleration utilizing Field Programmable Gate Arrays (FPGAs) possesses significant potential to enhance computational efficiency and minimize energy expenditures associated with computations involving CellML models incorporated within OpenMCSISS. Nevertheless, the presence of several hundred CellML models renders the manual implementation of hardware for each individual CellML model to be a complex and protracted endeavor. The benefits associated with FPGA designs will only be actualized if a comprehensive solution or instrument is available to automatically translate CellML models into hardware description languages [46].

Conversely, as the enthusiasm for the advancement of antibody-based therapeutics persists, the biopharmaceutical sector has commenced directing its attention toward intricate multi-specific antibodies as a nascent category of biologics that distinguish themselves from conventional monoclonal antibodies by virtue of their capability to interact with multiple antigenic entities [47].

As techniques to generate such molecules have evolved, so too have their formats and the necessity for standardized notation and nomenclature. Previous efforts to develop a notation language for macromolecule drugs have been insufficient, or too complex, for multi-specific antibodies [47]. In this sense, Antibody Markup Language (AbML[19]) is presented in [47]. It is a new notation language specifically for antibody formats that overcomes the limitations of existing languages and can annotate all current antibody formats, including fusions, fragments, standard antibodies and multi-specific antibodies, as well as all currently conceivable future formats. AbML V1.1 also provides explicit support for T-cell receptor domains. To assist users of this language, [47] has also developed a tool, abYdraw, that can draw antibody schematics from AbML strings or generate an AbML string from a drawn antibody schematic. AbML has the potential to become a standardized notation for describing new multi-specific antibody formats entering clinical trials [47].

Biological Pathway Exchange (BioPAX) constitutes an ontology, which is a formal framework for articulating knowledge that organizes biological pathway data, thereby enhancing its processability by computational software. It delineates biological semantics pertinent to metabolic, signaling, molecular, gene-regulatory, and genetic interaction networks. While SBML and CellML predominantly emphasize quantitative modeling and dynamic simulation, BioPAX is primarily oriented toward quantitative processes and their corresponding visualization. The ontology encompasses a singular superclass, referred to as Entity. Within this Entity superclass, two principal classes are delineated: PhysicalEntity and Interaction. The PhysicalEntity class encapsulates various molecules, including proteins, complexes, and deoxyribonucleic acid (DNA), whereas the Interaction class articulates the reactions and relationships existing between instances of the PhysicalEntity class.

---

[19] bioinf.org.uk - Prof. Andrew C.R. Martin's group at UCL.

Interactions are categorized into either Control or Conversion, both of which are further subdivided into a multitude of more specific subclasses. Similar to SBML, BioPAX is disseminated in a tiered manner, with level 1 encompassing interactions, level 2 facilitating the representation of signaling pathways and molecular interactions, and level 3 permitting the characterization of gene-regulatory networks and genetic interactions [15].

Due to the existence of numerous XML-based languages for the representation of biological system models, the initiative termed COmputational Modeling in BIology Network (COMBINE) has been established to facilitate the coordination of the diverse community standards and formats pertinent to computational models. This initiative aims to ensure that the collaborative projects will yield a comprehensive suite of interoperable and non-redundant standards that encompass all dimensions of biological modeling. Figure 2.7 illustrates an overview of the COMBINE standards and their related initiatives (this figure offers an extensive perspective on the COMBINE standards alongside associated endeavors. Special issues that concentrate on COMBINE standards have been periodically published since 2016, providing updates extending from 2015 to 2023) [48].

Reference [48] presents the latest specifications of all COMBINE standards. This special issue highlights the main updates, namely the releases of the following specifications: CellML 2.0.1, SBML Level 3 Package: Spatial Processes, Version 2 [49], and Synthetic Biology Open Language (SBOL, that is analyzed in Sect. 2.2.2) Version 3.1.0 [50–52]. Special issues on COMBINE standards have been published since 2016, and earlier editions provide updates for the years 2015–2021. In addition, this editorial provides a brief overview of the COMBINE 2022 meeting [48], and in [53] is presented the newest information regarding COMBINE in 2024.

In [48], the core standards are established as indicated in the following paragraphs.

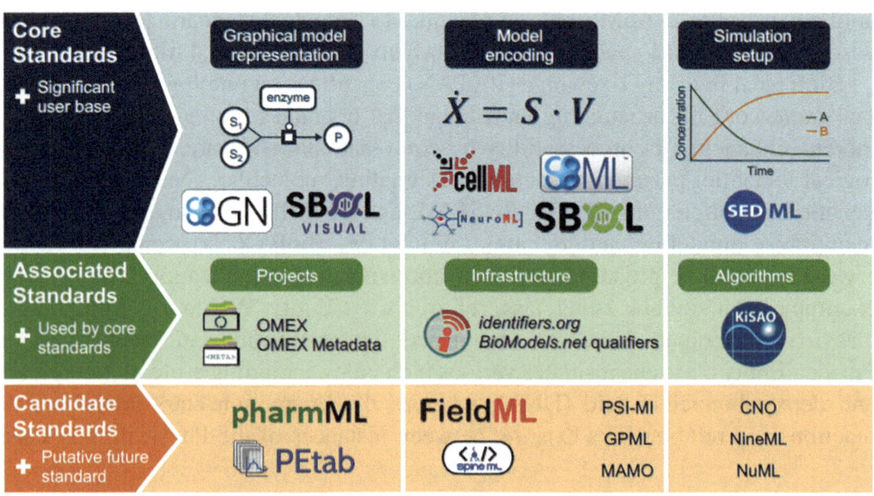

Fig. 2.7 COMBINE standards and associated efforts [48]

- BioPAX is a standard language for integration, exchange, and analysis of biological pathway data. It is expressed in OWL. The current specification is [48]:

Standard and specification
BioPAX

- The CellML language is an XML-based markup language designed for the storage and exchange of computer-based mathematical models. The current specifications are [48, 53]:

Standard	Specification	Version
CellML	CellML	2.0.1
	CellML Metadata Framework	2.0

- NeuroML: The Neural Open Markup Language (NeuroML) is an XML-based descriptive language that offers a standardized data format for developing and distributing representations of neuronal cell and network models. The present standard is [48, 53]:

Standard	Specification	Version
NeuroML	NeuroML	2.3

- Systems Biology Graphical Notation (SBGN[20]): SBGN is a standardized graphical language designed to graphically represent biological information. It comprises three languages that delineate Process Descriptions, Entity Relationships, and Activity Flows. Furthermore, SBGN-ML is an XML-based file format that delineates the geometry of SBGN maps while maintaining their intrinsic biological significance. The present specifications are [48, 53]:

Standard	Specification	Level	Version
SBGN	SBGN Process Description	1	2
	SBGN Entity Relationship	1	2.0
	SBGN Activity Flow	1	1.2
	SBGN Markup Language	–	0.3

- SBML represents an XML-based framework meticulously developed for the computational analysis of models pertaining to biological phenomena. Although it is particularly advantageous for models elucidating processes, the utilization of SBML extends beyond this specific realm. The prevailing specifications are: [48, 53]:

[20] We do not discuss SBGN because there are no works in literature using this ML for MCs applications explicitly.

Standard	Specification	Level	Version	Release
SBML	SBML	3 Core	2	2
	SBML	3 Package: Distributions	1	1
	SBML	3 Package: Flux Balance Constraints	3	1
	SBML	3 Package: Groups	1	1
	SBML	3 Package: Hierarchical Model Composition	1	3
	SBML	3 Package: Layout	1	1
	SBML	3 Package: Multistate, Multicomponent, Species	1	2
	SBML	3 Package: Spatial Processes	1	1
	SBML	3 Package: Qualitative Models	1	1
	SBML	3 Package: Render	1	1

Tools for SBML[21] include Complex Pathway Simulator (COPASI), roadrunner, CySBML,[22] and sbmlutils [53].

- SBOL represents a formalized linguistic framework designed for the precise articulation and dissemination of knowledge pertaining to synthetic biological constituents, apparatuses, and complex systems. SBOL Visual (SBOLv), an associated standard, provides a cohesive array of symbols and protocols for the graphical representation of genetic circuits. The prevailing specifications are delineated as follows [48, 53]:

Standard	Specification	Version
SBOL	SBOL	3.1.0
	SBOL Visual	2.3
	SBOL Visual Version	3.0.0

- The SED-ML constitutes a format predicated on XML, utilized for the elaboration of simulation experiments. It facilitates the articulation of the models to be employed, the experimental procedures to be undertaken, and the resultant outputs to be derived. SED-ML accommodates models encoded in a diverse array of programming languages. The recently developed SED-ML language specification for Level 1 Version 5 augments the functionalities available to modelers for delineating simulations within SED-ML via the KiSAO ontology. Although the specification of simulations utilizing KiSAO was achievable in Version 4, the latest iteration enhances this capability, allowing users to further leverage the ontology for the definition of tasks, modifications to models, parameter ranges, and outputs. The prevailing specification is delineated as follows: [48, 53]:

Standard	Specification	Level	Version
SED-ML	SED-ML Level Version	1	5

[21] A tool and model overview can be found at https://sbml.org/software/

[22] https://sbml4humans.de

2.2.1 Systems Biology Markup Language (SBML)

Computational models serve as invaluable instruments for researchers, facilitating the interpretation of data, elucidation of biological functions, and formulation of quantitative predictions. The SBML constitutes a specific file format designed for the representation of computational models in a declarative manner, thereby enabling interoperability among various software systems. SBML is specifically tailored to depict biological processes prevalent in research across a multitude of domains, encompassing metabolic pathways, cellular signaling pathways, gene regulation, among others. By endorsing SBML as a standard input/output format, diverse tools can function on a uniform representation of a model, thereby mitigating the potential for translation errors and ensuring a consistent foundation for analyses and simulations [54].

Biochemical modeling entails the formulation of a mathematical representation of a biochemical system. In the domains of systems biology and synthetic biology, considerable resources have been dedicated to the advancement of software applications designed for the biochemical modeling of both natural and engineered biological systems. These applications have empowered users to develop, simulate, and conduct comprehensive analyses of biochemical models of biological systems, with implications extending from hypothesis validation to engineering design [55]. Standards are key to the success of systems and synthetic biology since standards give biologists the ability to share models. The leading standard representation of biological systems is SBML [56].

SBML has been extensively employed for the purpose of modeling biological systems. While the SBML has demonstrated efficacy in encapsulating a diverse array of biochemical models, the foundational standard is deficient in its capacity to systematically represent large, complex regular systems, including whole-cell and cellular population models. These models necessitate the incorporation of a substantial number of variables to adequately depict specific elements inherent to these model types, such as the chromosome within the whole-cell model and the multitude of identical cell models present within a cellular population. Although the core SBML framework is not optimally designed to manage these particular models with efficiency, the proposed SBML arrays package offers a more straightforward means of representing such regular structures [56].

SBML constitutes a "free, open, XML-based schema for the representation of biochemical reaction networks." It serves as a standard that is independent of specific software, thereby ensuring interoperability across various tools employed in biological model simulations. This linguistic framework is utilized to formally represent diverse models prevalent in numerous domains of computational biology research [41].

The SBML provides a long-established and rigorously defined methodology for articulating annotated model structures, encompassing an explicit and systematic delineation of each component of a mechanistic model, including species, reactions, volumes, initial concentrations, parameters, rules, events, and equations. SBML constitutes an extensible, machine-readable markup language rather than a mere text file. Furthermore, SBML is compatible with interfaces and libraries in the majority of programming languages, such as Python, C++, and Perl, and can be utilized by a plethora of software applications, including Python, MATLAB, COPASI, Virtual Cell, among approximately 300 additional packages. Nonetheless, the task of authoring thousands of reactions in accordance with SBML standards, whether directly or via accessible Graphical User Interface (GUI-based) software, presents significant challenges. To address this issue, initiatives are underway to translate alternative model formats into SBML, such as Antimony. The Antimony format is articulated in a straightforward text format that is both human-readable and interpretable. Nevertheless, it is imperative that any constructed mechanistic model, irrespective of its adherence to the SBML format, is subjected to simulation within a reasonable timeframe dictated by central processing unit (CPU) constraints [7, 57, 58].

As SBML can be difficult to read and edit without additional tools and, in fact, was never designed for human consumption. Instead, various SBML editors have been developed to help users read and write SBML. In addition, at least two human-readable formats have been devised, SBML-shorthand[23]. A number of tools currently support Antimony include Tellurium and BioUML [59].

A software development kit, libAntimony[24], is provided in the form of a C/C++ library and associated optional Python bindings. The software kit provides a very simple Application Programming Interface (API) that allows the interconversion between Antimony and SBML. In order to make the software development kit available to web browsers there must be translated the C/C++ library into Web-assembly with associated JavaScript bindings [59]. A client-side web application called MakeSBML[25] is developed by [59] that allows a user to load, edit, and translate SBML and Antimony models from within the web browser. Because the application is a purely client-side application, MakeSBML is hosted by a GitHub page which makes the installation robust and relatively maintenance-free [59].

The SBML format describes models as a sequence of transformations affecting the biochemical entities within the network. There exist specified components that signify spatial compartments within the model, distinct biochemical entities, chemical transformations and reactions, as well as parameter values pertinent to these reactions. SBML establishes guidelines to impose constraints on the model and to execute general mathematical operations. The SBML format is not readily comprehensible to humans, necessitating that it be read, authored, and interpreted for the purposes of visualization or analysis through software that adheres to the standard. The Antimony scripting language enables modelers to articulate and comprehend models in a reaction scheme format via a straightforward text-based exposition. Tellurium serves as a modeling instrument that comprehensively supports the functionality of Antimony and facilitates the processes of importing, exporting, and converting SBML between formats that are human-readable and machine-readable.

---

[23] https://github.com/darrenjw/sbml-sh

[24] https://github.com/sys-bio/antimony

[25] https://sys-bio.github.io/makesbml

Alternative structural formats are available should SBML fail to provide the requisite functionality for a particular modeling investigation [60].

The key advantages of using a declarative language like SBML encompass the capacity to facilitate model exchange, promote reuse, enhance comprehension, and enable the re-targeting of models to accommodate alternative or novel simulation methodologies. The foundational development of SBML was markedly influenced by the necessity to enable model interchange among the limited array of modeling tools that were available during that period. The majority of those original tools have since become obsolete and have been supplanted by more contemporary iterations. Nevertheless, the current tools continue to utilize SBML as the standard exchange format, and SBML models that were created over a decade ago remain accessible for reading and simulation in the present day. Provided that tools persist in utilizing SBML, such models are anticipated to remain functional for decades into the future [60].

Models written in SBML format exhibit enhanced reusability due to the capacity of SBML to articulate the biological characteristics of the model, which can be readily scrutinized by the user. In instances where an SBML model is supplemented with additional metadata, it becomes feasible for subsequent users to attain a more comprehensive understanding of the model's composition. Engaging with annotations proves to be significantly more straightforward than attempting to analyze executable source code. Ultimately, a model formulated in SBML format can be repurposed within alternative modeling frameworks, including stochastic simulation or constraint-based modeling. The capacity to repurpose models across diverse frameworks addresses a fundamental challenge in the modeling of biological processes: the varying mathematical methodologies and modeling assumptions exert a considerable influence on the results of a simulation [60].

The SBML format facilitates researchers in the systematic comparison and contrast of various modeling methodologies across diverse modeling platforms. In disciplines such as multicellular modeling, where multiple distinct modeling strategies exist, the feasibility of model comparison is significantly hampered, necessitating extensive effort. Consequently, one cannot ascertain with certainty whether a model developed in one simulation environment will exhibit analogous behavior in another simulation environment that employs a somewhat different methodology [60].

The SBML is systematically organized into a hierarchy of upwardly compatible levels, whereby each successive level incorporates increasingly sophisticated features. The various versions delineate the progression and enhancement of these levels. Most recently, SBML level 3 has introduced modular architecture, which comprises a foundational set of fixed features, referred to as SBML level 3 core, alongside a framework for the integration of additional packages that enhance the core functionalities. This design facilitates extensive customization of the language while simultaneously promoting the reutilization of essential features. At present, there are eight packages that constitute the SBML 3 standard. These packages significantly expand the capabilities of SBML, enabling functionalities such as the representation of uncertainties through distributions, facilitating the encoding, exchange, and annotation of constraint-based models, and generating visual diagrams, among numerous other applications [44, 61].

Recent studies use SBML; for example, [7] analizes mechanistic models of how single cells respond to different perturbations, which can help integrate disparate big datasets or predict response to varied drug combinations. However, the construction and simulation of such models have proved challenging. Hence, in [7] is developed a python-based model creation and simulation pipeline that converts a few structured text files into an SBML standard and is high-performance- and cloud-computing ready. The authors in [7] applied this pipeline to their large-scale, mechanistic pan-cancer signaling model (named SPARCED) and demonstrated it by adding an IFNγ pathway submodel [7].

While numerous biological phenomena can be effectively represented by abstracting the spatial dimensions in which these phenomena transpire, certain modeling approaches (notably at the cellular level) necessitate the explicit incorporation of spatial dimensions, wherein processes occur not within homogeneously mixed compartments, but rather within spatially delineated compartments. The SBML Level 3 Core specification lacks a definitive mechanism for the representation of geometries and spatial processes within a model; however, it does furnish a framework through which SBML packages may enhance the Core specification by introducing supplementary syntactic constructs. The SBML Spatial Processes package for SBML Level 3 incorporates the requisite features that empower models to encapsulate geometries and other spatial attributes concerning the elements and processes they delineate [62, 63].

Moreover, the field of systems biology has witnessed significant expansion in terms of the quantity, scope, and intricacy of computational models. In order to replicate simulation outcomes and facilitate the reutilization of models, it is imperative for researchers to share precise and unequivocal descriptions of the models. The SBML has undergone modifications in its format specifically tailored for this objective. Over the preceding decade, a collective of modelers and software developers has contributed to the advancement of SBML Level 3. Its modular architecture comprises a foundational core that is adept at representing reaction-based models, along with packages that augment the core with functionalities pertinent to various other model types, including constraint-based models, reaction-diffusion models, logical network models, and rule-based models. This format capitalizes on two decades of SBML development and a robust software ecosystem that has fundamentally altered the methodologies employed by systems biologists in constructing and engaging with models. Recently, the emergence of multiscale models of entire cells and organs, alongside novel data sources such as single-cell measurements and live imaging, has prompted innovative approaches to the integration of data with models [64].

SBML's initial inspired by dialogs aimed at establishing a "metabolic model file format." Although SBML was originally conceived for the purpose of facilitating the exchange of compartmental models pertaining to biochemical reaction networks, which are predominantly articulated in the language of chemical kinetics, it has been consistently recognized that a broader spectrum of model types exists beyond the representational capabilities of the initial iteration of SBML. The achievement of SBML in functioning as an interchange format for fundamental categories of models has prompted various modeling communities to inquire whether it could be modified or expanded to accommodate additional types [64].

In conjunction with reaction-diffusion models, alternative modeling frameworks have gained significant traction in the last decade; consequently, researchers have encountered interoperability challenges among software tools designed for their implementation. These requirements catalyzed a substantial transformation in the architecture of SBML: An enhancement allowing the integration of the fundamental SBML framework with additional features tailored to accommodate a broader spectrum of models, along with a mechanism for individual models to specify which sets of extensions are requisite for accurate interpretation. The introduction of SBML Level 3 has established a novel foundation to facilitate the interchange of a more diverse array of models across various biological domains (Fig. 2.8) [64].

The subsequent paragraphs provide a comprehensive overview of the fundamental structure of SBML, followed by an examination of the modularity introduced in Level 3 and the extensive array of modeling formalisms that are supported by Level 3 packages. The foundational architecture of SBML primarily concentrates on the encoding of models wherein entities reside within containers and are influenced by processes that serve to modify, create, or eliminate these entities. It is noteworthy that the containers are not required to align with tangible structures; they may also

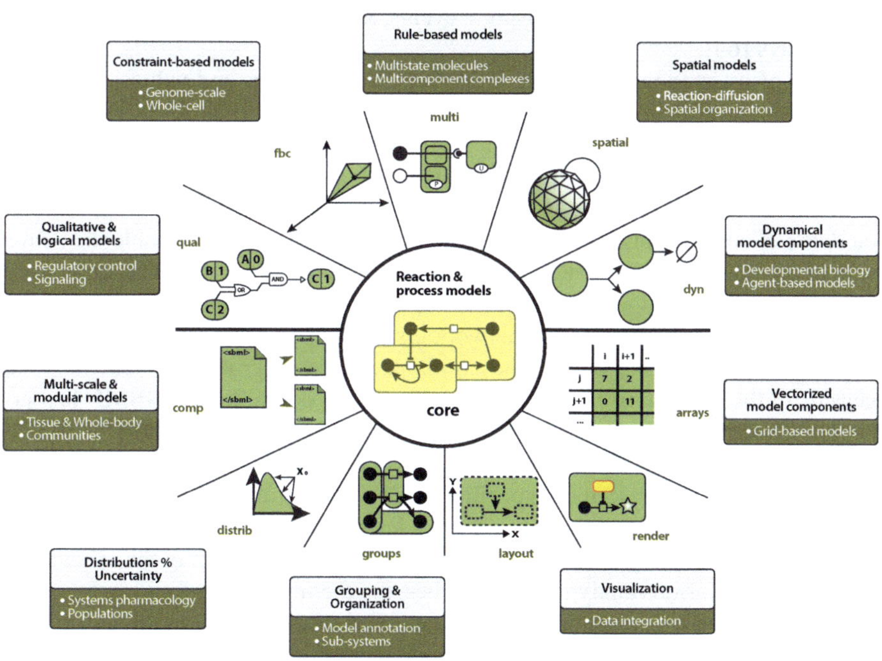

Fig. 2.8 SBML Level 3 consists of a core (center) and specialized SBML Level 3 packages, which provide syntactical constructs to support additional modeling approaches

be conceptual or abstract in nature. Furthermore, supplementary constructs permit the delineation of parameters, initial conditions, additional variables, and various mathematical relationships. In the predominant model type, the "entities" typically represent biochemical substances, the "containers" are characterized as well-mixed and spatially homogeneous, and the "processes" denote the biochemical reactions occurring within or between these containers [64].

This initial categorization resulted in the SBML constructs being designated as species, compartments, and reactions, respectively. Software applications possess the capability to associate these terminologies with alternative concepts to enhance their functionality. For example, the term "species" may be associated with populations of molecules, cellular entities, or even entire organisms [64].

Modelers and software developers are strongly advised to utilize the reaction construct inherent in SBML to delineate the dynamics of a model, as opposed to explicitly representing the model as a system of equations. This methodology affords users the flexibility to transform the model into their preferred final format, which is a more straightforward task compared to, for instance, deducing a reaction network from a set of differential equations. More significantly, this approach adeptly accommodates models in which reaction kinetics are either indeterminate or unnecessary, such as interaction maps, while also facilitating the enhancement of the reaction construct through the utilization of SBML packages. The values of model variables and their temporal variations may be either predetermined or derived from mathematical expressions, applicable prior to or during simulation, continuously or in response to discrete events, with or without the incorporation of time delays [64].

Units of measurement may be delineated for all entities and values; in addition to imparting a critical layer of fundamental physical understanding (after all, how can one accurately ascertain whether a temporal duration is measured in milliseconds or years?), the information regarding units can serve to corroborate the interrelations articulated within a model. Furthermore, units promote the reutilization of models and components, facilitate the interconnection of models, enable the conversion of models across disparate frameworks, and support the integration of data with models. Rule-based modeling possesses the capacity to characterize models that are untenable to convey as reaction networks, such as polymerization, or are merely impractical to delineate due to the combinatorial proliferation of reactions suggested by the governing rules. The storage of rule definitions in SBML has now become practicable through the implementation of the "multi" package, which permits the utilization of rule-based modeling tools [64].

SBML does not prescribe a specific framework that must be employed for the analysis or simulation of a model; in fact, it intentionally omits any explicit mechanisms to delineate the processes associated with a model, including the execution of simulations or various analytical approaches, the methodologies for their execution, or the modalities for presenting the resultant data, as externalizing this information promotes the reusability of models and facilitates independent innovation in distinct yet complementary formats. Among the most widely utilized techniques for time-course simulation are two prevalent methods: the first involves the numerical integration of differential equations derived from the reactions and other interrelations that impact model variables, while the second entails simulating the temporal evolution of the model as a stochastic system through algorithms such as that which was formulated by Gillespie (1977) [64].

Alternative methodologies are concurrently employed, especially when a model is augmented with SBML packages. Ultimately, software applications can also utilize annotations to encode tool-specific information in their proprietary formats, thereby establishing a mechanism to preserve data that may otherwise be neglected. Annotations consequently serve to enhance the semantic richness of model components, promote the comprehension and reutilization of models, and enable software to interact with SBML in a more adaptable manner [64].

The SBML community endeavored to tackle these challenges by instituting specific structural elements within the developmental framework of SBML. The initial component is the concept of Levels. A Level in SBML represents an endeavor to offer a particular array of features for the articulation of models, with elevated Levels furnishing more advanced capabilities. Collectively, these three components (Levels, Versions, and Packages) assist in mitigating the aforementioned challenges: they facilitate the navigation of methodological evolution by consolidating significant modifications into distinct phases (SBML Levels), they aid in managing the unavoidable necessity for amendments (Versions within Levels), and they empower developers to constrain the feature set they choose to implement [64].

The development of SBML has facilitated the application of a singular model representation throughout the entirety of a project's life cycle, even in instances where projects incorporate a diverse array of software tools. Presently, SBML-compatible software applications empower researchers to utilize SBML across all dimensions of a modeling endeavor, encompassing creation (whether manual or automated), annotation, comparison, merging, parameterization, simulation/analysis, result comparison, network motif discovery, system identification, omics data assimilation, visualization, among other activities. The adoption of a standardized format, in conjunction with standardized annotation frameworks and education in reproducible methodologies, enhances research workflows and is broadly acknowledged as fostering research reproducibility [64].

SBML has significantly facilitated the automated processing of models, transforming them into a distinct category of data within the life sciences. Currently, SBML serves as an import/export format utilized by numerous databases dedicated to mathematical models, as well as those focused on pathways and reactions. The adoption of SBML as the preferred format for model curation within the BioModels Database is attributable not only to its widespread acceptance but also to its capacity to accurately encode and annotate models, thereby promoting reproducible modeling practices. Furthermore, SBML is employed for the dissemination of models through more generalized data management platforms, such as SEEK, alongside comprehensive online simulation environments [64].

SBML serves as a formal representation for mathematical models that accurately depict biological phenomena, with a particular emphasis on biochemical reaction networks, gene-regulatory mechanisms, metabolic pathways, and signaling cascades. SBML encapsulates essential data regarding biological processes, which encompasses entities such as species, spatial compartments, reaction events, and various other attributes (including concentrations, volumes, stoichiometry, and kinetic laws) in a uniform and standardized framework. Furthermore, annotations pertinent to the biological context can be incorporated within the SBML structure. Given its endorsement by over two hundred external tools and its proficient capacity for conversion into alternative modeling formats, SBML has established itself as the preeminent language for models in the field of systems biology. Models constructed in SBML consist of various entities, such as species, which are situated within compartments that may be influenced by processes capable of creation, destruction, or modification of these entities. Additionally, supplementary elements facilitate the delineation of parameters, initial conditions, variables, and the mathematical relationships that interconnect these components [44].

In few terms, an SBML document delineates a singular model encapsulated within an SBML "container." The model delineates a system that encompasses zero or more of the six fundamental components, as a chemical reaction can be deconstructed into several conceptual elements: reactant species, product species, reactions, stoichiometries, rate laws, and parameters associated with the rate laws. To undertake an analysis or simulation of a network of reactions, it is imperative to explicitly identify additional components, including compartments for the species and units pertinent to the various quantities. Therefore, the definition of a model in SBML essentially comprises lists of one or more of these diverse components [41, 65]:

- Compartment: A container of finite volume for well-stirred substances where reactions take place [41, 65].
- Species: A chemical substance or entity that takes part in a reaction. Some example species are ions such as calcium ions and molecules such as ATP (adenosine triphosphate) [41, 65].
- Reaction: A statement describing some transformation, transport, or binding process that can change one or more species. Reactions have associated rate laws describing the manner in which they take place [41, 65].
- Parameter: A quantity that has a symbolic name. SBML provides the ability to define parameters that are global to a model, as well as parameters that are local to a single reaction [41, 65].
- Unit definition: A name for a unit used in the expression of quantities in a model. This is a facility for both setting default units and for allowing combinations of units to be given abbreviated names [41, 65].
- Rule: A mathematical expression that is added to the model equations constructed from the set of reactions. Rules can be used to set parameter values, establish constraints between quantities, etc. [41, 65].

Unfortunately, not all modeling initiatives within the domain of systems biology employ the SBML. There exist certain facets of the model that pose challenges when it comes to encoding them within the confines of SBML. For instance, the model extensively utilizes vectors to symbolize regular structures, such as chromosomes. In SBML, it is necessary to define a species for every individual locus within the chromosome. This approach does not exhibit scalability, as it demands the creation of hundreds of thousands of species, each with distinct identifiers and redundant common attributes. The discipline of synthetic biology, along with the Genetic Design Automation (GDA) tools that facilitate advancements in this area, necessitates the efficient modeling of cellular populations. For instance, a synthetic biologist may seek to model a population of cells that incorporate a specific genetic circuit design [56].

Recent work shows the applications sturdiness and flexibility of SBML software [9], including interactions with other programs, such as AMICI (Advanced Model Integration and Computational Intelligence), which facilitates the rapid simulation of models delineated by SBML, as well as PEtab and Datanator, which define data formats pertinent to parameter estimation. Additionally, formalisms are available that contribute to the unambiguous nomenclature of species, alongside compositional methodologies such as SBML merging, which serve to streamline the processes of model aggregation and expansion in manners that are congruent with efficacious large-scale simulation algorithms and promote ease of reuse [66].

In [66], authors focus on improving communication between different modules as a major impediment for computation speed in large-scale modeling (Fig. 2.9). The referenced study employs its SPARCED model as an illustrative case, representing a comprehensive mechanistic framework for the analysis of proliferation and apoptosis signaling in individual mammalian cells. This model is composed of 141 genes and 1196 distinct biochemical entities. It has been developed by converting a straightforward array of structured input text files into a module that adheres to the SBML standards, which encapsulates "protein biochemistry" (signaling) and is simulated utilizing AMICI, alongside a module that encapsulates "gene expression" implemented in Python. The findings indicated that 1.4% was attributed to stochastic gene expression, 15.8% was related to the resolution of ODEs via AMICI, and 81.9% pertained to the storage of solver outputs within a NumPy object to facilitate inter-module communication, with the residual 0.9% accounted for miscellaneous overhead. Thus, [66] focuses on inter-module communication as a rate-limiting step for simulation speed (Fig. 2.9a) [66].

Another example with SBML is presented in [67], for creating a structured approach to study information propagation processes in biological organisms and their relationship with information for their control, optimization, and exploitation. Hence, in [67] first it is modeled the overall functionality of a multicellular metabolic system, the human digestion, in terms of energy production from major nutrients in the food. Second, an examination of metabolic processes within an individual cell is conducted, along with their capacity for adaptation to the availability of nutrient information derived from the surrounding environment. Third, the characterization of the mechanisms that facilitate the transmission of information from the external environment to be assimilated by the cell is undertaken. Quantitative results are presented to offer an initial proof-of-concept characterization of these processes within the framework of communication theory [67].

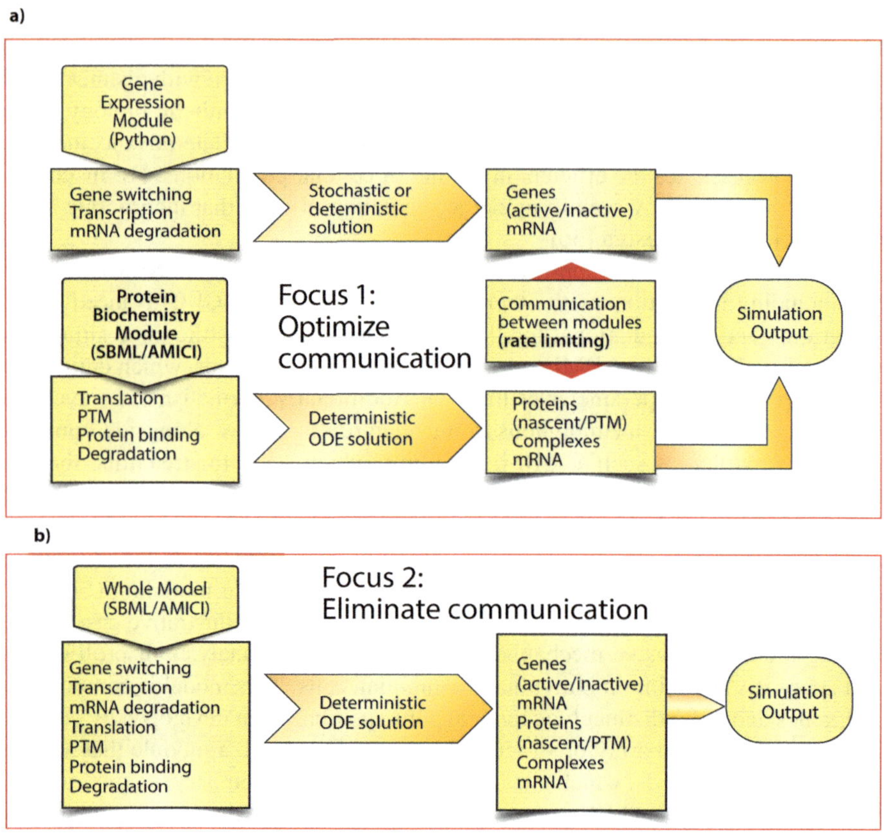

Fig. 2.9 Computational acceleration of the SPARCED model. (**a**) The simulation workflow of the original SPARCED model delineates the critical communication bottleneck between the gene expression module and the protein biochemistry module (One acceleration reported herein specifically addresses that bottleneck to facilitate more rapid stochastic simulations.) (**b**) A novel simulation workflow presented in this study integrates the gene expression module with the protein biochemistry module through the utilization of SBML, thereby allowing for significant computational acceleration in deterministic simulations (Currently, no solvers exist that can handle stochastic simulations at this magnitude.)

Figure 2.10 gives an overview of the system. In the model used in [67], the authors utilize JSBML (Java SBML) that is a Java library to parse SBML files. [67] employs the JAK-STAT model obtained from BioModels website to run ODE simulation [67].

In the presented ODE model, the chemical reactions represented as differential equations are accessible. The concentrations of each species may subsequently be determined by constructing the differential equations that delineate the variations in species concentration. The model sourced from BioModels is provided in the form of an SBML file. This SBML file was integrated into MATLAB, wherein the bespoke rate equations were altered to conform to mass action law equations. Such

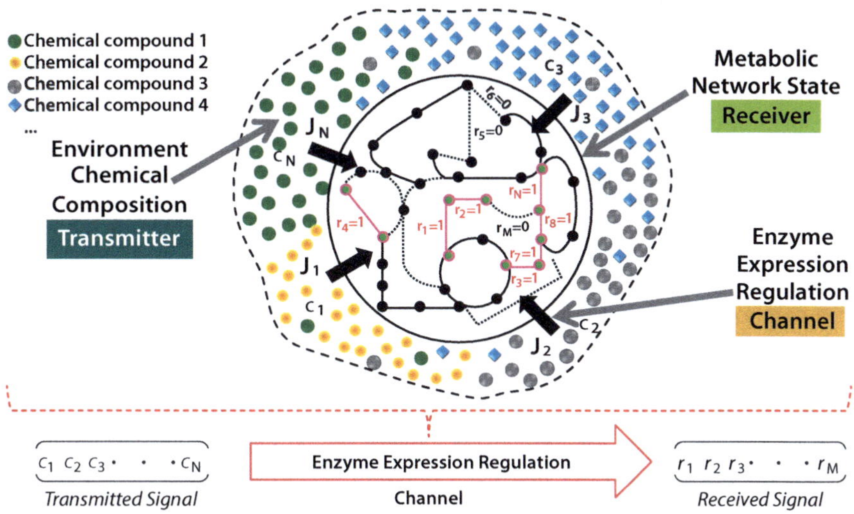

Fig. 2.10 Abstraction of the metabolic communication system, where J are pathway proteins

modifications facilitated the execution of Gillespie simulations within the MATLAB environment. The Janus kinase-signal transducer and activator of transcription (JAK-STAT) pathway constitutes a signal transduction pathway that regulates diverse cellular functions, including cellular proliferation. Alterations within the JAK-STAT pathway significantly contribute to the pathophysiology of diabetes and play a role in oncogenesis. Within the JAK-STAT pathway, receptors establish a complex with external ligands, such as cytokines. Subsequently, Janus Kinase (JAK) forms a complex with the receptor-cytokine assemblage. This resultant complex then recruits signal transducer and activator of transcription (STAT), which functions as a transcription factor that governs gene regulation [67].

Besides, in [67] MATLAB was employed for the execution of Gillespie simulations. Furthermore, COPASI offers an efficient platform for the formulation, modification, and simulation of pathway models. SBML files are capable of being imported to and exported from COPASI, thus facilitating the portability of models. Both COPASI and MATLAB incorporate algorithms specifically designed for conducting Gillespie simulations.

2.2.2 *Synthetic Biology Open Language[26] (SBOL)*

In the domain of synthetic biology, biological cells and associated processes are systematically deconstructed and reconstructed to create innovative systems that serve practical purposes. The designs are encoded within DNA; DNA synthesizes

[26] The Synthetic Biology Open Language (sbolstandard.org).

biological bio-parts; these bio-parts are amalgamated to construct devices; and these devices are integrated into biological systems. Computational systems are employed throughout all phases of the design-build-test-learn cycle, encompassing mathematical modeling and extending to the utilization of robotic systems for the automation of assembly and experimentation [68].

Synthetic biology integrates engineering, concepts such as standardization, modularity, and abstraction, thereby facilitating rapid prototyping and the seamless interchange of designs among synthetic biologists globally. Analogous to construction toys, compatible modular frameworks allow for the straightforward combination and optimization of biological components; specifications for these bio-components are disseminated through open registries. The emergence of synthetic biology is attributable to significant advancements in DNA sequencing and synthesis technologies, alongside insights derived from the discipline of systems biology. The objective of systems biology is to comprehend biological processes across various scales, ranging from molecular and cellular levels to tissues and whole organisms, positing that cells function as intricate information-processing entities. In contrast, synthetic biology endeavors to engineer and construct its own systems. The applications of synthetic biology are diverse, encompassing enhancements in healthcare aimed at improving diagnostic capabilities and developing superior therapeutic interventions for various diseases; it also aspires to ameliorate environmental conditions by devising innovative methods for pollution remediation, promoting sustainability in chemical synthesis industrial processes, and mitigating the detrimental effects of conventional agricultural practices through the development of enhanced fertilizers. The transformative potential of synthetic biology could profoundly influence human lifestyles and contribute to the safeguarding of our planet's future [68].

Thus, synthetic biology is predicated on the foundational disciplines of genetics, molecular biology, and metabolic engineering, utilizing engineering principles to facilitate the design of biological systems. In the process of constructing a synthetic system, practitioners in synthetic biology must engage in the exchange of information pertaining to various molecular types, the anticipated functionality of the system, as well as empirical experimental data. SBOL has been established as a standardized framework to enhance the specification and interchange of biological design information within the field of synthetic biology, developed through a collaborative process that encompasses both experimental scientists and computational modelers and software developers across academic, industrial, and other institutional contexts [69–89]. To facilitate the sharing of designs and data throughout the synthetic biology community, a data standard known as SBOL has been established. This data standard offers a means to express biological designs in silico. The objective is to enhance the efficiency of data sharing among scientists. SBOL data models may be articulated in several computer languages, including C++, Java, or Python [68].

SBOL facilitates a more explicit encapsulation of not only structural data encompassing DNA, ribonucleic acid (RNA), proteins, and other chemical constituents within a given design but also information pertinent to the functional interactions among these elements. The significance of functional information is paramount for the automation of model development, and in the ambit of whole-colony modeling (WCM), it empowers the explicit incorporation of our understanding of the interconnections within biological processes. In the realm of biosafety, the documentation of this information guarantees that any individual evaluating a design can more transparently discern potential issues, as less domain-specific knowledge remains obscured, thereby promoting more comprehensive testing of a model for potentially undesirable phenotypes. Furthermore, the process of testing is augmented by the streamlined exchange of this information [90].

SBOL is a standard to support the specification and exchange of biological design information in synthetic biology, following an open community process involving both "wet" bench scientists and "dry" scientific modelers and software developers across academia, industry, and other institutions. A principal objective driving the advancement of SBOL is the imperative to render the knowledge embedded in the synthetic biology life cycle computationally manageable, thus permitting the automation of associated processes [91].

The inquiry into the manner in which domain-specific knowledge can be systematically deconstructed into a format amenable to computational methodologies is a well-established concept within the field of computer science. The Resource Description Framework (RDF) constitutes a data model that has been formally ratified by the World Wide Web Consortium (W3C) for the purpose of delineating named attributes and their corresponding values, which is already extensively employed within the bioinformatics sector, with several of the most substantial biological datasets, including UniProt and PubChem, disseminating officially sanctioned RDF versions. The SBOL is constructed upon the principles of RDF and is further supported by a rigorously defined ontology, thereby enabling design-related data to be navigable by machines in the capacity of a knowledge graph [91].

Since its inception in 2011, the SBOL has emerged as the preferred format for engineered nucleic acid constructs within the American Chemical Society (ACS) Synthetic Biology domain, receiving endorsement from numerous biological design tools. For example, platforms such as Eugene, GEC, Cello, GenoCAD, ShortBOL, and GeneTech offer computational frameworks that facilitate the exploration of combinatorial design spaces, allowing users to delineate structural, functional, and performance parameters. The outputs produced by these aforementioned tools in SBOL format can subsequently be utilized by various DNA assembly planning software applications, including BOOST, Raven, j5, and DeviceEditor, to streamline the process of physically constructing DNA entities. Additionally, applications such as iBioSim, MoSeC, and SBOLDesigner are compatible with the SBOL data format and contribute to the modeling, analysis, and simulation of biological systems [91].

A variety of data repositories, registries, and databases exist that facilitate the storage and management of data in the SBOL format, including but not limited to SynBioHub, SBOLme, JBEI-ICE, and the Virtual Parts Repository (VPR). Furthermore, the SBOL community has innovatively developed a graphical language aimed at the visualization of biological designs, which has been effectively utilized alongside the data standard in various tools, such as Pigeon, DNAPlotlib,

VisBOL, Constellation, and SBOLCanvas. These tools are instrumental in the visualization of constructs within the realm of computational synthetic biology, encompassing genetic circuits, biochemical components, and potential design spaces governed by structural or functional constraints [91].

Numerous additional instances underscore the efficacy of the SBOL data exchange format in facilitating the connection and integration of data to establish a cohesive computational workflow. For example, Cello has embraced the notion of a User Constraint File (UCF), which is employed in digital logic design, to delineate the library of genetic gates along with the requisite properties and metadata essential for the synthesis of combinational Boolean logic circuits. Complementing this UCF file, the identical library is also accessible in SBOL format, thereby enabling the utilization of Cello's data in alternative tools and workflows, as emphasized in a recent initiative that leveraged the Cello library and VPR API to construct computational models encoded in SBML, which could subsequently be simulated utilizing iBioSim [91].

The first version of SBOL defined a simple data model for the description of engineered DNA components and their sequences. Since then, SBOL has evolved to support the capture of information at many different levels of representation across entire synthetic biology workflows (Fig. 2.11). In particular, the preceding significant amendment, SBOL version 2 (SBOL2), broadened the data model to permit designs to encompass not solely DNA components, but also various molecular entities, such as RNAs, proteins, larger constituents of a system, including entire cells, and connections to models articulated using complementary standards, such as SBML. The standard was further augmented through numerous minor revisions to accurately capture information regarding combinatorial design libraries, external file attachments, sequence construction, experimental evaluations, and measurements [91].

Moreover, by utilizing the Provenance Ontology (PROV-O), SBOL2 is capable of capturing provenance data that facilitates the linkage and tracing of information and processes throughout the comprehensive design-build-test-learn continuum. The progressive enlargement of the SBOL2 scope in recent years has engendered a marked enhancement in the intricacy of the SBOL data model and has illuminated facets of the representation that have constrained prospective advancements. While the SBOL Enhancement Proposals (SEPs) aimed at ameliorating this complexity have garnered acceptance from the academic community, they were deemed excessively substantial for inclusion in a 2.x release, thereby underscoring the necessity for a new major iteration of SBOL [91].

SBOL version 3 (SBOL3) represents a significantly refined standard that seeks to mitigate existing limitations, drawing upon the collective experiences of the SBOL community in implementing SBOL within various scientific and industrial contexts. This novel standard facilitates a more straightforward and sophisticated articulation of the myriad forms of biological design information currently employed, while concurrently diminishing the intricacy of the data model, thereby aiding in the simplification of the development of supporting libraries and the exchange of data with compatible tools. SBOL3 embodies an endeavor to glean

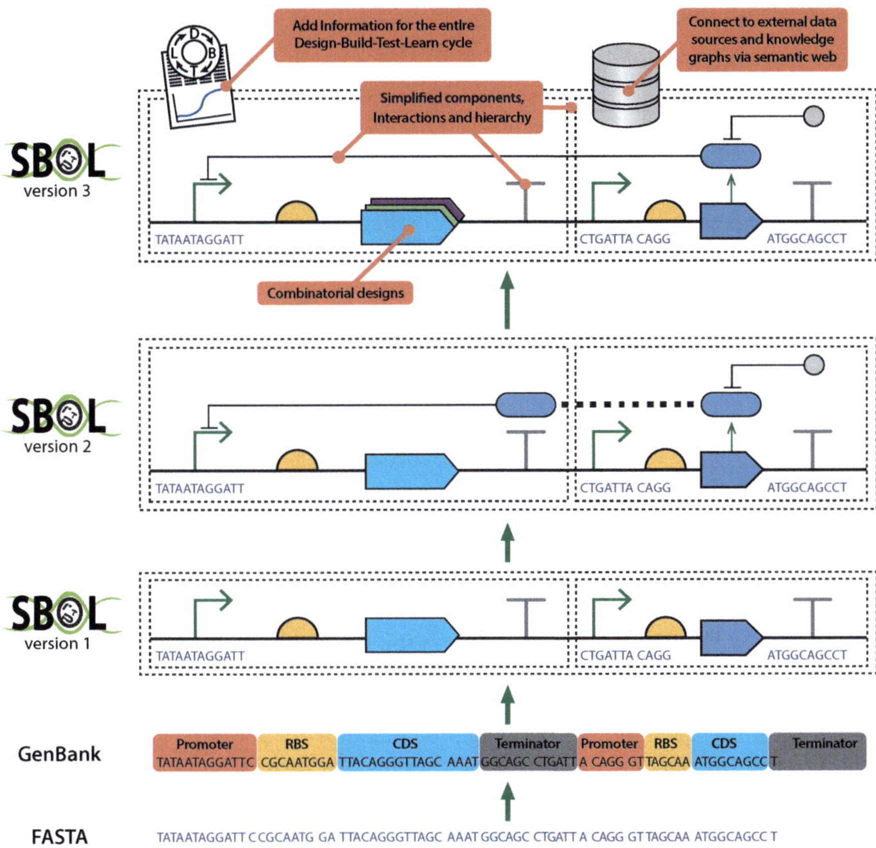

Fig. 2.11 The progression of the SBOL from its antecedent formats, namely Fast-All (FASTA) and GenBank, is noteworthy (FASTA was conceived to encapsulate unadulterated sequence data. Conversely, GenBank enhances this foundational concept by permitting the inclusion of annotations, thus facilitating the capture of certain structural and functional attributes. SBOL1 introduces the capability to employ hierarchical composition in the articulation of a design while also allowing for the partial specification of sequences. The associated SBOL Visual standard provides a means for the unequivocal visual depiction of biological design data (designs from SBOL version 1, 2, and 3 are all represented utilizing SBOL Visual). SBOL2 further expands this framework by incorporating the capacity to delineate modules and the functional interactions among various components. Ultimately, SBOL3 streamlines the data model established in SBOL2 and significantly enhances interoperability with other computational systems through the implementation of a standardized knowledge graph representation.)

insights from the utilization of prior SBOL standards, assess recent advancements and trajectories in the discipline, and establish a robust foundation for enhanced data interchange and computational accessibility within the realm of synthetic biology [91].

SBOL3 encompasses ten principal top-level classes that facilitate the diverse elements of the design-build-test-learn workflow (Fig. 2.12). Specifically, designs may

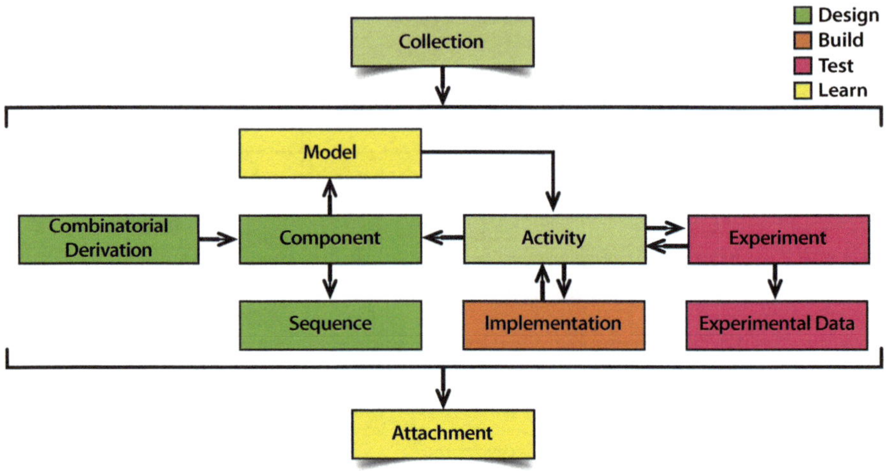

Fig. 2.12 Main top-level classes of SBOL3 and their relationships. The color of each class corresponds to its role in the design (green), build (orange), test (pink), and learn (yellow) stages of the synthetic biology workflow. Additional utility classes are represented by gray boxes

be articulated through the Component, Sequence, and Combinatorial Derivation classes. The Component class is designed to be extensively applicable across various dimensions of biodesign, serving to delineate not only genetic constructs but also other biological entities, including proteins, functional RNAs, strains, multicellular systems, media, and experimental samples. For those Components possessing a well-defined primary structure, such as nucleic acids and proteins, an instance of the Sequence class may be allocated. A Combinatorial Derivation permits the specification of a design pattern wherein individual Subcomponents can be selected from a range of variants. In addition to the design aspect, the Implementation class pertains to the build phase of the synthetic biology life cycle and is utilized to represent tangible entities such as a sample of plasmid, a stab of transformed bacteria, or an aliquot of liquid culture [91].

The Experiment and Experimental Data classes facilitate the testing phase, enabling the integration of data produced during an experimental procedure. The Model class correlates acquired knowledge with a specific design framework. All this data can be interconnected through the Activity class derived from PROV-O. For instance, a design Activity may elucidate the process by which a Component is formulated from a Model description. A build Activity delineates the methodology by which an Implementation is realized according to the specifications of a Component description. A test Activity explicates the manner in which an Experiment is executed utilizing an Implementation artifact. Lastly, a learning Activity may articulate how a Model is revised employing information obtained from an Experiment. The Collection class encompasses members that may consist of any of these types or even other Collections. Ultimately, every one of these objects has the capability to reference instances of the Attachment class, which is employed for establishing connections to external data sources (images, spreadsheets, textual documents, and outputs from experimental instruments, among others) [91].

SBOL 3.0.0 condenses and simplifies previous versions of SBOL based on experiences in deployment across a variety of scientific and industrial settings. In particular, SBOL 3.0.0 has the following characteristics [69]:

- Separates sequence features from part/sub-part relationships.
- Renames Component Definition/Component to Component/Sub-Component.
- Merges Component and Module classes.
- Ensures consistency between data model and ontology terms.
- Extends the means to define and reference subcomponents.
- Refines requirements on object Uniform Resource Identifiers (URIs).
- Enables graph-based serialization, moves for Component types.
- Makes all sequence associations explicit.
- Makes interfaces explicit.
- Generalizes Sequences.
- Constraints into a general structural Constraint class.
- Expands the set of allowed constraints.

As synthetic biology designs can be described using [69]:

- Structural terms, e.g., a set of annotated sequences or information about the chemical makeup of components.
- Functional terms, e.g., the way that components might interact with each other.

In SBOL 3, both the structural and functional aspects are described using a class called *Component*, as depicted in Fig. 2.13. Namely, to represent structural aspects, a *Component* can include *Features*, some of which may be at some *Location* within a *Sequence*. A *Component* can also include *Constraints* between these features. To represent functional aspects, a *Component* can include *Interactions* that can refer to relationships between participating *Features*. Finally, a *Component* can have its behavior described using a *Model* [69].

Consequently, SBOL facilitates the depiction of abstraction hierarchies across diverse dimensions of bioengineering, encompassing individual molecules, multicellular assemblies, and comprehensive synthetic genomes. The SBOL data model accommodates a broad spectrum of significant applications pertinent to synthetic biology and bioengineering, which include visualization, automation of sequence design, dissemination of genetic design information, metabolic engineering, and the derivation of dynamic models from sequence representations [91].

Furthermore, the SBOL can be utilized to encapsulate data regarding the workflows employed in the engineering of biological systems, thereby enhancing the reproducibility and automation associated with these methodologies. As articulated in this manuscript, the SBOL community has leveraged several years of empirical insights from the practical application of SBOL within both scientific and industrial contexts to formulate a specification for SBOL3 that is concurrently more straightforward and more expressive. Enhancements to the standard in SBOL3

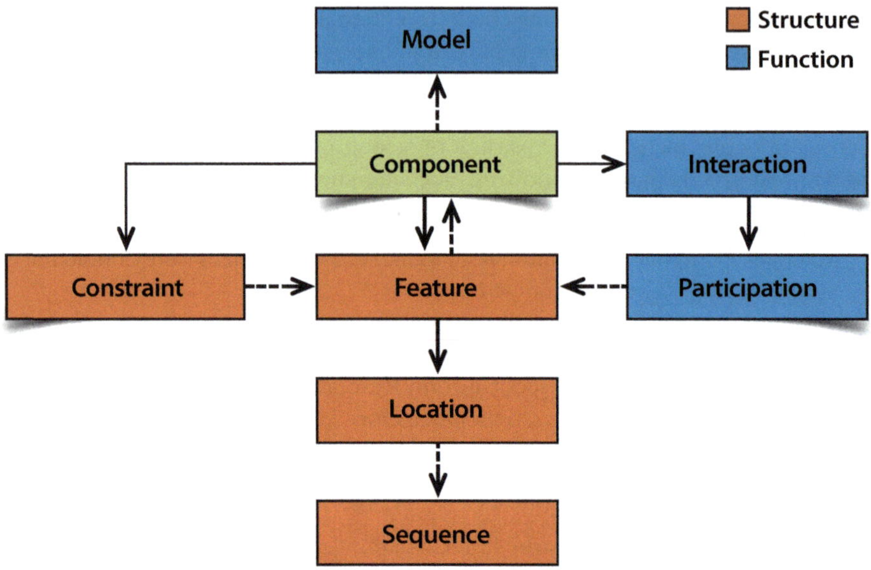

Fig. 2.13 The SBOL *Component* object and related objects (Solid arrows indicate ownership, whereas a dashed arrow represents a reference to an object of another class. Orange boxes represent structural objects, while blue boxes represent functional objects. To represent structural aspects, a Component can include Features, which may refer to Locations within a Sequence. A Component can also include Constraints between these features. To represent functional aspects, a Component can include Interactions that can refer to relationships between participating Features. Finally, a Component can have its behavior described using a Model) [69]

predominantly categorize into two distinct areas: the simplification of the data model or the alignment with best practices in the Semantic Web. Significant simplifications within the data model encompass the integration of structural and functional compositions into a unified component hierarchy; the streamlining of the descriptions of subcomponents and sequence features; as well as the simplification of connections between inputs and outputs across modular interfaces (for instance, transcriptional logic gates). The alternative category of advancements in SBOL3 modifies the standard to more effectively leverage Semantic Web technologies [91].

By incorporating established advancements, this transformation will facilitate the expedited advancement of SBOL tools and libraries while streamlining their upkeep. Furthermore, it will allow SBOL users to more readily assimilate biological knowledge within their tools by utilizing ontologies, which are extensively employed in the life sciences to explicitly characterize biological entities and their interrelations [91].

In conjunction with the enhancement of pre-existing ontologies wherever feasible, such as the Sequence Ontology and Systems Biology Ontology, the SBOL is now articulated as a machine-readable ontology, designated as SBOL-OWL. In a manner analogous to the construction of ontologies upon the RDF layer to establish the semantics of RDF graphs, SBOL-OWL delineates entities of the data model that are employed in the formulation of SBOL graphs. The formal articulation of the data model as an ontology facilitates the potential for utilizing diverse Semantic [91].

SBOL 3.1.0, detailed in [92], enhances version 3.0.0 by incorporating several corrections, clarifications, and other updates and improvements. This version offers a comprehensive array of validation procedures to ascertain the validity of documents, such as SBOL 3. Second, it incorporates updates derived from six community-sanctioned upgrade suggestions. Two improvement requests pertain to the depiction of an object's namespace. The *Namespace* class has been eliminated and substituted with a *namespace* property in each class. Another improvement is the expansion of the *Combinatorial Derivation* class to facilitate the direct utilization of *Features* and *Measures*. The *Participation* class now permits *Interactions* to serve as participants in order to delineate higher-order interactions. A further modification is the application of Sequence Ontology terminology for *Feature orientation*. This iteration of SBOL has transitioned from utilizing URIs to Internationalized Resource Identifiers (IRIs) to accommodate foreign character sets [92].

Web-based tools, such as the utilization of pre-existing reasoners to deduce information, or the validation of SBOL data against a defined schema, are employed. Logical axioms are subsequently implemented to delineate the permissible interactions among various SBOL entities. SBOL-OWL has been incorporated into the SBOL Visual Ontology, which has been established as a machine-readable repository of glyphs. This integration further enhances the capability to search for standardized SBOL glyphs through the use of ontological terminology, while a web service layer permits the retrieval of these glyphs via the Internet. Collectively, these enhancements yield a revised iteration of SBOL that allows for a more direct and sophisticated articulation of an extensive array of bioengineering data, simultaneously minimizing the complexity of classes and regulations to an essential threshold, thereby offering a significantly enhanced mechanism for data interchange. These advancements will consequently promote a more seamless adoption by novice users and expedite the development of software tools and datasets that leverage the standard [91, 93].

Beyond the foundational data, a complementary standard referred to as SBOL Visual additionally provides a methodology for the visualization of biological designs in a delineated manner, thereby facilitating a clearer transmission of information inherent within the SBOL data files. By enhancing the expressiveness of intent in engineered biological designs, it becomes feasible for prospective complications to be more readily identified as a reduced amount of domain-specific knowledge remains with the designer [90].

In this context, for instance, there exists a visual representation of SBOL, namely, VisBOL, which is a web-based application facilitating the rendering of genetic circuit designs, thereby enabling synthetic biologists to effectively communicate designs in the SBOL visual format. Designs created within VisBOL can be exported into various formats, including Portable Network Graphics (PNG) and Scalable Vector Graphics (SVG) images, suitable for incorporation into web pages, presentations, and academic publications. The VisBOL tool supports the automated generation of visual representations from designs articulated using SBOL, as well as a diverse array of established bioinformatics formats, including GenBank and Pigeoncad notation. VisBOL is available both as a publicly accessible website and as an open-source BSD JavaScript library, which can be utilized to integrate diagrams within other content and software applications [94].

Additionally, SBOLCanvas constitutes a web-based platform that facilitates the creation and modification of genetic constructs through the utilization of SBOL data and visual standards. SBOLCanvas enables users to visually and structurally develop a genetic design comprehensively from inception to completion. Furthermore, it permits users to integrate pre-existing SBOL data sourced from a SynBioHub repository. By virtue of its web-based nature, SBOLCanvas is characterized by its accessibility and user-friendliness [95].

Furthermore, people engaged in the engineering of biological organisms frequently find it advantageous to utilize diagrams for the purpose of conveying both the structural characteristics of the nucleic acid sequences they are manipulating and the functional interrelationships among sequence attributes and various molecular entities. A number of standard practices and conventions have started to materialize for the creation of such diagrams. SBOL Visual seeks to organize and systematize these conventions to establish a coherent lexicon for articulating the structure and function of genetic designs [96]. Hence, [96] details version 3.0 of SBOL Visual, a new major revision of the standard. The primary distinction between SBOL Visual 3 and SBOL Visual 2 lies in the fact that the diagrams and glyph representations are articulated according to the SBOL 3 data model, as opposed to the SBOL 2 data model. An ancillary consequence of this modification is the elimination of the utilization of dashed undirected lines for subsystem mappings, which is subject to future deliberation regarding the representation of general SBOL 3 constraints; during this interim period, such annotations may still be employed as a form of annotation. Ultimately, obsolete materials have been excised from the compilation of glyphs: the outdated "insulator" glyph and its "macromolecule" alternative glyphs have been eliminated, along with the deprecated BioPAX alternatives to SBO terms [96].

Diagrams formulated from standardized glyphs are pivotal in the dissemination of intricate design information across numerous engineering disciplines. For instance, circuit diagrams are ubiquitous in the field of electronics and facilitate an appropriate abstraction of the physical system that aids in the design process. With the advent of the Synthetic Biology Open Language Visual (SBOLv), bioengineers are now equipped to more effectively articulate and disseminate their biological designs in a visual format. Nonetheless, the advancement of computational tools to facilitate the generation of these diagrams is presently hindered by an overwhelming maintenance burden resulting from the considerable and ever-increasing number of glyphs encompassed within the standard [97].

Therefore, in [97] is presented a Python package called paraSBOLv that enables access to the full suite of SBOLv glyphs through the use of machine-readable parametric glyph definitions. These significantly streamline the rendering procedure while facilitating extensive personalization of the resultant diagrams. It is illustrated how the implementation of paraSBOLv can expedite the advancement of highly specialized biodesign visualization instruments or even establish the foundation for more intricate software by alleviating the obligation of maintaining glyph-specific rendering algorithms. Looking ahead, it is proposed that the incorporation of machine-readable parametric glyph definitions into the SBOLv standard could further enhance the development of tools aimed at generating standard-compliant diagrams and promote the integration of visual standards across various disciplines [97].

Figure 2.14 establishes an example of SBOLv. The standard glyphs (or symbols) for transcription of a bacterial gene, described using SBOLv, are shown in Fig. 2.14. As in circuit diagrams used in electronic engineering, here elements of the genetic machinery have been simplified and abstracted [68].

Another illustration of SBOL Visual pertains to synthetic regulatory circuits, which constitute bio-parts engineered to execute logical functions, analogous to the operational principles of electronic circuits. In the domain of "Biological Systems and Systems Biology," logic gates are employed in cellular signal processing to conceptualize and construct a genetic toggle switch utilizing two repressor motifs (Fig. 2.15) [68].

Each repressor motif has its own promoter, which is a DNA sequence upstream of the gene to which the enzyme RNA polymerase binds to initiate gene transcription. In this synthetic circuit, a repressor (R1) produced by the first repressor motif (*r1*) blocks transcription from a promoter (P1), which would otherwise drive expression of a second repressor motif (*r2*). Likewise, repressor R2 blocks production of R1 by binding to P2. Expression of GFP downstream of *r1* enabled the state of the system to be ascertained at any one time—if *r1* was active and *r2* repressed, GFP was synthesized and the cells fluoresced green; if *r2* was active and *r1* repressed, the cells would not fluoresce. As anticipated from the computational modeling, the system demonstrated bistability: activation was observed in either one promoter or the other, but not concurrently, and the cells were incapable of transitioning between the two states in a stochastic manner. Rather, the circuit necessitated an external perturbation to facilitate this transition and to allow for a shift to the alternate stable configuration, exemplified by the introduction of a chemical agent into the cellular environment or a variation in temperature exceeding a critical threshold [68].

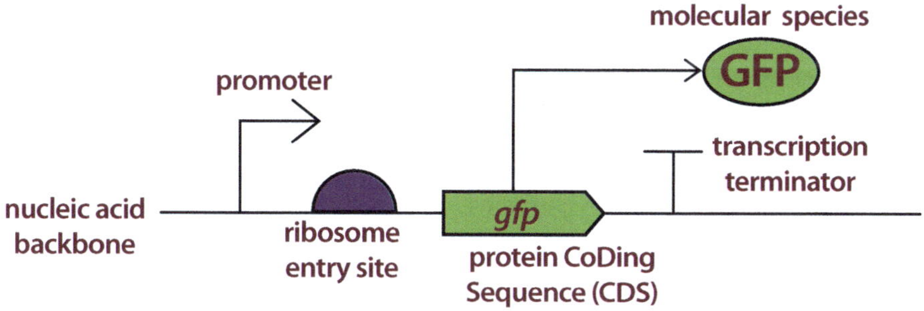

Fig. 2.14 SBOL Visual. It is a standard description of a bacterial protein-coding gene made using SBOL Visual where GPF stands for Green fluorescent protein

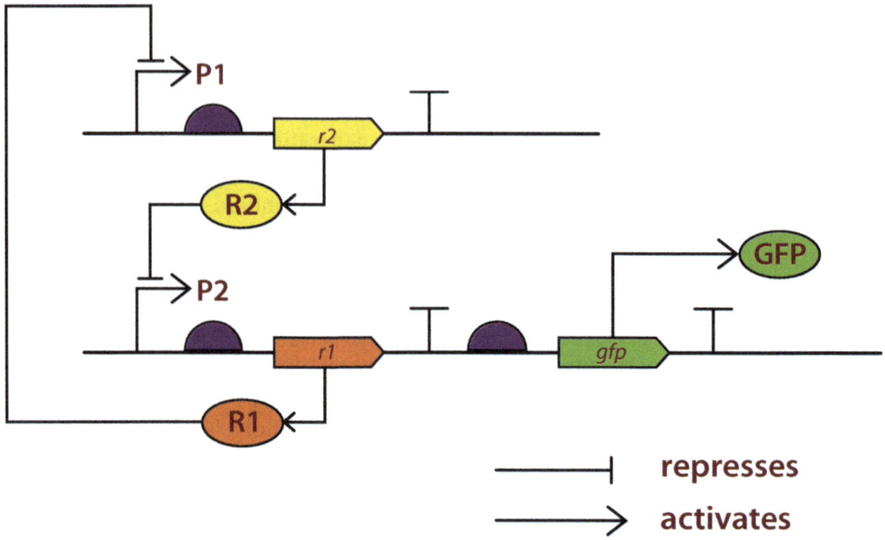

Fig. 2.15 The genetic toggle switch. (The genetic toggle switch designed and built by Gardner, Cantor, and Collins, drawn using SBOLv. This synthetic circuit combines two repressor motifs, as discussed in the main text. P1/P2, promoters; *r1*/*r2*, genes encoding repressor proteins; R1/R2, repressor proteins; P2 drives expression of *r1* and also gfp, the latter of which is translated into GFP, the system read-out. Both repressor motifs reside on opposing strands of the same double-stranded DNA plasmid, but they have been drawn here separately for clarity.)

The effective implementation of the principles of FAIR data management in research necessitates a combination of specialized knowledge, adequate temporal resources, adherence to (meta)data standards and formats, the utilization of appropriate software tools, and the establishment of public repositories. The SBOL2 metadata standard facilitates the FAIR sharing of designs pertaining to synthetic biology constructs, particularly within the SynBioHub repository. The production of extensive libraries comprising such constructs is becoming increasingly feasible in practical scenarios, exemplified by advancements in DNA foundries. Nonetheless, the manual curation of equivalent design libraries continues to pose significant challenges for the average laboratory researcher, thereby impeding the process of data sharing [98].

In [98], a straightforward tool named SynBio2Easy is introduced, which facilitates and automates processes on various SBOL designs utilizing *Microsoft Excel*® tables as metadata inputs. The tool offers various utilities for the manipulation of SBOL documents and interaction with the SynBioHub repository, including the generation of a plasmid library based on an original design template, bulk deposition into SynBioHub, and the annotation of existing SBOL component definitions with notes and authorship details. The instrument was employed to produce and submit a compilation of 3661 cyanobacterium *Synechocystis* plasmids to the public SynBioHub repository. During the software development and data upload process, some features of the SynBioHub platform and SBOL ecosystem are examined.

Software like SynBio2Easy enables the provision of user-centric tools to actualize FAIR principles throughout all phases of the project life cycle in synthetic biology research [98]. Figure 2.16 illustrates an example of this platform.

Additionally, an Excel-SBOL Converter exists that facilitates the reciprocal transformation of SBOL2 into Excel Templates and vice versa. This conversion of metadata provides researchers with the adaptability to operate within Excel while preserving the advantages of SBOL2 and ensuring interoperability with the SBOL2 tool suite. Furthermore, the converter enhances comprehension of the metadata interrelations, management practices, and structural components encapsulated within the SBOL2 XML file. The converter has been developed and disseminated as two Python packages, Excel2SBOL and SBOL2Excel, under a permissive and open license, and has been made available on PyPI for straightforward pip installation and integration with additional projects [99].

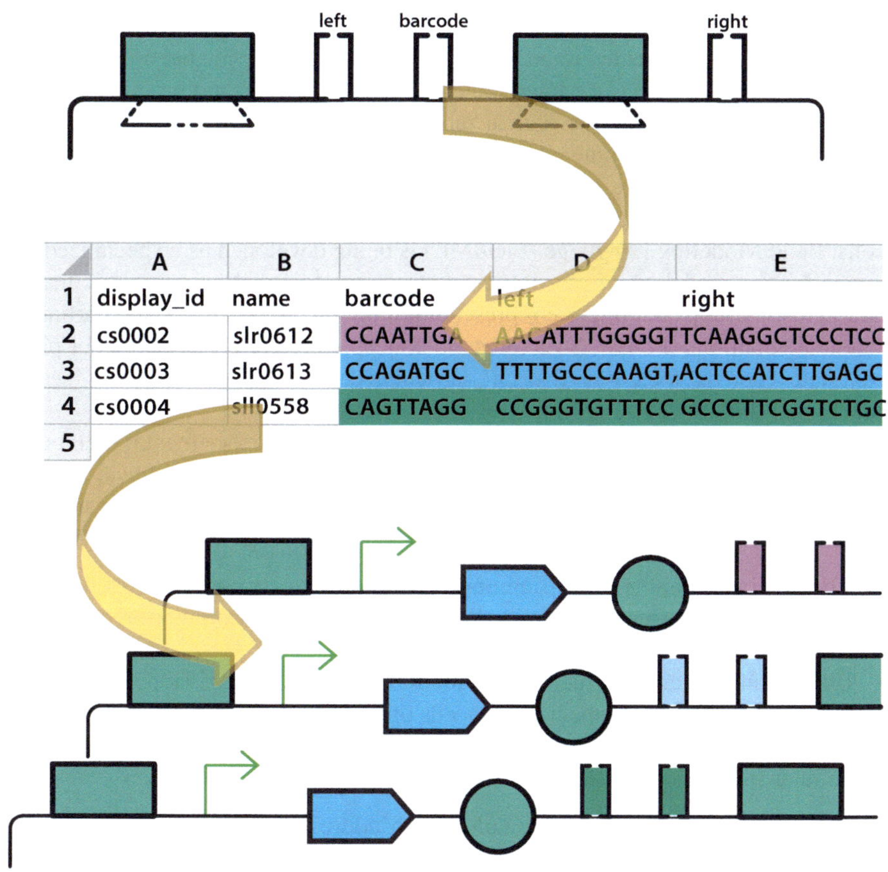

Fig. 2.16 SynBio2Easy application platform

2.2.3 Cellular Modeling Markup Language (CellML)

CellML, which emerged from the domain of physiological modeling, constitutes a markup language designed for modeling that aspires to encompass a diverse array of biological phenomena, with a primary focus on cellular functionality. The inception of CellML was motivated by the necessity to rectify the absence of standardized protocols for the depiction of cellular processes and to furnish unequivocal representations of various models. Due to the deficiencies in rigor and the lack of established standards within the publication framework, the validation of models has proven to be a formidable challenge. The introduction of errors is frequently observed when models are disseminated through platforms specifically tailored for particular software frameworks and tools, thereby diminishing the portability of the models themselves. This situation presents significant obstacles when engaging in collaborative research with individuals who may lack familiarity with the modeling methodologies, frameworks, and tools that others may have employed [100].

Similar to SBML, CellML employs Content Mathematical Markup Language (MathML) for the representation of systems modeled through mathematical equations. CellML is architected to be modular, permitting encapsulated models (potentially of varying scales) to be interconnected via both public and private interfaces. This capability facilitates the logical interlinking of multiple models whose variables may reference the same entity. This component-based methodology promotes the reutilization of entire models or segments of models delineated with CellML markup. In order to augment CellML's functional characterization of biological cells, Field Modeling Language (FieldML) is being developed as a specialized language for the modeling of physiological structures founded on geometric meshes and fields, thereby enabling the representation of spatial variability and partial differential equations (PDEs). Structures are conceptualized as abstractions of physical states across locations that are articulated and approximated by sets of functions [100].

Moreover, progress in the fields of biotechnology and experimental methodologies has resulted in the clarification of extensive quantities of biological information. Mathematical models offer a framework for the comprehensive analysis of this information; nevertheless, there exist two concerns that require attention [101]:

- The need for standards for defining cell models so they can, for example, be exchanged across the World Wide Web and also read into simulation software in a consistent format [101].
- Eliminating the errors which arise with the current method of model publication. CellML has evolved to meet these needs of the modeling community. CellML is a free, open-source, XML-based standard for defining mathematical models of cellular function [101].

CellML has emerged as a viable solution to address the discrepancies observed between computational and published models, while concurrently seeking to overcome the constraints associated with alternative computational cell biology modeling languages. CellML provides a precise methodology for articulating models of cellular function in a format that is comprehensible to both human users and machine processes. It facilitates the exchange of models via the Internet and promotes collaboration among modelers, irrespective of the modeling software utilized. The scope of models that can be represented using CellML exhibits considerable overlap with that of several other biological model specification languages [101].

CellML is an XML-based language. XML was selected as the underlying language for describing CellML for three principal reasons: it is a widely adopted standard for describing domain-specific structured data [102]; XML provides well-defined methods for integrating multiple domain-specific information into a single document; and many existing XML-based standards for describing domain-specific data are already available, for instance, MathML for describing mathematical expressions [101]. CellML is a standard format to store and exchange reusable, modular computer-based mathematical models [103]. From the start, CellML has been meticulously engineered to facilitate extensive modeling initiatives: it possesses a modular architecture, allowing for the independent construction of models and their constituent parts, which can subsequently be amalgamated into a comprehensive and cohesive model; additionally, CellML offers mechanisms for information concealment, ensuring that the intricacies of lower-level model components do not obstruct the current stage of model advancement [101, 104].

CellML employs Content MathML to deliver a representation that is comprehensible to both humans and computers, characterized by unambiguous semantics regarding the mathematical interrelations of biophysical models. Furthermore, CellML encompasses additional functionalities that enhance the model creation process. Notably, the stipulation that each variable possesses corresponding physical units permits CellML processors to systematically verify the dimensional consistency of equations. The modular architecture of CellML, achieved through the integration of components, encapsulation, and import mechanisms, streamlines the model-building process by utilizing previously validated models as foundations for new developments. Again, the insistence on the presence of physical units guarantees that models interact appropriately concerning dimensional consistency [101].

A substantial proportion of the inaccuracies that are presently engendered throughout the conventional processes of creation, dissemination, and testing can be markedly diminished through the utilization of CellML as the standardized exchange format for the biophysical model [101].

In [105], CellML 2.0 is analyzed, which is an XML-based language for describing and exchanging mathematical models of physiological systems. To define the underlying mathematics of models, MathML is embedded in CellML documents. Models are comprised of an interconnected system of reusable components, each characterized by variables and equations that delineate the relationships between these variables. Furthermore, models possess the capability to incorporate other models, thereby enabling the construction of systems with escalating complexity. The specifications for CellML 2.0 are delineated by the normative framework

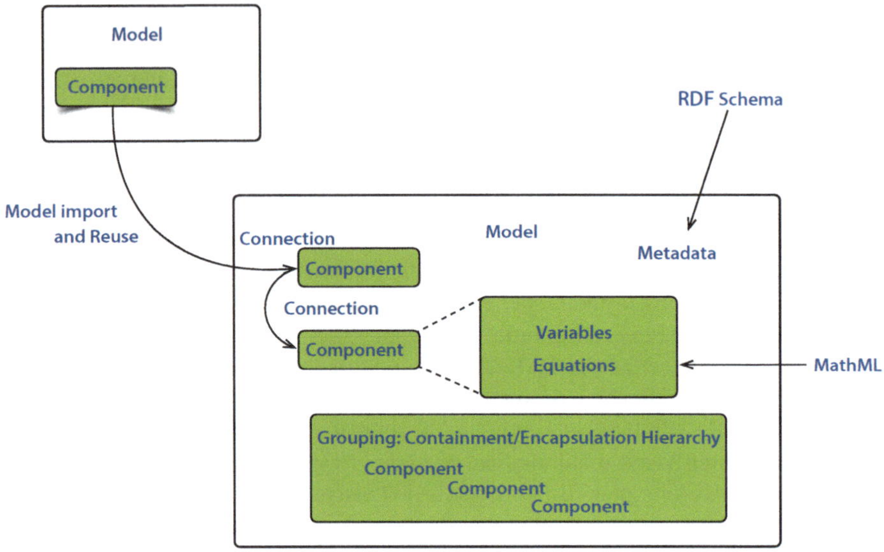

Fig. 2.17 A schematic diagram showing the general structure of a CellML model

detailed in [105][27] which extends the normative specification with explanations of the rules combined with examples of their usage [105].

In [106], CellML 2.0.1 is presented and this version improves the identification of rule statements and corrects errata present in the CellML 2.0 specification. For CellML format, the process of translating a biological process into CellML document capable of simulating this procedure in a computer system follows a specific template. This template describes the different sections required as follows [41]:

- Metadata: It contains information about the model's name and author and creation date, among others in RDF format [41].
- Units: It defines all different units of time, volume, and quantity [41].
- Components, variables, and equations: Always the model begins with a component, named Environment, and it contains all global variables such as time. Additionally, components also include kinetic laws giving information on how every compound takes part in reactions in the unit of time [41].
- Interfaces, groups, and connections: Each variable can have a public and/or private interface, with a value of either out, in or none. Interfaces describe the direction in which the value of a variable is passed between components, i.e., variable export, import, or no movement [41].

Models and components: CellML models are conceptualized as a network of intricately interconnected components (Fig. 2.17). A component is defined as the most fundamental functional unit of a model (a solitary component has the capacity to

[27] https://cellml.org/specifications/cellml_2.0/

represent a valid CellML model). Each component is required to encompass a minimum of one variable, and it may additionally incorporate mathematical equations to elucidate the behavior of that component within the model. These equations are articulated in MathML, an XML-based language which is integrated within the CellML framework [101].

A CellML variable represents a designated entity that is affiliated with a singular component. A variable possesses the potential to have various attributes. Mathematical equations are articulated through content MathML, an XML-based linguistic framework that is integrated within the CellML architecture. Within a CellML model, mathematical expressions are delineated within components under the auspices of the MathML namespace. In addition to its comprehensive array of predefined containers and operators, MathML also furnishes a mechanism for the amalgamation of these elements within mathematical expressions. Although a CellML document may incorporate any MathML content markup elements, the intent is to promote interoperability among software applications [101].

While a variable may be instantiated and defined within a singular component, it remains feasible to associate its value with other components within a given model. Nevertheless, it is imperative to acknowledge that a variable declared with either a public interface or a private interface of in cannot undergo mathematical modification within the current component, as its value is derived from a variable situated within an alternative component. The inter-component mapping of variables is facilitated through connections. These connections enable the transmission of information between components within the network. A connection is characterized by a set of variable mappings linking two components, and it is stipulated that there can exist only a singular connection between any two components within a model. The validation of this principle is crucial in order to prevent the establishment of inconsistent or redundant variable mappings between any two components [101].

The Units CellML specification mandates that all variables and numerical values must be articulated with a designated set of units. This characteristic of the CellML framework guarantees the integrity and reusability of the CellML components and models, as it permits the interconnection of components and models that encompass variables with disparate units yet identical dimensions (for instance, ounces and kilograms). CellML offers a compendium of standard units that is grounded in the International System of Units (SI) [101].

Groups: Grouping adds structure to a model by defining named relationships between components [101].

Analogous to SBML yet encompassing a wider range of applications, CellML constitutes an XML-based framework designed for the representation of mathematical models across various domains, including biochemical reaction networks. The architecture of CellML models is composed of multiple interrelated components. Each component is mandated to incorporate a minimum of one variable, which must be attributed with physical units. This requirement facilitates the capability of CellML processors to conduct automated verifications of equations for dimensional congruence. Furthermore, mathematical equations elucidating the operational dynamics of the component within the overarching model may be incorporated. For instance, a specific component may embody a cell membrane and incorporate the variable indicative of the potential difference across the cell membrane, expressed in terms of voltage [107].

Both CellML and SBML employ nearly indistinguishable mathematical representations in MathML, which serves as a globally recognized standard for the encoding of mathematical expressions utilizing XML. CellML explicitly delineates all mathematical constructs, including ODEs. It demonstrates greater versatility compared to SBML, as it possesses the capability to articulate any variety of mathematical models [15]. Then, CellML constitutes a standardized framework employed within the bioengineering discipline to delineate ordinary differential equation (ODE)-based models at the (sub)cellular scale. It establishes a format that is amenable to machine interpretation for the representation of differential-algebraic systems of equations. Furthermore, it is feasible to develop novel solvers for supplementary model components and integrate them with the pre-existing ones [107].

CellML has been devised to enhance the storage and interchange of computer-generated mathematical models. CellML constitutes a textual descriptive language. By employing CellML, researchers are required to specify solely the variables and the mathematical equations that articulate the dynamics of the system, and the associated solution tools for CellML can be invoked to directly assess the system's responses. The text-based nature of the model description further enables researchers to disseminate the models across diverse operating systems and various model-building software. CellML employs a modular configuration for model description, which facilitates the reutilization of components defined for one model in another, thereby augmenting the efficiency of the modeling process [108].

To assist researchers in modeling different systems using CellML, the CellML language developer also provided an integrated development environment called OpenCell[28], which encapsulates the various necessary functionalities including model construction, editing, validation, executing, and results displayed in a single computer program. For the model solution, OpenCell provides the options of using different algorithms including the Backwards Euler scheme, the Adams-Moulton scheme, and the two-stage implicit Runge-Kutta scheme, for the solution of the governing ODEs [108].

The results derived from the simulation can be exhibited directly within the OpenCell environment, or alternatively, they can be exported in Comma Separated Values (CSV) format for subsequent analysis in Microsoft Excel or other data visualization software. Furthermore, the CellML working group offers an additional modeling environment known as OpenCOR, which may be utilized interchangeably with OpenCell. Additionally, both CellML and the associated modeling tools, OpenCell and OpenCOR, are accessible to the research community for complimentary downloading [108].

[28] http://www.cellml.org/tools/downloads/opencell

As an example of the model implementation, Figs. 2.18, 2.19, and 2.20 show the schematics of the CellML model structure for the left heart and the systemic loop [108].

While numerous modeling methodologies have been employed, the Bond Graph (BG) remains the singular paradigm for representing physical systems, wherein "causality" is depicted in a graphical format. Furthermore, the BG methodology, with its inherent characteristics, facilitates the modular development of models and enables the algorithmic verification of mass and energy conservation. The BG methodology has been extensively utilized in the field of engineering and has, in recent times, witnessed an increasing application in biological and physiological systems. The significance of BGs is paramount in elucidating cellular mechanisms, biochemical interactions, as well as musculoskeletal and cardiovascular systems. In this context, CellML serves as an open standard language grounded in XML, which

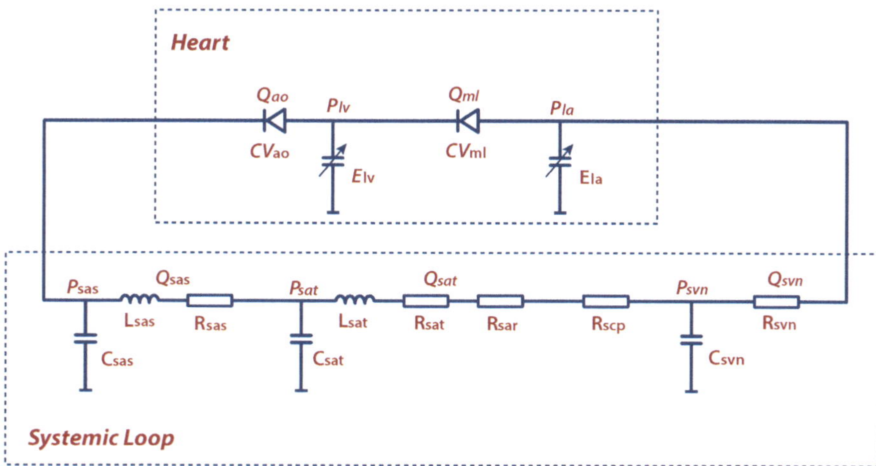

Fig. 2.18 A series of cardiovascular models developed. Systemic model with multi-compartment afterload

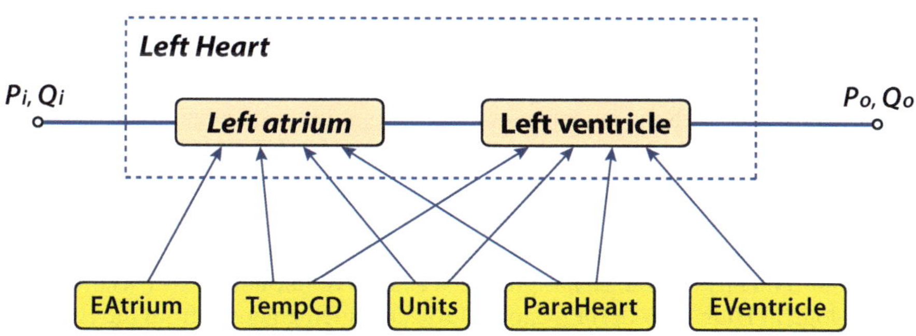

Fig. 2.19 Schematic of the CellML model structure for the left heart

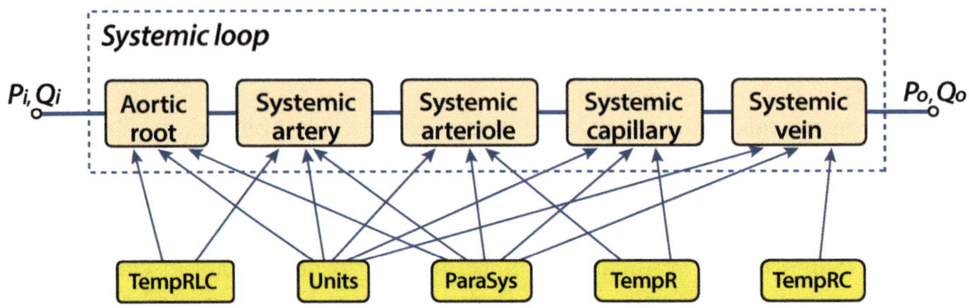

Fig. 2.20 Schematic of the CellML model structure for the systemic loop

facilitates the modeling process by transforming them into corresponding analogous BG representations [109].

The Auckland Bioengineering Institute, situated within the University of Auckland, oversees the Physiome Model Repository, which presently houses an extensive array of over 800 CellML models, alongside models and simulation protocols represented in diverse additional formats. The annotation of CellML models remains somewhat constrained; however, a compilation of metadata specifications is available, offering guidelines and exemplary practices for the effective annotation of models. While the CellML metadata specification articulates that semantic annotations ought to be serialized externally, the prevailing tools utilized within the CellML community, such as OpenCOR, incorporate RDF/XML annotations directly into the CellML documents, employing *identifiers.org* URI formatting and *BioModels.net* qualifiers. [110].

The CellML format prioritizes the representation of the mathematical dimensions of models, excluding biological constructs that are present in SBML models. Therefore, the variables within a model must be associated with terms from knowledge resources to accurately convey the specific implications of what a CellML model simulates. This presents challenges, as these variables represent concepts that can be very fine-grained (e.g., *concentration of cytosolic glucose in a pancreatic beta cell*), and publicly available knowledge resources do not provide adequate coverage for such concepts. Thus, the CellML group has been collaborating with the creators of the Semantic Simulation architecture, described next, to develop an annotation approach that uses *composite* annotations to describe such fine-grained concepts [110].

The Physiome Project was established by the International Union of Physiological Sciences in 1997 with the objective of employing multiscale engineering modeling methodologies to facilitate the physiological analysis of the extensive data, spanning from the molecular to the organismal level, that was progressively becoming accessible. Its primary goal is to furnish a quantitative, multiscale characterization of physiological dynamics and the functional behaviors exhibited by the intact organism. The Physiome idea has produced a range of initiatives, such as the Virtual Physiological Human (VPH) Physiome Project, which emerged from the European Commission VPH Project, in combination with the US Interagency Modeling and

Analysis Group (IMAG) initiative and others (for more background on different Physiome Projects and related initiatives[29]) [111].

The objective of the project is to delineate the Physiome through the utilization of ordinary PDEs, specifically denoted as ODEs and PDEs, which are rendered compatible, for instance, by the imposition of energy and mass fluxes via Bond-graph theory, and by adopting standardized nomenclature, ontologies, and formats, such as CellML [112]. While CellML models overlook spatial gradients, spatial data can be addressed through the complementary standard FieldML, which facilitates broader applications, including the implementation of constitutive material laws. CellML is meticulously crafted to assist in the design and specification of biological processes, while concurrently serving as a repository for validated models [111].

In [113] is presented the Cardiac Electrophysiology Web Lab, a user-friendly web interface that allows modelers to characterize their (and others') cardiac electrophysiology models, provided they are encoded using CellML, and to compare a model's behavior against that of any other model under a wide range of simulated experimental conditions. It is imperative to underscore that the Web Lab refrains from rendering any evaluations regarding the appropriateness of the models' behavior within a specific experimental context, nor does it determine which model is superior. Rather, it offers a framework that facilitates meticulous comparison and examination of the models' behavior across numerous virtual experiments, as well as enables the juxtaposition of a model's performance against that of any alternative model across an extensive array of simulated experimental conditions [113].

The main features are the use of annotations indicating the physiological significance of model variables, thereby mitigating potential ambiguities pertaining to nomenclature (these annotations can be integrated into the CellML files within a matter of minutes in the Web Lab), along with automated unit conversions that guarantee mathematically coherent simulations. These simulation instruments are constructed upon the foundational Chaste libraries tailored for computational biology [113].

The adaptable neuromuscular system, which comprises skeletal musculature and the nervous system, facilitates the execution of essential daily activities by humans. To examine its operational dynamics and pathological conditions through computational simulations, intricately detailed, multiscale models are advantageous, as their numerical resolutions necessitate high-performance computing capabilities. In [107] is analyzed OpenDiHu, a versatile, high-performance computing, open-source software framework for detailed, systemic simulations of skeletal muscles and their recruitment mechanisms.

OpenDiHu facilitates the resolution of a diverse array of multiscale models, encompassing three-dimensional muscle mechanics, quantifiable dysfunction signals, the propagation of APs within muscle tissue, subcellular bio-chemo-electrical

[29] Background to the VPH-Physiome project—OpenCOR Tutorial 0.18.0-alpha documentation (tutorial-on-cellml-opencor-and-pmr.readthedocs.io)

phenomena, and the neural stimulation of muscle fibers. These elements can be integrated with an extensive assortment of numerical solution methodologies to construct comprehensive simulation frameworks for the entire system. Empirical investigations utilizing nearly 27,000 processing cores illustrate the effectiveness and parallel scalability of OpenDiHu. This capability permits in silico experiments to be conducted at exceptionally high spatial and temporal resolutions [107].

An OpenDiHu scenario comprises modular solvers designated for submodels and model components, organized within a hierarchical tree structure. The framework facilitates the essential infrastructure for both data and control flow among the solvers. In this framework, a solver module can encompass any operation performed on specified data; for instance, it may involve a time-stepping method or an operator-splitting technique to resolve ODEs, a linear or nonlinear solver, or an interface to a CellML model [107].

2.2.4 Neuroscience Markup Language (NeuroML)

An essential component of computational modeling is the implementation of standardized model interchange languages that facilitate a simulation-independent characterization of the computational model. Model interchange languages, exemplified by NeuroML, are gaining increasing acceptance and enable researchers to more effectively modify pre-existing computational models and employ a diverse array of simulator backends. Notwithstanding the inclusion of features that support the dynamic creation and dissolution of elements, as observed in reaction networks, the explicit endorsement of neurogenesis within model interchange languages such as NeuroML remains constrained [114].

NeuroML is characterized by its simplicity and ease of use; however, it possesses significant power by enabling the definition of intricate models and their associated components in an independent format, which can be utilized across a multitude of simulators and stored in a standardized format for archival purposes. Despite the numerous advantages that NeuroML currently provides, its primary objective was to systematically address the diversity of neurological systems in a comprehensive manner. Furthermore, entities are categorized solely based on biological characteristics derived from biological foundations, such as chemical elements and cellular types, among others [115].

NeuroML-based neural simulator achieves both high model flexibility and high computational performance, through an innovative model-analysis and code-generation technique [116]. Furthermore, NeuroML offers a standardized data format for the delineation of neuronal cells and network architecture. It constitutes a simulator-agnostic framework for the articulation of neuronal model descriptions. This framework possesses the capacity to characterize data-driven models of neurons alongside their intricate networks, encompassing substantial biological intricacies. This descriptive language facilitates the representation of the elaborate branching configurations of dendritic trees and axonal projections, as well as their respective biophysical characteristics [117].

NeuroML constitutes an XML-based descriptive language that furnishes a standardized data format for the delineation and interchange of descriptions pertaining to neuronal cellular and network models [118]. NeuroML represents an independently developed methodology aimed at facilitating the standardization of neuronal modeling, thus permitting the application of models across various simulation platforms. NeuroML adopts a declarative modeling paradigm, systematically articulating the model in a structured format utilizing XML. One example is MorphoML [119], which is a part of NeuroML. For example, defining soma and a dendrite can be written as it is indicated in Fig. 2.21 [120]:

This XML-oriented neuromorphological specification is amenable to verification through specialized software, in addition to being transmutable into formats compatible with platforms for simulating neural networks, such as GEneral NEural SImulation System (GENESIS), NEURON, or Multiscale Object-Oriented Simulation Environment (MOOSE) [116, 119, 121–123].

Schematics of hierarchy-based representations of neuromorphological data are illustrated in Fig. 2.22. NEURON, one of the dominant players in computational neuroscience, has a dedicated file type termed “HOC” (*Hoc has* an object-oriented syntax addition which can be used to implement abstract data *types*, encapsulation of data, and polymorphism (but no inheritance)). It has C-like syntax with an additional object-oriented expressability. One of the uses for “HOC” is defining a

```
<cells>
  <cell name = "Example">
    <meta:notes>A Simple cell</meta:notes>
    <segments>
      &lt;segment id="0" name = "Soma" cable = "0"&gt;
        <proximal x = "0.0" y = "0.0" z = "0.0" diameter = "16.0"/>
        <distal x = "0.0" y = "0.0" z = "0.0" diameter = "16.0"/>
      &lt;/segment&gt;
      &lt;segment id= "1" name = "Dend" parent = "0" cable = "1"&gt;
        <proximal x = "8.0" y = "0.0" z = "0.0" diameter = "5.0"/>
        <distal x = "28.0" y = "2.0" z = "0.0" diameter = "6.0"/>
    &lt;/segment&gt;
    &lt;/segment&gt;
    <cables>
      <cable id = "0" name = "SomaCable"/>
      <cable id = "1" name = "DendriteCable"/>
    </cables>
  </cell>
</cells>
```

Fig. 2.21 Definition of a dendrite in MorphoML

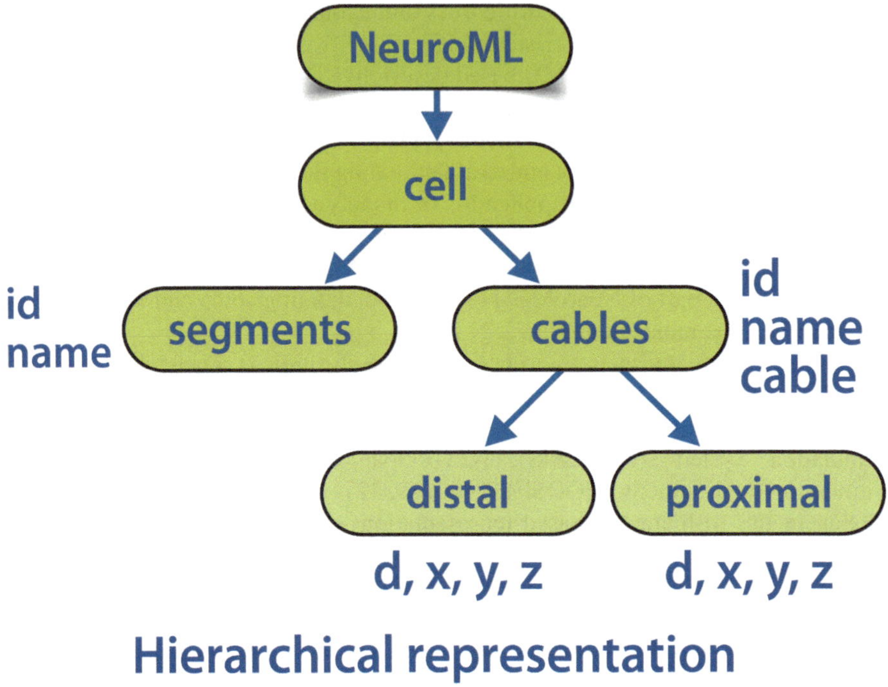

Fig. 2.22 Hierarchical and object-based representation of neuromorphological data

neuronal morphology by constructing an array of "section" objects, each defined by a series of four points (using neuron's "*pt3dadd*" function): three coordinates and a radius. Sections can be connected to one another (using neuron's "*connect*" function) [120].

Moreover, data-driven models of neurons and neural circuits are crucial for elucidating how the characteristics of membrane conductances, synaptic connections, dendritic structures, and the anatomical interconnection among neurons contribute to the intricate dynamic behaviors exhibited by brain circuits in both normative and pathological states. Nevertheless, the intrinsic complexity associated with these biological phenomena renders the development and reutilization of biologically precise models a formidable challenge. A diverse array of tools has been engineered to facilitate their construction and simulation; however, disparities in design and internal representation pose significant technical obstacles for researchers seeking to incorporate data-driven models into their investigative workflows [124].

Consequently, NeuroML, a formalized language for the description of models within the realm of computational neuroscience, was conceived to remediate this disjunction in modeling instruments. From its initial development, NeuroML has matured into a widely accepted standard within the community, incorporating a diverse array of model classifications and methodologies pertinent to computational neuroscience [124].

The advancement of a comprehensive and mechanistic comprehension of cerebral functionality in both normative and pathological conditions necessitates the employment of diverse scientific methodologies that traverse multiple scales, ranging from genetic expression to behavioral manifestations. While "wet" experimental methodologies are indispensable for delineating the characteristics of neural systems and for empirically testing theoretical propositions, theoretical frameworks and computational modeling are vital for investigating the behaviors of these intricate systems across a broader spectrum of conditions, as well as for formulating novel experimentally verifiable and physically plausible hypotheses. Furthermore, theoretical constructs and modeling facilitate the synthesis of a multitude of experimentally acquired parameters, functional attributes, and responses to disturbances into a physio-chemically coherent schema that accurately reflects the properties of the neural system under investigation [124].

The inaugural iteration of the NeuroML standard, designated as version 1 (NeuroMLv1), was initially conceptualized as a framework for model description and materialized as a three-tiered, declarative, modular, simulator-independent language. NeuroMLv1 possessed the capability to articulate intricate neuronal morphologies alongside their biophysical characteristics, as well as particular instantiations of neural networks. It facilitated the preservation of models within a uniform format and addressed the challenge of simulator fragmentation by serving as the universal language for model interchange among established simulation platforms (e.g., NEURON, GENESIS, and MOOSE) [124].

In the course of addressing numerous enduring challenges within the domain of computational neuroscience, NeuroMLv1 exhibited several significant constraints. The most prohibitive of these was the absence of a formal description of the dynamical behavior of model components within the standard itself, rendering it only partially amenable to machine interpretation. The information regarding the dynamics of these components (i.e., the temporal evolution of state variables) was solely available in the form of documentation intended for human comprehension, necessitating that developers of each new simulation environment replicate the behavior of these components in their respective programming languages. Furthermore, the incorporation of novel model components mandated revisions to both the standard and all associated simulators; the reliance on XML as the principal interface language constrained usability, as applications typically had to implement additional code to facilitate the reading and writing of XML files [124].

In order to mitigate these constraints, NeuroML underwent a comprehensive redesign in its second iteration (NeuroMLv2) utilizing the Low Entropy Modeling Specification (LEMS) language. LEMS was specifically formulated to articulate a diverse array of physio-chemical systems, facilitating the development of entirely machine-readable, formal representations of the architecture and dynamics of any model constituents. The modeling components within NeuroMLv2 (such as cells, ion channels, and synapses) possess their mathematical and structural characterizations delineated in LEMS (for instance, the requisite parameters and the temporal evolution of state variables). Consequently, NeuroMLv2 preserves all the attributes of NeuroMLv1—it remains modular, declarative, and continues to accommodate

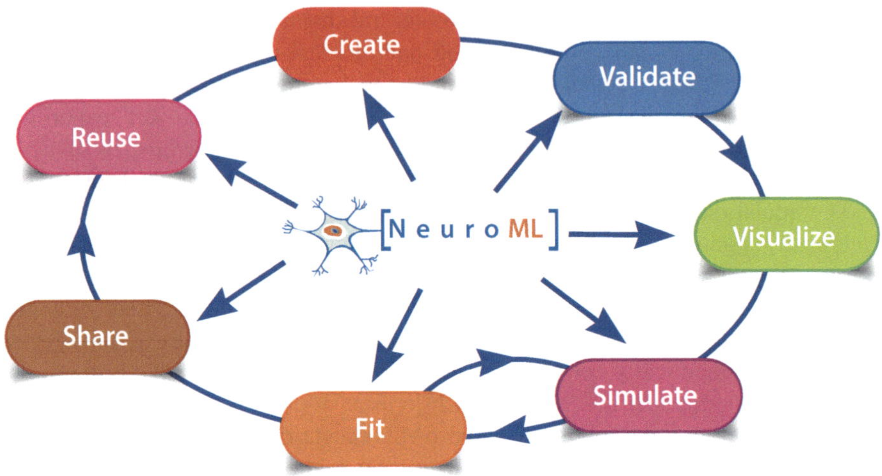

Fig. 2.23 The NeuroML software ecosystem supports all stages of the model development life cycle

various simulation engines—but in contrast to version 1, it exhibits extensibility, and all specifications are entirely machine-readable. Furthermore, NeuroMLv2 transitioned to Python as its primary interface language and offers an extensive suite of Python libraries to enhance usability, while XML continues to serve as a machine-readable serialization format (that is, the manner in which the model files are stored and disseminated) [124].

Since its inception in 2014, the NeuroMLv2 standard, along with its associated software ecosystem and community, has experienced consistent growth. A governance framework characterized by open, community-driven principles has been placed by an elected Editorial Board, which under the auspices of an independent Scientific Committee is responsible for the maintenance of the standard and the core software tools, which include APIs, reference simulators, and various utilities. While these tools were originally designed to facilitate the simulation of models across diverse platforms, their functionality has since been broadened to encompass all phases of the model life cycle (Fig. 2.23). Researchers and modelers are afforded the capability to utilize these tools to seamlessly create, inspect, visualize, validate, simulate, fit, optimize, share, and disseminate NeuroMLv2 models and their resultant outputs [125, 126].

NeuroML documentation has been significantly expanded and re-deployed using modern web technologies[30], providing clear, concise, searchable information for both users and developers. Increased community-wide collaborations have also extended the software ecosystem well beyond the NeuroMLv2 tools developed by the NeuroML team: additional simulators such as Brian, NetPyNE [125, 126],

[30] https://docs.neuroml.org

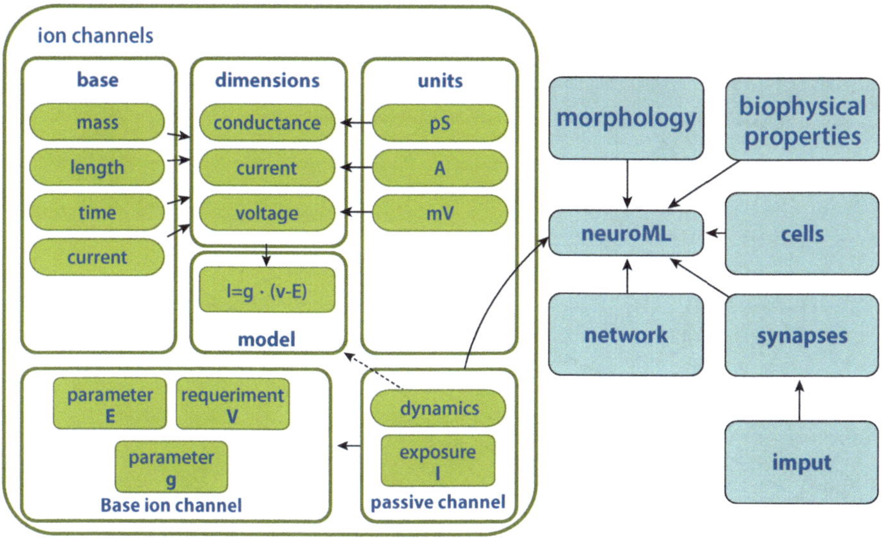

Fig. 2.24 NeuroMLv2 hybrid approach to morphological, biophysical, and network modeling

Arbor, and EDEN all support NeuroMLv2. As discussed in [116, 124], NeuroML works to ensure interoperability with other structured formats for model development in neuroscience, such as PyNN and SONATA[31] [127]. Platforms for collaboratively developing, visualizing, and sharing NeuroML models Open-Source Brain (OSB) as well as a searchable database of NeuroML model components [NeuroML Database (NeuroML-DB)] have been developed [116, 117, 128–130]. These enhancements, driven by an ever-expanding community, have helped NeuroMLv2 grow into a standard that has been officially endorsed by international organizations such as the INCF and COMBINE, and that is now sufficiently mature to be incorporated into a wide range of research workflows [124].

Furthermore, in the subsequent iteration of NeuroML, an innovative integrative methodology is being formulated for the purposes of modeling, referred to as low entropy model specification (LEMS). LEMS constitutes a hierarchical, XML-oriented language through which ion channels, synapses, neurons, and networks can be collectively delineated. It amalgamates a hybrid hierarchical object-oriented strategy for the purposes of modeling. An exemplification is provided in Fig. 2.24 [120, 131].

NeuroML offers a comprehensive and readily accessible collection of meticulously curated model components. The primary objective of the NeuroML initiative is to facilitate and promote the implementation of multiscale, biophysically intricate models of neurons and neuronal circuits within the realm of neuroscience research. The initiative adopts a variety of strategies to fulfill this objective. NeuroML supplies users with a curated repository of model components that constitute NeuroML

[31] https://github.com/AllenInstitute/sonata

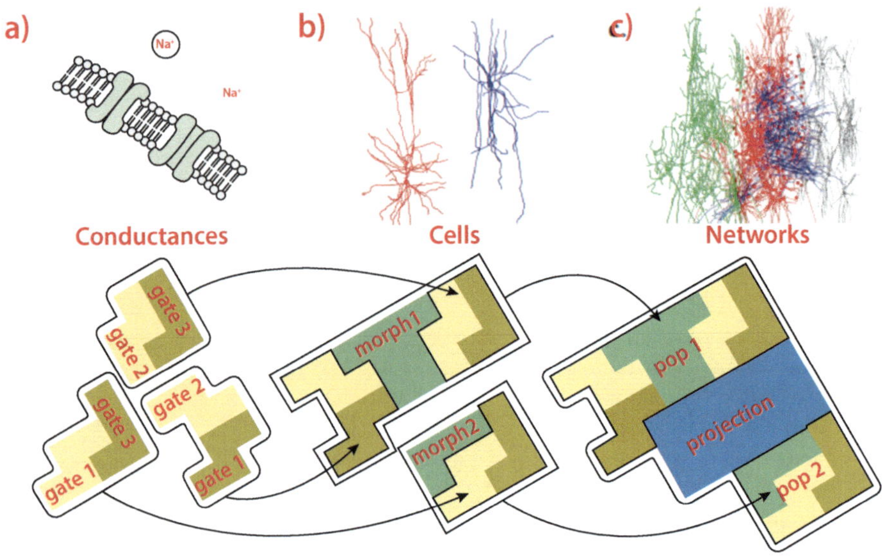

Fig. 2.25 NeuroML represents a modular and hierarchical framework that facilitates multiscale modeling endeavors. (The components within NeuroML are rigorously delineated, autonomous, and self-sufficient entities that possess hierarchical interrelations.) (**a**) The models encapsulating ionic conductances may be articulated as an assemblage of gates, each characterized by particular voltage (and potentially [Ca^{2+}]) dependencies that govern the conductance. (**b**) Neuronal models with morphological intricacies delineate the three-dimensional architecture of the cells, alongside their passive electrical characteristics, and reference ion channels that impart membrane conductances. (**c**) Network models encompass collections of these cellular entities interconnected through synaptic projections

standard2 (refer to Figs. 2.25 and 2.26). This standard is overseen by the NeuroML Editorial Board, which has delineated a fundamental array of model types to endorse, thereby ensuring that a substantial proportion of frequently utilized neurobiological modeling entities can be articulated using this framework. This encompasses (but is not confined to): active membrane conductances (employing Hodgkin-Huxley style or kinetic-scheme-based ionic conductances), a multitude of synaptic and plasticity mechanisms, detailed multi-compartmental neuron models characterized by distinct morphologies and biophysical properties, abstract point neuron models, as well as networks of such neurons spatially arranged in populations, interconnected by targeted projections, and receiving inputs in the form of spikes and currents [124].

The NeuroMLv2 standard consists of two levels that are designed to enable users to easily create their models without worrying about simulator-specific details. The first level defines a formal "schema" for the standard model elements, their attributes/parameters (e.g., an integrate and fire cell model and its necessary attributes: a threshold parameter, a reset parameter, etc.), and the relationships between them (e.g., a network contains populations; a multi-compartmental cell morphology contains segments). This allows the validation of the completeness of the description of individual NeuroML model elements and models, *prior to simulation*. The second level defines the underlying dynamical behavior of the model elements (e.g., how the time-varying membrane potential of a cell model is to be calculated). Most users do not need to interact with this level (which is enabled by LEMS), which, among other features, enables the automated translation of *simulator-independent* NeuroML models into *simulator-specific* code [124].

Therefore, modelers have the capability to utilize the standardized elements of NeuroML to effectively construct models that are independent of any specific simulator, while simultaneously being able to scrutinize and enhance the fundamental implementations of those models. As a language that is independent of simulators, NeuroML fosters interoperability among various computational modeling tools, thereby resulting in the standard library being augmented by a substantial and meticulously maintained ecosystem of software tools that facilitate all phases of the model life cycle—from creation, analysis, simulation, and fitting, to dissemination and reutilization. Ultimately, as elaborated in subsequent sections, for sophisticated use cases in which the pre-existing components of NeuroML are inadequate, NeuroML also encompasses a framework for the creation and integration of novel model elements [124].

Furthermore, NeuroMLv2 constitutes a modular, systematic, hierarchical format that is independent of any particular simulator. Each element within NeuroML is rigorously defined, autonomous, and self-sufficient, exhibiting hierarchical interrelations among themselves. For instance, an "ionic conductance" model element in NeuroML may encompass zero, one, or multiple "gates" and can be incorporated into a "cell" model element alongside a "morphology" element, which in turn can be integrated into a "population" within a "network" (Figs. 2.25 and 2.26) [124].

To facilitate the diverse array of electrical characteristics exhibited by biological neurons, ionic conductances characterized by specific ionic selectivities and dynamic behaviors can be formulated in NeuroML through the incorporation of various types of gates (e.g., activation and inactivation), their correlation with parameters such as voltage and [Ca^{2+}], and their respective reversal potentials. Distinct cell types exhibiting varied functional and biophysical attributes can subsequently be produced by assigning specific combinations of ionic conductances to their membranes. The density of conductance may be fine-tuned to replicate the electrophysiological attributes observed in authentic neuronal cells. In practical applications, numerous instances of ionic conductances that form the basis of neuronal electrical activity are readily accessible in NeuroMLv2 and can be seamlessly integrated into a cellular membrane (Figs. 2.25 and 2.26) [124].

Indeed, a model element, once defined in NeuroML, acts as a building block that may be reused any number of times within or across models. Elements such as ionic conductances, cell biophysics, cell morphologies, and cell definitions that incorporate them can be serialized in separate files and "included" in other models (e.g., morphologies[32]). Such reuse of model components speeds model construction and prototyping irrespective of the simulation engine used [124].

[32] https://docs.neuroml.org/Userdocs/ImportingMorphologyFiles.html#neuroml2

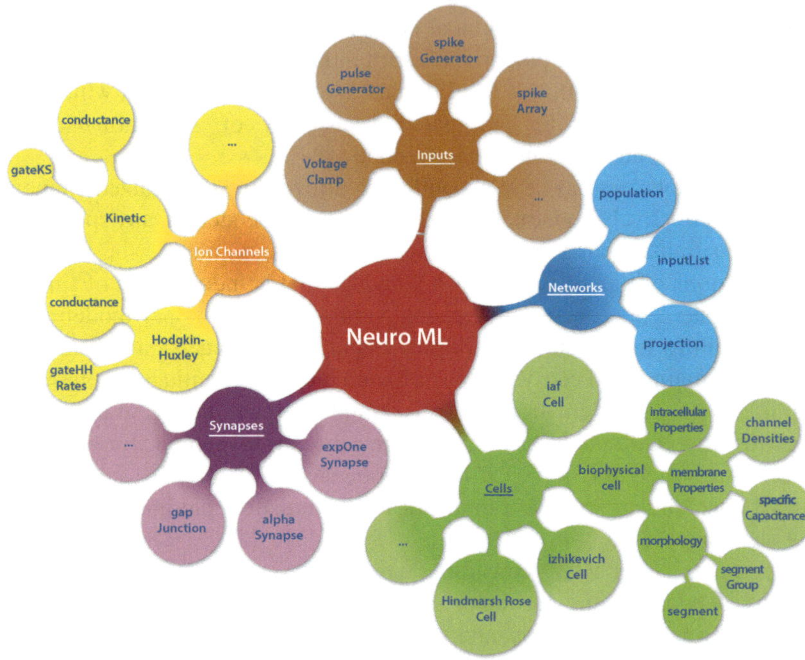

Fig. 2.26 A condensed depiction of the primary categories of the NeuroMLv2 standard components and their hierarchical organization. The standard encompasses frequently utilized model components/building blocks that have been pre-established for users, namely, Cells: neuronal models that range from basic spiking point neurons to biophysically intricate cells characterized by multi-compartmental morphologies and active membrane conductances; Synapses and ionic conductance models: widely adopted chemical and electrical synapse models (including gap junctions), alongside various representations for ionic conductances; Inputs: designed to stimulate cellular and network activity, for instance, current or voltage clamp, as well as spiking background inputs; Networks: composed of populations (which may include any of the previously mentioned cell types), along with their projections

The delineated framework of each model component, along with the interrelations among them, provides users with precise guidance on the methodologies for the creation and integration of model elements. This fosters the development of rigorously organized models, minimizes the incidence of errors and redundancies, and guarantees that the principles of FAIR are intricately woven into both NeuroML models and the associated tool ecosystem. Furthermore, NeuroML's structured formalism facilitates functionalities such as the validation of models prior to simulation, the conversion into formats tailored for specific simulations, and the utilization of NeuroML as a standardized language for interaction among various tools [124].

NeuroML facilitates a comprehensive ecosystem of software applications that encompasses all phases of the model life cycle. The process of constructing models and extracting scientific insights through simulation and analysis is a complex, iterative endeavor that necessitates a diverse array of software tools. NeuroML effectively supports every phase of the model development life cycle (refer to Fig. 2.23) by offering a unified model description format that interfaces with numerous tools throughout the entirety of the process. Typically, researchers compile improvised collections of scripts, applications, and methodologies to assist them in their inquiries. In the absence of established standards, they are compelled to operate within the specific model formats and Application Programming Interfaces (APIs) mandated by each tool they employ, necessitating the conversion of model descriptions when utilizing multiple applications within a toolchain [124].

NeuroML effectively resolves this challenge by establishing a standardized linguistic framework that facilitates the utilization and interchange of models and their respective components among various simulation engines and modeling instruments. The NeuroML ecosystem encompasses an extensive array of software tools, both of which are developed and sustained by the primary contributors of NeuroML (the "core NeuroML tools and libraries": jNeuroML, pyNeuroML, APIs) and those external applications that have integrated NeuroML support (Figs. 2.27 and 2.28) [124].

The fundamental NeuroML tools and libraries encompass APIs across various programming languages, including Python, Java, C++, and MATLAB. These instrumental tools deliver essential functionalities that facilitate user interaction with NeuroML components and the construction of models. Through the utilization of these resources, researchers are empowered to develop models from the ground up, as well as to read, alter, analyze, visualize, and simulate pre-existing NeuroML models on compatible simulation platforms. Moreover, developers are afforded the opportunity to leverage the core tools, libraries, and APIs to integrate NeuroML capabilities within their own software applications. The simulation platforms (for instance, EDEN), in conjunction with other independently developed tools, constitute the subsequent tier of the software ecosystem, offering enhanced functionalities, such as interactive model construction (e.g., neuro-Construct and NetPyNE), supplementary visualization (e.g., OSB), analytical tools (e.g., NeuroML-DB), data-driven validation (e.g., SciUnit), and archival/sharing mechanisms (e.g., OSB and NeuroML-DB). Indeed, OSB and NeuroML-DB exemplify how sophisticated neuroinformatics resources can be constructed upon established standards such as NeuroML [124].

NeuroML possesses the capability for extensibility. Although the foundational NeuroML components offer an extensive array of meticulously curated model categories suitable for simulation-based inquiries, NeuroML allows for expansion (through the utilization of LEMS) to integrate innovative model components and classifications when they are not (as yet) accessible within the standard framework. LEMS serves as a versatile model specification language designed for the generation of entirely machine-readable definitions pertaining to the architecture and functionality of model elements [124].

The dynamic behavior of NeuroML components is articulated within the framework of LEMS. The inherent hierarchical structure of LEMS facilitates the development of novel components that can leverage existing elements within modular NeuroML architecture. For instance, an innovative ionic conductance element may

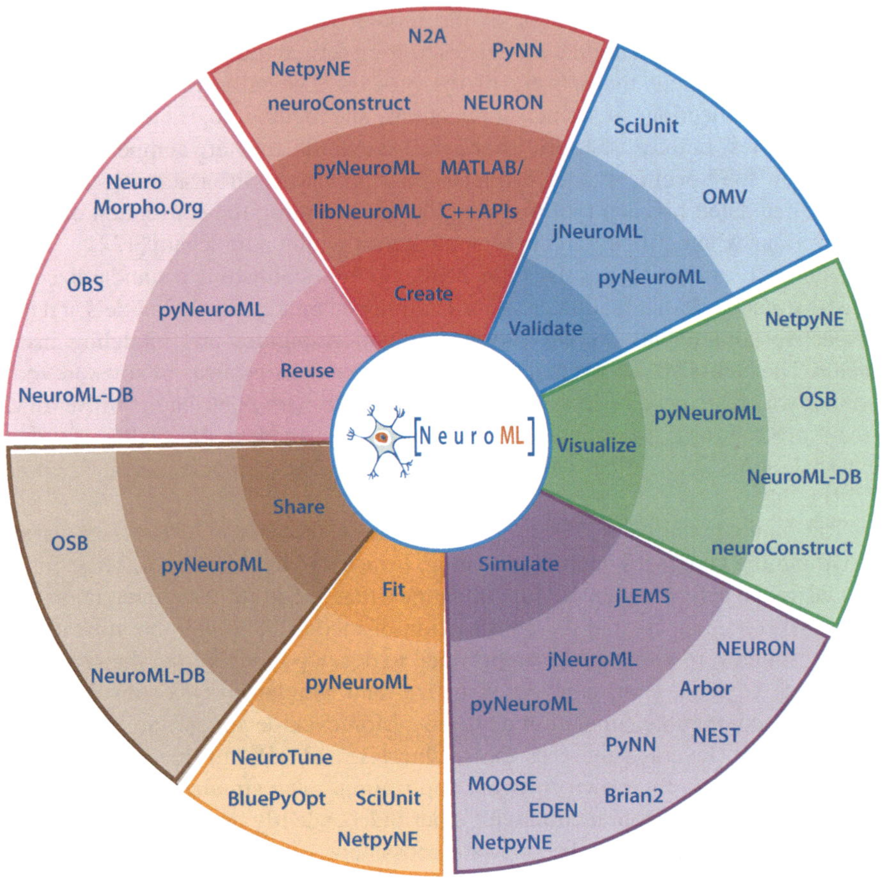

Fig. 2.27 The ecosystem comprising NeuroML compliant tools and their association with the model life cycle. (The innermost circle delineates the fundamental NeuroML tools and libraries that are curated by the NeuroML development team. These tools offer functionalities for the reading, modification, or generation of new NeuroML models, in addition to enabling the validation, analysis, visualization, and simulation of these models. The peripheral layer illustrates the NeuroML compliant tools that have been independently devised to facilitate diverse interactions with NeuroML models. These tools serve to enhance the core offerings by promoting model creation, validation, visualization, simulation, fitting/optimization, sharing, and reuse)

augment the "ionChannelHH" component, which subsequently builds upon the "baseIonChannel." Consequently, this newly introduced element will be recognized within the NeuroML ecosystem as reliant on an external voltage and generating a conductance, attributes that are derived from the "baseIonChannel." Additional components, such as cellular structures, are capable of assimilating this new type without necessitating any supplementary knowledge regarding its internal mechanisms [124].

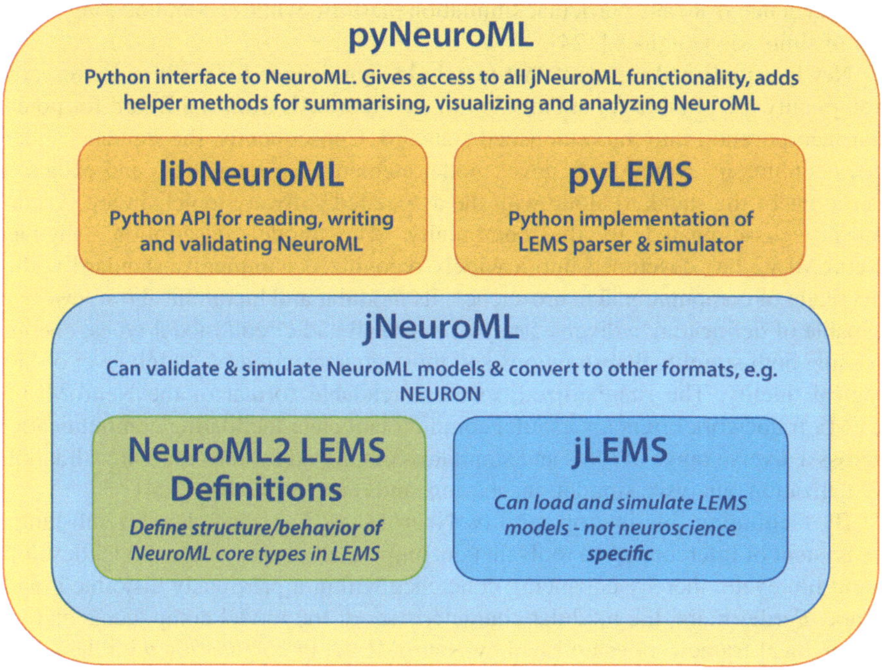

Fig. 2.28 The fundamental NeuroML software architecture, along with a representative NeuroML model constructed utilizing the Python NeuroML tools, is presented. (The fundamental NeuroML software architecture comprises Java (indicated in blue) and Python (denoted in orange) based applications and libraries, alongside the LEMS model Component Type definitions (illustrated in green), all encapsulated within a singular package, pyNeuroML. Each of these modules possesses the capacity for independent utilization, or alternatively, the entire architecture can be acquired through the installation of pyNeuroML via the default Python package manager, Pip: pip install pyNeuroML)

LEMS, along with its associated framework NeuroML, provides definitions for elements (termed "ComponentTypes") that delineate the dynamic properties of model elements through a set of parameters that are yet to be specified. Upon establishing the generic behavioral characteristics of the model, researchers are merely required to input the pertinent values for the necessary parameters (such as conductance density, reversal potential, etc.) to instantiate new entities (referred to as "Components") of the element. Consequently, users are enabled to construct arbitrary and reusable model elements within LEMS, which can be utilized in a manner analogous to that of standard model elements. A significant benefit of NeuroML's incorporation of the LEMS language lies in its ability to be translated. Given that LEMS is entirely machine-readable, its fundamental components (including state variables and their dynamics, which are articulated as ODEs, can be efficiently converted into other programming languages. Consequently, simulator-specific code can be generated from NeuroML models and their corresponding LEMS extensions (as illustrated in Fig. 2.28), thereby allowing NeuroML to maintain its independence from any particular simulation platform while accommodating a variety of simulation engines [124].

Newly established elements that may hold significance for the broader research community are eligible for submission to the NeuroML Editorial Board for potential incorporation into the established standard. Consequently, the standard undergoes continuous evolution as novel model elements are integrated, and enhanced iterations of the standard along with the associated software toolchain are systematically disseminated to the community. The model description language NeuroMLv2 has developed into a widely recognized community standard within the realm of computational neuroscience. Its modular and hierarchical framework is capable of delineating a diverse array of neuronal and circuit model types, encompassing both simplified abstractions and those characterized by a high level of biological fidelity. The standardized, machine-readable format of the NeuroMLv2/LEMS framework offers a versatile, common language facilitating communication across a diverse range of tools and simulators employed for the creation, validation, visualization, analysis, simulation, sharing, and reuse of models [124].

By facilitating this interoperability, NeuroMLv2 has engendered a substantial ecosystem of interconnected tools that encompass all phases of the model development life cycle, thereby enhancing coherence within a previously disjointed landscape. Furthermore, the modular characteristics of the model components and the hierarchical framework established by NeuroMLv2, in conjunction with the adaptability afforded by programming in Python, have culminated in a robust "building block" methodology for the systematic construction of standardized models from inception. Consequently, NeuroML has transitioned from a standardized archival format to a sophisticated language that underpins an ecosystem of tools dedicated to the formulation and execution of models that adhere to the FAIR principles and advance open, transparent, and reproducible science [124].

Furthermore, software engineers aiming to verify, validate, and contribute to intricate software projects depend on collections of straightforward executable tests known as "unit tests". Consequently, inspired by this methodology, SciUnit has been developed in [132] as an accessible framework for creating data-driven "model validation tests," which are executable functions written in Python. Each test creates and statistically evaluates predictions from a model against a pertinent aspect of empirical data to get a score reflecting the concordance between the model and the data. Validation test suites may effectively delineate the advantages and drawbacks of current models and the advancement of new models. NeuronUnit, as described in [132], is a library that extends SciUnit and incorporates several neuroinformatics resources to facilitate the validation of single-neuron models utilizing data collected by neurophysiologists and neuroanatomists. NeuronUnit is compatible with established technologies, such as NeuroML [132].

2.3 PART II: Most Used Simulators for Molecular Communications

The utilization of simulators across various domains is pivotal for advancing research and facilitating corresponding scientific progress. Simulations culminate in the emulation of operational processes that occur in the real world. Within the realm of MCs, numerous simulation platforms have been developed to investigate the propagation and detection mechanisms of specific species, alongside the examination of the intrinsic characteristics of MCs. There exist several constraints in this context, one of which pertains to the preference for nano- and microscale analyses over macroscale due to the limited availability of experimental evaluations. Consequently, simulations have emerged as a critical component in the investigation and assessment of a multitude of parameters within the field of MCs.

2.3.1 Biological and Nano-Scale Communication Simulator (BiNS)

An extensive simulation framework, which encompasses a Markov chain model of the receiver, was developed for biological communications within the Networking Laboratory at the University of Perugia, Italy, and is referred to as Biological and Nano-Scale Communication Simulator (BiNS). One of the primary aims of this simulator is fundamentally medical, entailing the implementation of a simulation-based approach to evaluate, for example, the risk of cardiovascular diseases on an individual patient basis. This area of research also aids in the elucidation of mechanisms that facilitate generic communications among nanomachines within such a context. BiNS is specifically engineered to simulate nanoscale communications, characterized by its meticulous granularity, whereby the position of each constituent is assessed at each simulation interval, and collisions are addressed in accordance with an elastic model. In this model, receptors are allocated across the surface of each stationary or mobile nanomachine, with their quantity ranging from hundreds to thousands. A nanomachine is capable of receiving a carrier when the latter interacts with one of the carrier-compliant receptors [133, 134].

The software library associated with our simulation framework was engineered utilizing the Java programming language. Its primary objective is to furnish a comprehensive toolkit for the emulation of various categories of nanonetworks. The preliminary phase of its development entailed the delineation of the fundamental characteristics of the communication protocols employed by nanomachines across a majority of scenarios. The requirements analysis was conducted by scrutinizing the MCs posited in certain nanonetwork models, notwithstanding the fact that this methodology does not preclude the capability to simulate a diverse array of nanocommunications. The salient features of the package are [28]:

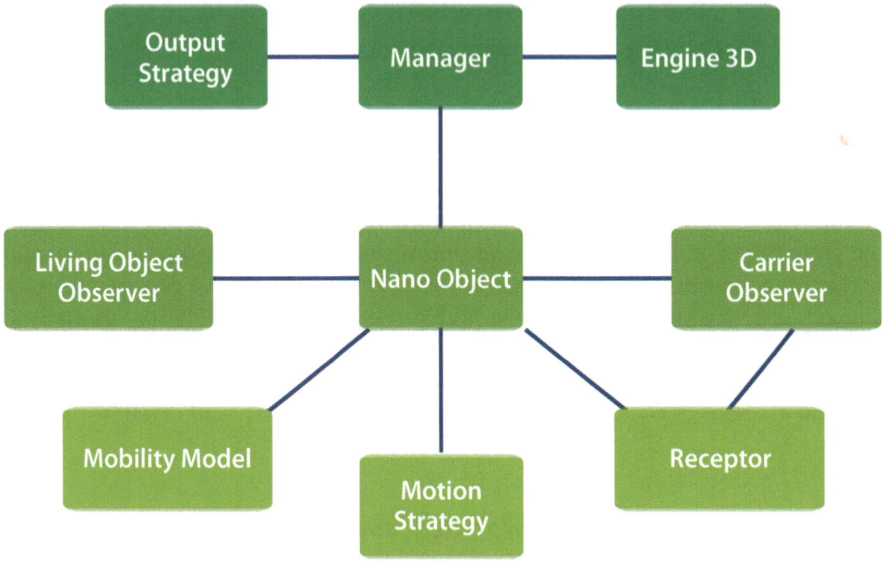

Fig. 2.29 A simplified picture of the class diagram

- The possibility for different entities (nanomachines) to send, receive, and process information. Each nanomachine has well-defined physical properties: mass, size, shape, motion pattern, and lifetime [28].
- The signal transduction is done by specific structures placed on the outer surface of nanomachines, referred to as receptors. Their size and affinity make them specialized for particular types of received objects [28].
- The information is transferred by carriers, which can be modeled through different attributes, such as shape, size, motion pattern, and lifetime, similarly to nanomachines [28].
- Carriers can be bound to compatible receptors, which may transduce the associated signal and transfer information from/to the interior of a nanomachine [28].

The implementation of the simulator necessitates the delineation of the environmental attributes within which information dissemination occurs, including the factors that may influence such dissemination. The library additionally furnishes the overarching principles governing the specific interactions among nanomachines and carriers, carriers and carriers, as well as nanomachines and nanomachines. These interactions facilitate a generalized behavioral framework and do not constrain the formulation of intricate or tailored scenarios [28].

The software package, whose simplified design is depicted in Fig. 2.29, is structured into several classes. Several components, including *NanoObject, Receptor, MobilityModel, MotionStrategy, CarrierObserver,* and *LivingObjectObserver,* exemplify the standard aspects of a nanonetwork, while the other components, *Manager, OutputStrategy,* and *Engine3D*, oversee the actions occurring at each time interval. The software design employs a multithreaded framework to enhance the efficiency of computationally intensive processes, including nano-object propagation and collision control. Nano-object propagation is executed in a meticulous manner. This indicates that the position and velocity of the Nano Object are computed at each time interval. The collision management model relies on the mass and velocity of each Nano Object, whereas the propagation strategy is founded on the Brownian molecular diffusion process [28].

Therefore, the BiNS simulator offers functionalities for the simulation of molecular signaling and communication dynamics within confined environments. The BiNS simulator has been meticulously developed with the intention of ensuring a high degree of customization. A comprehensive suite of tools is accessible. This suite enables the development of computational entities that replicate the behaviors of biological organisms, categorized as either nodes (transmitters, receivers, or both) or carriers, which are emitted by transmitters and serve as the signaling mechanism in molecular-based communications. Moreover, it is possible to modify the attributes of the simulated communication medium (such as the bloodstream or the conditions of an in vitro experiment) with the necessary accuracy. Every object is classified as a generic sort of software entity known as a *Nano Object*. Nodes and carriers are particular implementations of the *Nano Object*, while they have common characteristics, they may offer significantly diverse functionalities [28, 134].

In particular, within any given scenario, it is feasible to distinguish the type of node object at any moment, thereby resulting in a diverse array of distinct node objects. The identical properties are applicable for the modeling of carriers like *Nano Objects*. The simulation is structured in discrete temporal increments. Each increment comprises a series of phases, during which software objects are activated to perform the operations pertinent to their specific behavioral characteristics. The primary phases are [28, 134]:

- Transmission phase, i.e., emission of carriers.
- Reception phase, i.e., carriers' assimilation.
- Information-processing phase, which is executed depending on the received signal intensity, which corresponds to the number of assimilated carriers.
- Motion phase, during which all mobile objects (node and carriers) are moved according to the rules specified for them in the simulation configuration (mobility model).
- Object destruction phase, during which objects are removed due to lifetime expiration or because they exited from the area of interest.
- Collision management phase, which, in turn, consists of a collision-check phase, during which the objects under collision are identified, and a relocation phase, during which those objects are moved as a result of the (in)elastic collision.

The BiNS2 simulator, which is an upgraded version of BiNS, has been meticulously crafted with an emphasis on modularity and customizability. The six primary phases previously delineated are also executed in BiNS2, with respect to collision management, which encompasses a collision detection phase, during which colliding objects are discerned, and a relocation phase, wherein these objects are repositioned as a consequence of the inelastic collision (elastic collisions were managed by the original BiNS). It is evident that certain phases are exclusive to specific object types. The simulator adopts a granular methodology for addressing collisions between objects. A collision may result in either a rebound or an assimilation. The latter occurs exclusively when a carrier collides with a compliant receptor situated on the surface of a node [135].

In order to model different scenarios of MCs more accurately, in [135] has proposed a generalized methodology for addressing the complexities of spatial domains. The proposed approach is hierarchical and splits the overall simulated environment into smaller volumes of well-defined geometrical shapes. Four distinct types of domains can be identified, each differing in geometric shape: spherical, cubic, cylindrical, and unbounded. Each domain possesses comprehensive knowledge of all nanostructures contained within it, implements a mobility model that delineates the motion protocols applicable within its governed space, monitors conditions necessitating the destruction of an object, and oversees the management of collision events. Moreover, the collision management system is adaptable and can be modified in real-time to facilitate the transition between various collision sensing and handling strategies appropriate for different environmental contexts within a singular simulation [134].

Therefore, each domain possesses the capability to independently manage its own spatial parameters and all associated embedded entities. This characteristic proves advantageous for the simulation of distinct components within an identical experimental framework, which may encompass various regions characterized by specific propagation laws. Consequently, the incorporation of domains facilitates a delineation between spatial attributes and management features, such as the initialization of the simulation parameters, the administration of discrete temporal intervals, and the aggregation of outcome data. The domains that can be modeled encompass even dynamic objects; for example, to investigate the internal processes of a nanomachine or a cellular structure, if deemed necessary. In this context, the domain would oversee the positioning of (potentially) subdomains and nano-entities integrated within the modeled construct [134].

When domains are incorporated, a more comprehensive elucidation is warranted, as it constitutes one of the principal innovations of BiNS2. Indeed, the advent of domains necessitates the differentiation of collision types involving nano-objects. Consequently, it is imperative to treat collisions between two nano-objects (Fig. 2.30a) distinctly from those occurring between a nano-object and a domain boundary, irrespective of the nature of the collision, which may involve the inner domain surface (Fig. 2.30b) or the outer domain surface (Fig. 2.30c). In the scenarios depicted in Fig. 2.30b, c, it becomes essential to consider the geometry of the domain [134, 135].

Each domain is hierarchically attached to its parent domain; thus, its position in the 3D space is defined as a function of the coordinates of its parent domain, as shown in Fig. 2.31 [134].

At the highest level of the hierarchy, there is the Root domain, which constitutes a unique form of an unbounded domain. Each subordinate domain is evaluated against any nano-object that is affiliated with its parent domain by employing an

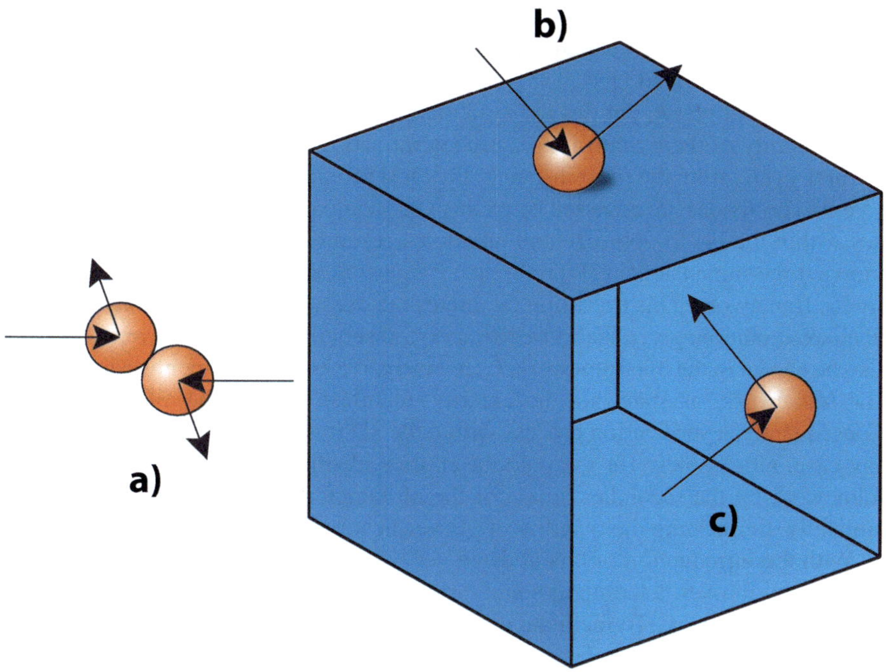

Fig. 2.30 Different types of collisions. (**a**) Between nano-objects, (**b**) external collision, and (**c**) internal collision

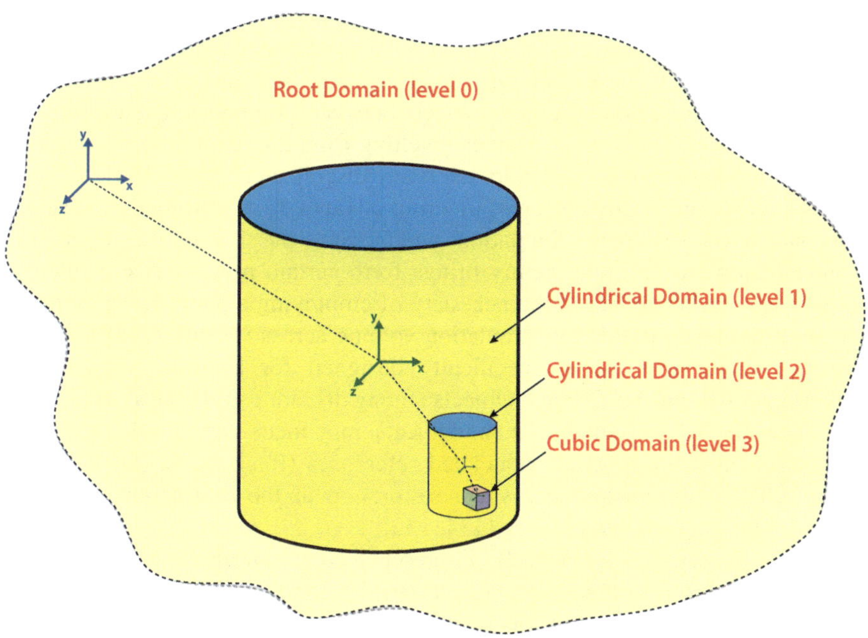

Fig. 2.31 Hierarchical management of system of coordinates in nested domains

algorithm that has been adapted from the methodology utilized to detect collisions among nano-objects. To enhance comprehension, the two principal phases of this algorithm are reiterated [134, 135].

In the initial phase, entities are systematically arranged in a sequence based on the proximity of their centroid to a predetermined reference point in a descending hierarchy (i.e., from the most distant to the most proximal). In the subsequent phase, potential collisions are assessed by exclusively comparing only the contiguous entities within the list to optimize computational resources. This indicates a transition from a three-dimensional (3D) reference framework to a one-dimensional (1D) reference framework. This transition facilitates the execution of a preliminary, streamlined examination, given that if collisions are absent in the 1D framework, they will also be absent in the 3D framework. Conversely, if a collision is identified within the 1D framework, it may not necessarily manifest within the 3D framework. Consequently, a verification process within the 3D framework must be conducted in any case, albeit solely for the pairs of entities identified as "colliding" in the 1D framework by the secondary phase of the aforementioned algorithm, thereby significantly diminishing the quantity of 3D verifications required [134, 135].

With the introduction of this updated version of the simulator, it is now feasible to execute it over a computational grid infrastructure utilizing the renowned GridGain framework. To facilitate this capability, the entire simulation domain has been subdivided into several smaller volumes, each overseen by a distinct grid node, which may be either physical or virtual in nature. Consequently, the simulation is fragmented into multiple parallel simulations, with each simulation responsible for managing a segment of the divided volume. These sub-simulations independently complete all phases associated with each discrete time step. However, it is imperative that they attain synchronization prior to advancing to the subsequent time step. More specifically, each grid node must await a synchronization signal before progressing to the next phase. This approach enables the accurate management of potential exchanges of moving nano-objects between grid nodes that are responsible for adjacent volumes (i.e., volumes resulting from the partitioning process that share a common boundary) and helps avert any discrepancies [134, 135].

This novel functionality presents numerous advantages, including the capacity to both conduct more extensive simulations and facilitate the sharing of computational resources. However, it concurrently brings forth certain performance challenges. The primary challenge lies in the necessity of employing a partitioning algorithm that can proficiently divide the simulation volume across various simulation environments, rather than being specifically designed for a singular case study. Furthermore, the transfer of Nano Objects among disparate grid nodes, whether due to collisions or the movement of Nano Objects, may incur substantial overhead, as each object must transport all associated references (for instance, to the current domain). These references necessitate restoration at the destination grid node, thereby addressing any redundancy [134, 135].

In [136], the same designers of BiNS present a communication protocol between a pair of biological nanomachines, i.e., a transmitter and a receiver, built upon MCs in an aqueous environment. In this scenery, the receiver, acting as a control node, sends a connection setup signal to the transmitter, which stokes molecules, to start molecule transmission. The molecules emitted by the transmitter disseminate throughout the environment and are subsequently absorbed by the receiver via its receptors. Upon the absorption of the requisite quantity of molecules by the receiver, a tear-down signal is generated to inform the transmitter to cease transmission. The proposed protocol establishes a bidirectional communication mechanism through the application of several techniques that were originally devised for the Transmission Control Protocol (TCP). Indeed, the proposed protocol is characterized as connection-oriented and employs TCP-like probing methodologies to ascertain an appropriate transmission rate between the transmitter and the receiver, thereby mitigating the risk of receiver congestion. In contrast to TCP, however, explicit acknowledgments are eschewed, as their inclusion would adversely affect communication throughput due to the substantial delays inherent in mobile communications (MCs). Consequently, the proposed protocol resorts to implicit acknowledgments, wherein feedback signals are dispatched by the receiver to regulate the transmission rate at the transmitter, effectively serving as explicit negative feedback. The findings derived from an extensive simulation utilizing BiNS2 are employed to substantiate the proposed protocol and to accurately calibrate the principal protocol parameters [136].

Reference [137] demonstrates a straightforward yet dependable receiver model for diffusion-based MCs systems. The intricate nature of MCs systems, characterized by a vast array of interacting elements, renders scalability an essential attribute of simulators and modeling tools. An example scenario entails targeted drug delivery systems that employ biological nanomachines situated near a biological target, capable of releasing therapeutic chemicals in a diseased area. The suggested framework meets the temporal requirements for system analysis by introducing an analogous Markovian queuing model that simulates the collective behavior of several receptors scattered on the receiver's surface. The findings in [137] indicate that the proposed method significantly aligns with simulation outcomes obtained from extensive simulations involving numerous receivers using the BiNS2 simulator, despite the result acquisition time being several orders of magnitude shorter than the simulation duration.

In [138], diffusion-based MC systems including interacting carrier molecules facilitate the exploration of several communication paradigms. Simulating these interactions on a network scale is challenging from an analytical perspective, and the ongoing identification of which pairs of particles may interact is a computationally intensive endeavor. [138] presents several techniques for identifying interacting particles. Hierarchical grid-based methods are examined and developed to maximize the parallel processing potential provided by Multiple-Graphics Processing Units (Multi-GPU) designs. A comparative analysis of the implemented algorithm is conducted against the prior simulation attributes executed by the BiNS2 simulator within a particular domain, as well as against a leading-edge collision detection library and nanosimulator [138].

Diffusion-based MCs are defined by the Brownian motion of signal molecules during their transit from the transmitting nanomachine to the receiving one, which impedes the directionality of the sent signal [139]. To enhance the reliability of diffusion-based MCs, [139] presents an improvement on the receiver side to augment the capacity for receiving biological signals. Drawing inspiration from intelligent antennas in traditional wireless communications, [139] presents a concept with a directed receiver nanomachine, intended to enhance the average concentration of signal molecules near its surface to augment the received signal intensity.

In greater specificity, it is regarded as a spherical receptor encompassing a finite array of receptors and equipped with a purely reflective shell positioned at a variable distance from the receptor surface, whose configuration is that of a spherical cap or a cylinder. The existence of the shell facilitates the retention of numerous signal molecules in proximity to the receptor surface for a limited duration. This occurrence enhances the likelihood of a signal molecule being assimilated by one of the receptors positioned on the receptor surface. Through a comprehensive simulation campaign, the intensity of the received signal acquired via the smart antennas is juxtaposed against that which can be attained under conventional settings. From the resultant analysis, the most appropriate configurations are identified concerning the aperture of the shell and the distance of the shell from the receptor [139].

Given that MCs represent a novel multidisciplinary field, including medicine, biology, and communication engineering, they are employed for the modeling of communication in nano/micro-size devices [140]. Consequently, the study of MCs systems is proposed in [140], utilizing chemical signals that convey information in gaseous or liquid mediums. Like other communication systems, minimizing errors in the transmission of information from transmitter to receiver is a primary objective in MCs.

In MCs systems, the physical features of the medium result in elevated levels of Inter Symbol Interference (ISI) and noise, hence increasing the chance of mistake. The authors in [140] assert that the acquired graphics of receiver mechanisms and signal detecting algorithms are the primary elements to be adjusted in order to reduce mistake probability. In this area, several receiver models have been presented, including reversible adsorption and desorption, the protrusion approach, ligand-receptor interactions, and linear catalytic receiver models. This study analyzes these models and offers a concise evaluation of the results obtained from their implementation using BiNS2 [140].

Cyber-physical systems (CPS) represent an emergent category of engineered systems predicated upon the interplay between cyber and physical elements, achieved through the integration of three fundamental components: communication, control, and computing. The transition of these systems to the nanoscale engenders a variety of design and implementation challenges. The elevated complexity inherent in a CPS can be attributed to the incorporation of biological components, such as engineered cells, which may function as sensors, actuators, or even controllers [141].

The study in [141] examines the efficacy of control solutions applied via the utilization of MCs in a biological nanoscale cyber-physical system (BioNanoCPS), whereby a biological nanomachine functions as an actuator to release medicament molecules, while others serve as both sensor and controller. The objective of the idea is to regulate the release rate, ensuring that target cells receive the requisite dosage of the medicine within a certain timeframe, while minimizing unwanted adverse effects. The objective is to mitigate congestion that may occur when substantial quantities of molecules are directed toward a target. To this end, [141] provides a straightforward congestion detection technique and analyzes several rate control algorithms employed to regulate the molecular release rate at the transmitter upon receiving a feedback signal from the receiver. [141] assesses the suggested methodologies concerning distribution efficiency and delivery time of molecules with a comprehensive simulation exercise.

The most recent endeavor undertaken by the designers of BiNS2 is delineated in [142], which encompasses an analysis of the complexity inherent in MCs systems characterized by a substantial number of interacting entities; thus, scalability emerges as a quintessential attribute of the BiNS2 simulator. A prototypical scenario pertains to targeted drug delivery systems, which employ biological nanomachines in proximity to a biological target, capable of releasing therapeutic molecules within the affected area [142].

Consequently, in [142], a straightforward yet dependable receiver model for diffusion-based MC (DMC) systems addresses the time required to examine such a system. The proposed model employs an analogous Markovian queuing framework that simulates the collective behavior of many receptors distributed throughout the receiver surface. It considers that molecular absorption occurs solely via receptors and that this process is not immediate, necessitating a considerable duration during which the receptor is unavailable for binding with additional molecules. The findings in [142], quantified by the number of absorbed molecules and the average count of occupied receptors, indicate that the proposed methodology aligns well with outcomes derived from particle-based simulations involving a substantial quantity of receptors, despite the proposed model yielding results in significantly less time than the simulation duration [142].

2.3.2 *Simulation Framework for Diffusion-Based Molecular Communication Nanonetworks (N3Sim)*

N3Sim[33] is a simulation framework for nanonetworks with transmitter, receiver, and harvester nodes using DMC. In DMC, transmitters encode the information by releasing molecules into the medium, thus varying their local concentration. N3Sim models the movement of these molecules according to Brownian dynamics, and it also takes into account their inertia and the interactions among them. Harvesters collect molecules from the environment to reuse them for later transmissions. Receivers decode the information by sensing the particle concentration in their

[33] N3Sim is available for download at www.n3cat.upc.edu/n3sim and related information can be found at: http://www.n3cat.upc.edu

neighborhood [143]. N3Sim may become a model of a superior application. Another possibility is to develop N3Sim as a stand-alone application with a graphical interface. Or it could be developed as a library that can be used by other developers [144].

N3Sim constitutes a comprehensive simulation framework for DMC, facilitating the assessment of the communication efficacy of molecular networks comprising multiple transmitters and receivers that emulate an infinite spatial domain characterized by a specified concentration of molecules. The transmitters convey information through the release of particles into the surrounding medium, thereby altering the concentration gradient in their immediate vicinity. The diffusion process of particles within the medium is represented as Brownian motion, incorporating the factors of particle inertia and interactions among particles. Ultimately, the receivers interpret the conveyed information by detecting the local concentration variations within their immediate environment [145].

The advantages conferred by such a simulator are manifold: the corroboration of extant channel models for MCs and the assessment of innovative modulation techniques represent merely two illustrations. It employs a tri-layered architecture. The user interface layer engages with the user to assimilate the input data requisite for the simulation, while the data layer records the simulation outcomes to files. The domain layer encompasses the MC model. N3Sim is an application developed in Java; consequently, it is operable across various operating systems equipped with the requisite Java virtual machine. The parameters for simulation are ascertained via a text-based configuration file that is amendable by the user, which delineates the values of all conceivable variables pertinent to the simulation. These variables are categorized into four distinct groups for enhanced clarity: simulation, spatial parameters, transmitters, and receivers. Furthermore, N3Sim facilitates the automation of the execution of multiple simulations in an uncomplicated manner through the utilization of user-defined scripts [145].

N3Sim has been meticulously engineered to replicate a cohort of nanomachines that engage in communication via molecular diffusion within a fluidic environment. Subsequently, the data transmitted by the sender nanomachines influences the frequency at which they discharge molecules, conceptualized as particles, into the molecular channel. For example, the conveyance of a logical bit "1" may be exemplified by the release of a specific set of particles, whereas the conveyance of a bit "0" is characterized by the lack of such emission. These discharges induce fluctuations in the local particle concentration, which disseminate throughout the medium as a result of Brownian motion and inter-particle interactions. The receiving nanomachines possess the capability to estimate the particle concentration within their vicinity by quantifying the number of particles present in a specified volume surrounding their location. Based on this quantification, they are able to decode the information that has been transmitted [143].

Figure 2.32 illustrates a schematic representation of the sequential processes requisite for executing a simulation. Initially, the researcher delineates the parameters pertinent to the simulation within a configuration document. These parameters encompass the quantity and positioning of transmitters and receivers, the signal intended for transmission, the dimensions of the emitted particles, and the diffusion

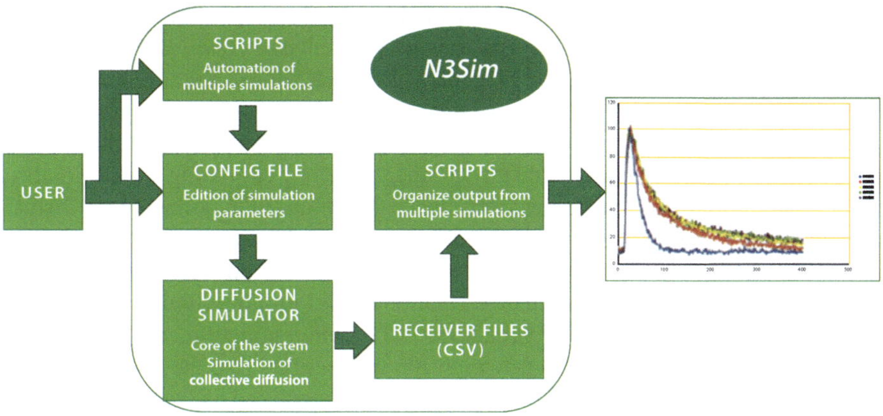

Fig. 2.32 Block diagram of N3Sim

coefficient of the medium, among other variables. A script file facilitates the execution of multiple simulations in an automated fashion utilizing a singular configuration file, thereby enabling a streamlined assessment of the impact of a particular parameter (for instance, the quantity of transmitted particles) on the system's output [143].

Next, the diffusion simulator takes the configuration file and the automation scripts as input and performs the actual simulation of the DMC scenario. The diffusion simulator meticulously calculates the position and velocity of each individual particle at every discrete time step throughout the simulation process. Upon the completion of the simulation, the resulting outputs are systematically stored in designated receiver files (one corresponding to each receiver), which encapsulate the concentration data recorded by each receiver as a function of time. Finally, an additional suite of scripts may be employed to aggregate the results obtained from multiple receivers and to visually represent them in a consolidated graphical format [143].

A fundamental component within the simulation framework is the node. A node represents a generic agent that can be specialized into various categories based on its operational function within the system. As previously indicated, N3Sim currently encompasses three principal node types as part of its MCs architecture, namely transmitters, receivers, and harvesters. The transmitter is characterized by its spatial location within the simulation environment and its dimensions, which dictate its area of influence (i.e., the zone within which it is capable of emitting particles). Each transmitter encodes the information intended for transmission into a corresponding waveform, which delineates its particle emission pattern. This waveform can be selected from a variety of predefined options, including a square pulse, a Gaussian pulse, or a pulse train. Additionally, a bespoke waveform may be established by the user as well [143].

The detection apparatus can be conceptualized as either a spherical or cubic entity, capable of quantifying the immediate count of particles within its operational detection range, thus facilitating the estimation of local concentration levels. This representation serves as an abstraction of the ligand–receptor binding dynamics observed in biological systems. Numerous parameters associated with the receiver are subject to customization; for example, upon detecting a specific ensemble of particles, the receiver may either assimilate these particles or remain entirely transparent to them. N3Sim further enables the incorporation of multiple transmitters and receivers within the simulation environment, thereby emulating a multipoint-to-multipoint communication framework. The user is required to delineate the distinctive attributes of each transmitter and receiver, specifically, their spatial coordinates, geometric configuration, and dimensions, which may vary among the different entities. This capability facilitates the investigation of various communication phenomena, such as the viability of broadcast MC communication, as well as the characteristics of interference when multiple transmitters operate simultaneously. Lastly, the third category of nodes within N3Sim is identified as the harvester node. Harvesters possess the proficiency to extract molecules from their immediate surroundings. These molecules are subsequently accumulated in a reservoir with a predetermined capacity and may be utilized in subsequent transmission events [143].

The main characteristics regarding MCs systems with NS3 are [143]:

- Particle model: The emitted particles are modeled as indivisible spheres. The main reason for this choice is the simplicity of the collision detection algorithm for this shape. Collisions between the emitted particles and the fluid molecules cause them to diffuse throughout the medium. It would be computationally infeasible to model each of these collisions individually, since the number of collisions between each particle and the fluid molecules is in the order of 10^{20} per second. Fortunately, the seemingly random movement of the suspended particles caused by collisions between the particles and the smaller fluid molecules can be mathematically modeled as Brownian motion. Assuming that the particles have no inertia, Brownian motion allows them to statistically calculate their movement, which can be modeled as a Gaussian random variable with zero mean and whose root-mean-square displacement in each dimension after a time t is $\sqrt{2Dt}$, where D is the diffusion coefficient of the medium [143].

  The high-level effect of the particles moving with a Brownian pattern is their diffusion throughout the medium, according to Fick's laws of diffusion. However, Fick's laws do not take into account the influence of interactions among the particles themselves (such as collisions and electrostatic forces), and thus, they are only valid when the particle concentration is very low and these interactions are neglected. In order to simulate environments with anomalous diffusion, which cannot be modeled by Fick's laws of diffusion, N3Sim is able to account for the inertia of the emitted particles and the interactions among them. The most relevant interactions which affect the particle diffusion process are collisions and electrostatic forces. For simplicity, collisions among particles are considered elastic (i.e., the total kinetic energy is conserved). Electrostatic forces appear when the emitted particles are ions, which have a non-zero electric charge [143].
- Simulation space: N3Sim possesses the capability to be configured for the simulation of either a two-dimensional or a three-dimensional spatial framework. A

salient parameter of the simulation space pertains to the initial concentration of particles within the medium. In instances where this concentration is nil, both bounded and unbounded spatial configurations can be effectively simulated. Conversely, when the initial concentration of particles exceeds zero, it becomes imperative for the simulation space to be bounded, thereby preventing the occurrence of an infinite particle population [143].

When a confined spatial domain is modeled, a rectangular simulation environment is postulated wherein particles reflect off the boundaries of the spatial confines. N3Sim employs this framework as the most accurate representation for potential implementations of MC. For example, within a system of communicating nanomachines situated in a blood vessel, each time a particle makes contact with the vessel wall (termed tunica intima), it undergoes a rebound. To mitigate the risk of the particles emitted by the transmitters leading to an incremental rise in background concentration over time, N3Sim integrates a discretionary mechanism that permits certain particles to vanish upon reaching the boundaries of the simulation domain, in accordance with the principles of diffusion [143].

Furthermore, entities may be strategically positioned within the experimental framework to emulate impediments between the transmitting and receiving apparatuses. This facilitates the replication of scenarios, such as one in which a consortium of bacteria traverses the medium, thereby obstructing the pathway between the transceivers and the receivers [143].

- Collision detection: The challenge of identifying instances of collision among entities exhibiting stochastic motion is well-known and extensively examined issue within the realm of algorithms. Owing to their substantial computational overhead, these algorithms constitute a significant impediment in the majority of applications that necessitate their implementation. Besides dealing with the n-body interaction, which is in itself a problem with a computational cost of $O(n^2)$, the problem of collision detection has an additional requirement. Collisions must be solved sequentially in time, since each collision changes the trajectory of the particles involved and may cause new potential collisions. In consequence, the collision detection process is mostly sequential. Since it is not possible to find a collision until the previous one has been solved, it is difficult to parallelize any significant part of the algorithm [143].

  The algorithm utilized within N3Sim represents an adaptation of the Baraff's algorithm (commonly referred to as sweep and prune, which identifies collisions a posteriori, that is, subsequent to their occurrence), thereby facilitating its application as an a priori algorithm to guarantee that no potential collisions are overlooked. The modification of the Baraff's algorithm consists of two principal components. First, rather than examining the overlap between two discrete objects, the analysis will focus on the overlap between their respective trajectories. Second, it will be imperative to document all potential collisions identified, organized chronologically, within a data structure to leverage the benefits of temporal coherence [143].

  Figure 2.33 illustrates that for two particles to undergo a collision, it is essential (although not sufficient) that the projections of their respective trajectories

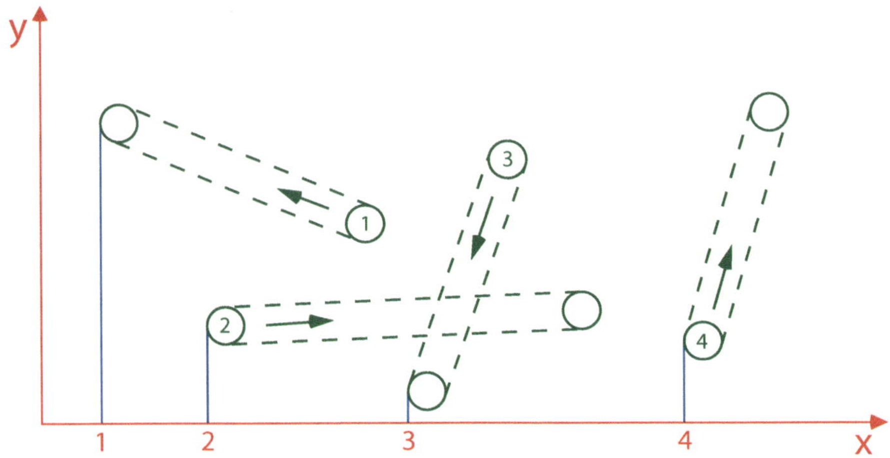

Fig. 2.33 Schematic description of the algorithm used for collision detection in N3Sim. Objects are sorted by the minimum x coordinate of their trajectories

intersect on each axis. This reasoning parallels that of Baraff's algorithm, wherein objects are replaced by their trajectories. In accordance with this analogy, the initial procedure entails sorting the list of objects based on the minimum x dimension of their trajectories. Similar to the method employed in Baraff's algorithm, each object is evaluated against the subsequent objects in the list, eliminating all subsequent entries when the trajectories of two objects do not exhibit overlap. Instances of collision are cataloged within a data structure referred to as the collisions queue, which is organized according to the temporal sequence of collisions. This way, the cost to obtain the next collision is always $O(1)$ [143].

Once the queue designated for collision events has been established, the algorithm entails the process of identifying the subsequent collision and addressing it iteratively until the queue is devoid of entries. Upon the resolution of a collision, it is imperative to acknowledge that the trajectories of the entities engaged in the collision will have undergone alterations. Consequently, two significant ramifications must be considered: first, the inventory of objects will no longer maintain its sorted order, and second, the prior collisions involving the entities implicated in the recently resolved collision will be rendered invalid [143].

To address the initial issue, it is apparent that the list of objects necessitates re-sorting. In alignment with Baraff's algorithm, an insertion sort algorithm is employed due to the near-sorted nature of the list (with merely two objects being out of order and situated nearly in their correct positions). Concerning the subsequent issue, it is imperative that all collisions involving the objects associated with the resolved collision be removed from the collisions queue. Navigating through the collisions queue to eliminate such collisions would incur significant temporal costs; consequently, a tagging system is implemented. Each object is assigned a tag; specifically, an integer that is incremented each time the object is involved in a resolved collision. Conversely, each collision possesses two tags, one corresponding to each object. These tags adopt the value of the object tag at the moment the collision is instantiated. Subsequently, when a collision is retrieved from the collision queue for resolution, if the collision tags do not align with the object tag, this indicates that the objects have previously engaged in a collision; thus, the current collision must be disregarded, and the subsequent one must be addressed. Following this procedure, the cost of discarding collisions is $O(1)$ [143].

Finding the new collisions for the objects involved in a collision means comparing the object with the rest of the objects in the sorted list (as it is done in the first part of the algorithm) and saving the new collisions into the collisions queue. The cost of this step is $O(n \log(n))$, $O(n)$ to compare the object with the object in the list and $O(\log(n))$ to save each collision in the collision's queue [143].

Finally, the data structures that store collisions have been modified to improve the memory costs. The problem that led to these changes was that in simulations of large time steps and high number of collisions, there were too many false collisions. This produced that the collisions queue grew too much. To avoid this problem, a list of collisions is associated with each object. This list is deleted and recalculated when a particle has a collision. This way, the memory cost is kept constant [143].

In subsequent paragraphs are cited some cases of N3Sim simulations. Thus, in [143] N3Sim is used for analyzing the influence of neighboring absorbing receivers in a point-to-point DMC link, following a simulation-driven approach. It is shown that the distance from the transmitter-receiver link, the distance between receivers, and their radius have a noticeable impact upon both amplitude and signal detection. The system model has been described through the probability when no absorption of hitting molecules has been modeled by [143]:

$$F = \left(t \middle| r_n, D, D_f\right) = \left(\frac{r_n}{r_n + D}\right) \operatorname{ercf}\left(\frac{D}{\sqrt{4 D_f t}}\right) \tag{2.1}$$

where D_f denotes the diffusion coefficient and t the time. There is no expression modeling the probability where there is absorption of hitting molecules. Hence, [146] characterizes this situation with nearby receivers. For that it is considered a point-to-point DMC link in Fig. 2.34, which is passively affected by an additional nearby absorbing receiver. The two receivers are located at the same distance D from the transmitter and maintaining a distance d between them. The radius of the receivers is denoted by $r1$ and $r2$ [146].

In [147] is studied the ISI caused by the heavy tail of the impulse response in DMC systems, because these systems produced free diffusion when particles are emitted to carry the message information between the transmitter and the receivers. Then, in [147] analyzes the influence of neighboring absorbing receivers upon ISI of a DMC point-to-point link. The authors in [147] conclude (through simulation results with N3Sim) that neighboring absorbing receivers have a noticeable impact

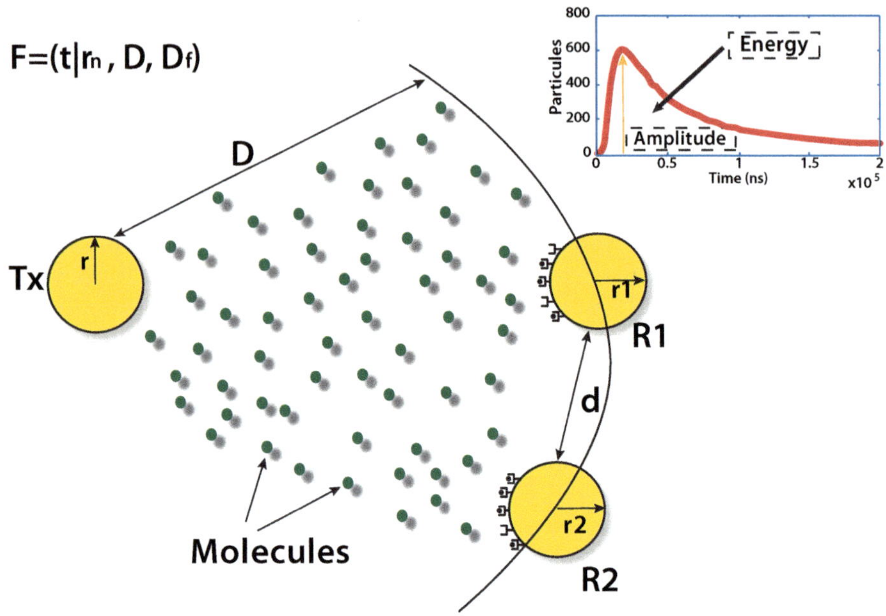

Fig. 2.34 Considered point-to-point DMC link with interfering absorbing receiver

on reducing the tail of the detected pulse-shape and, hence, higher achievable throughput is reachable; besides, the optimal number of neighboring absorbing receivers and the optimal distance between the receivers lead first to a steady state in the maximum amplitude of the impulse response and cancel the effect of ISI by reducing the amplitude of the tail proper of the diffusion systems in the molecular transmission [147].

Another case of the use of N3Sim is shown in [148], the analyzed topology is exhibited in Fig. 2.35, in which environment with changeable concentration of nanodevices was created. There, it has been defined the communication area bounds to 2500 nm × 2000 nm. The simulations were performed in aqueous environment with diffusion coefficient $D = 0.1\frac{\text{nm}}{\text{ns}^2}$. For the better possibility to set up each simulation scenario, the settings for the emitter and receiver were divided into two separate sections in the code. In all simulation scenarios, one emitter and one receiver were defined. The emitter had radius of the influence area set up to 100 *nm*. [148] concludes (through Figs. 2.35 and 2.36) that with active collisions between the emitted particles the number of the successfully received particles on the side of the receiver is slightly higher in comparison with the blue curve (simulation without the collisions between the particles) [148].

The authors in [148] determine that the dynamic distance between the emitter and the receiver and the values of the successfully received particles are lower till the distance between nanodevices is lower than $d = 1.5$ *mm*. This behavior is caused by the character of simulated space (area) which influences the results for this scenario [148].

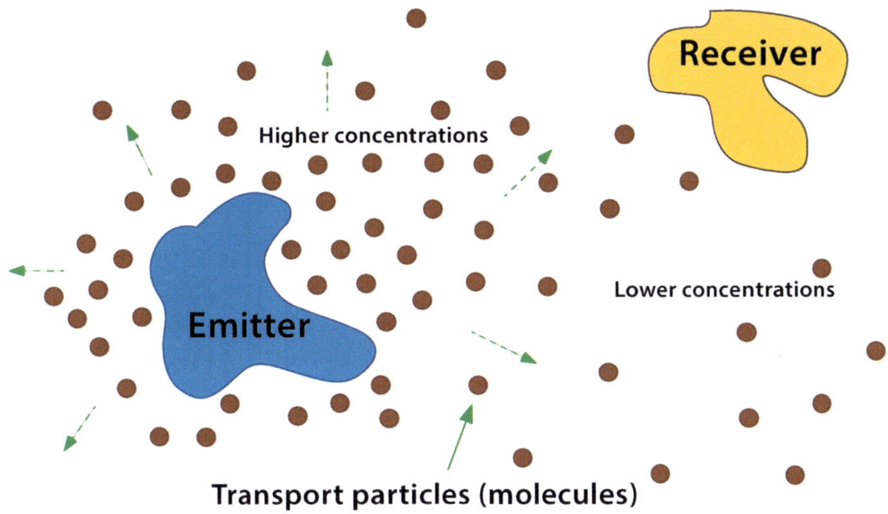

Fig. 2.35 Generic nanonetwork's topology

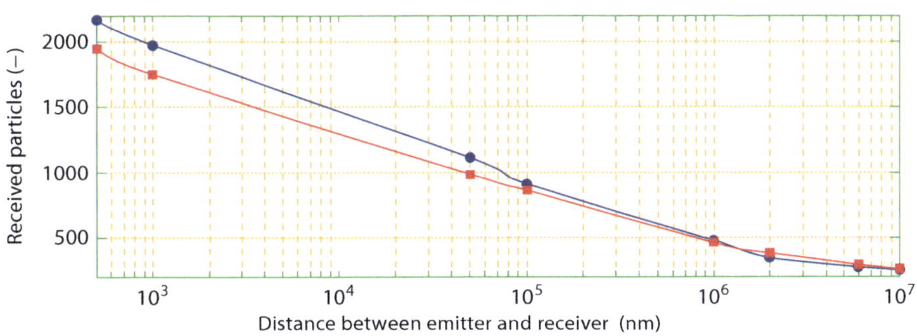

Fig. 2.36 Obtained results in N3Sim for the topology in Fig. 2.35

Also, N3Sim is utilized in [149] to analyze an MC system that is illustrated in Fig. 2.37. This model has a cooperative communication system with two-hop communication. The number of hops will be larger with more relay nanomachines. In each hop, the unguided DMC is used for realizing wireless communication between nanomachines [149].

The system in [149] comprises components that operate as information molecules, which embody the data intended for transmission, transmitter nanomachines that facilitate the release of these information molecules, receiver nanomachines that function to detect the presence of information molecules, relay nanomachines endowed with the capability to emit and receive molecules, and the surrounding environment through which the information molecules propagate from the transmitter nanomachine to the receiver nanomachine. It is posited that each nanomachine

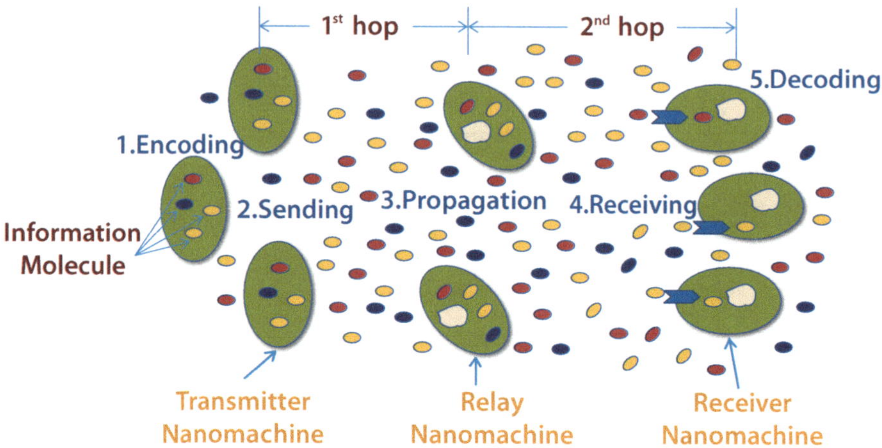

Fig. 2.37 Molecular communication and network model

possesses a self-identifying label that can be affixed to the transmitted molecules, thereby establishing an addressing scheme [149].

The informational molecule encompasses the sender's address, the data content, and the recipient's address. The communication processes at each intermediary step consist of: encoding the data into an informational molecule by transmitting nanomachine, transmitting the informational molecule into the surrounding environment, the propagation of the informational molecule through the environmental medium, reception of the informational molecule by the receiving nanomachine, and ultimately decoding the informational molecule into a chemical reaction at the receiving nanomachine. Simulation results indicate that the BER is significantly reduced when employing the cooperative technique (relay) [149].

In another case, the application in N3Sim in [149] uses a multiple transmitter local drug delivery system associated with encapsulated drug transmitters. One of the limitations of drug delivery systems is the reservoir capacity. In order to enhance the operational longevity of drug-delivering nanomachines, and consequently, the duration of drug administration protocols, the system incorporates encapsulated pharmaceutical transmitters. Encapsulated pharmaceuticals exhibit a lack of reactivity with their external environment unless they are released within a drug transmitter nanomachine. Consequently, long-range transmitters do not exert deleterious effects on the healthy tissues of the organism. The merit of this methodology lies in its capacity to prolong the interval between successive drug administrations while simultaneously mitigating toxicity levels. As a consequence, this approach fosters improvements in the mental well-being of patients and diminishes the financial burden of treatment. The operational lifespan of this drug delivery mechanism is contingent upon the spatial distribution and topology of encapsulated drug transmitters rather than their individual release rates. Furthermore, a lower bound is established concerning the anticipated lifespan of a Poisson-distributed random network of nanomachines [150].

For affecting the mentioned research in [150], three protocols are designed to increase the lifetime of the drug delivery system. The initial two protocols represent a direct augmentation of the drug delivery system, whereas the third protocol facilitates the further activation of inactive nanomachines to liberate encapsulated pharmaceuticals. The duration of the drug delivery system is characterized as the span of efficacy or validity, which refers to the interval during which the drug transmitters dispense the pharmaceutical agents at their optimal rates that fulfill the minimum effective concentration at the designated site. Consequently, the operational lifespan of the system concludes immediately upon the depletion of the first or the least durable drug transmitter. It is established that drug transmitter nanomachines exhibit a continuous emission pattern with a constant release rate as long as their reservoirs contain a sufficient quantity of molecules. These encapsulated pharmaceuticals remain inert with respect to environmental interactions until they are encountered by an active drug transmitter. The drug transmitter subsequently unpacks the encapsulated pharmaceutical and releases it in an uncapsulated form [150].

Continuing with drug delivery systems analysis, this topic is also investigated in [151], where a multiple transmitter drug delivery system is proposed which releases drug molecules at an optimal rate to maintain the minimum effective concentration inside the tumor. Both single and multiple transmitter scenarios are examined. In the initial scenario, the optimal rate of transmitter release, along with the optimal density of deployed transmitters, is derived for a specified tumor size through mathematical formulations, and the findings are corroborated through simulations conducted in N3Sim and MATLAB, which serves as an interface to execute N3Sim simulations with appropriate parameters in an automated fashion, as well as to analyze the resultant data. In the subsequent case involving multiple transmitters, the optimal allocation rate for each transmitter is computed in order to attain the minimum cumulative release rate. The influence of tumor size and the density of the deployed transmitters are also scrutinized in [151].

Moreover, the medical domain utilizing N3Sim simulations in [152] examines a molecular scenario including nanomachine applications for targeted drug administration, wherein drug molecules released by nanomachines disseminate and chemically interact with tumor cells in an aqueous environment. If the nanomachines discharge drug molecules more rapidly than the tumor cells can respond, it will lead to the loss and wastage of drug molecules. This is a possible issue related to the interplay between reaction rate, release rate, and efficiency. [152] seeks to investigate this architecture using two drug reception models, with the authors intending to provide a framework for developing a drug release control mechanism. Two analytical approaches have been adopted: one focuses on the drug reception process through tumor contact, while the other is based on Michaelis-Menten enzymatic kinetics. The analytical findings about the correlation between reaction rate, release rate, and efficiency, as shown in [152] align well with the numerical simulation results in a three-dimensional context. Utilizing two drug reception models, their findings would aid in the development of a control mechanism for nanomachine-mediated drug release [152].

Regarding another medical application with N3Sim, in [153] is proposed Alzheimer's disease research with MCs. Alzheimer's disease affects multiple brain functions; symptoms usually develop gradually and worsen with time. Alzheimer's disease is driven by the production and deposition of the Amyloid *βeta* plaques into the synaptic cleft. In [153], the physical channel is characterized with an MC system between glutamate vesicles and AMPA receptors, as well as the effect of increasing the number of *A βeta* on this channel. In addition, [153] introduced a new closed-loop system vision which leverages such characterization to infer *A βeta* concentration for fighting Alzheimer's disease while using serotonin which, as is widely accepted, can reduce the production of *A βeta* proteins. The characterization of the physical channel of an MC between glutamate vesicles and AMPA receptors shows that the aggregation of *A βeta* proteins is an indicative of the channel loss [153].

2.3.3 Calcium Signaling Molecular Communication Sim (CalComSim)

CalComSim or Calcium Signaling-based MC System Simulator is a calcium simulator, designed for both synthetic and natural communications found inside a 3D human tissue. An integration of the biological models (that are incorporated into the simulator are based on real experimental data [145]) for the signaling process governed by stochastic solvers is used to handle multiple and parallel reactions in each cell of the tissue. Also, it implements three different types of tissues. They can be categorized into three distinct classifications: non-excitable (tissues incapable of propagating electrical impulses), excitable (tissues proficient in the propagation of electrical impulses), and hybrid (tissues that possess the capacity to communicate with both excitable and non-excitable tissues). The various cell types modeled within the simulation are delineated as follows: epithelial cells (non-excitable), smooth muscle cells (excitable), and astrocytes (hybrid). The biological models integrated into the simulator are predicated upon empirical experimental data. Furthermore, the simulator encompasses the stochastic dynamics pertaining to the opening and closing mechanisms of gap junctions for each respective cell type [154]. The simulator can be used not only by telecommunication engineers and biological scientists but also by pharmaceutical researchers that can design new drugs and treatments for diseases that emerge from impaired calcium signaling [22]. CalComSim establishes the integration of the biological models for the signaling process governed by stochastic solvers and is used to handle multiple and parallel reactions in each cell of the tissue.

CalComSim has been implemented in Python [155] and requires the following dependencies: NumPy, SciPy, PyPy, and CPython. NumPy and SciPy: They are used to access mathematical and scientific functions, while the PyPy is an alternative interpreter for the simulator. The CPython is used to run C libraries that can be used in python codes. The Gillespie stochastic algorithm was used to solve partial differential equations that model intracellular and intercellular calcium signaling. The simulator is controlled by a command line interface [22].

In [156], CalComSim is explicitly analyzed; this simulator models biochemical reactions and diffusion processes within a 3D tissue-like environment. The core of the simulation is based on the Gillespie Stochastic Simulation Algorithm (SSA), which is a method for simulating the time evolution of a set of coupled chemical reactions. CalComSim is particularly relevant for studying MCs in biological tissues, which has been explored in the context of calcium signaling and cellular tissue deformation. Hence, the Gillespie SSA is a stochastic method used to simulate the time evolution of chemically reacting systems. Unlike deterministic models, SSA accounts for the intrinsic randomness in chemical reactions [156]. For Gillespie SSA, CalComSim establishes the characteristics defined in the following paragraphs [156].

- Reactions and propensities: Consider a system of N chemical species $\{S_1, S_2, \ldots, S_N\}$ interacting through M reaction channels $\{R, R_2, \ldots, R_M\}$. Each reaction R_j has an associated propensity function $a_j(x)$ and a state-change vector v_j [156].
  - $x(t) = (x_1(t), x_2(t), \ldots, x_N(t)\ )$ represents the number of molecules of each species at time t [156].
  - The propensity function $a_j(x)$ gives the probability per unit time that reaction R_j will occur, given the current state x [156].
  - The state-change vector v_j specifies the net change in the system state if reaction R_j occurs [156].
- SSA algorithm steps: The Gillespie SSA proceeds as follows [156]:
  - Initialization: Initialize the system state $x(t_0)$ and set the initial time $t = t_0$ [156].
  - Calculate propensities: For each reaction R_j, calculate the propensity $a_j(x)$ [156].
  - Calculate time to next reaction: Calculate the time τ to the next reaction through Eq. (2.2) [156].

$$\tau = \frac{1}{a_0(x)} \ln\left(\frac{1}{r_1}\right) \tag{2.2}$$

    where $a_0(x) = \sum_{J=1}^{M} a_j(x)$ is the total propensity, and r_1 is a uniform random number in (0,1) [156].
  - Select the reaction: Select the reaction R_j that will occur using the cumulative sum approach, through j, as Eq. (2.3) indicates [156]:

$$\sum_{k=1}^{j-1} a_k(x) < r_2 a_0(x) \le \sum_{k=1}^{k} a_k(x) \tag{2.3}$$

    where r_2 is another uniform random number in (0,1).

– Update the system state: Update the state $x \rightarrow x + v_j$ and advance the time $t \rightarrow t + \tau$ [156].
– Iteration: Repeat the steps from the propensity calculation to update the system state. This must be done until the desired simulation time is reached [156].

CalComSim models a series of biochemical reactions, each with a specific rate and propensity. The general form of the propensity function for a reaction is [156]:

$$a_j(x) = c_j h_j(x) \tag{2.4}$$

where c_j is the rate constant and $h_j(x)$ is a function of the current state x that represents the number of distinct molecular reactant combinations available.

As an example, reaction in [156] is considered a simple reaction as follows [156]:

$$S_1 + S_2 \xrightarrow{k} S_3 \tag{2.5}$$

The propensity function for this reaction would be [156]:

$$a(x) = k x_1 x_2 \tag{2.6}$$

where x_1 and x_2 are the molecule counts of species S_1 and S_2, respectively.

Regarding the diffusion process, CalComSim models molecules in the 3D tissue as a set of reactions where molecules move between neighboring compartments. The diffusion rate is determined by Fick's first law [156]:

$$J = -\frac{\partial C}{\partial x} \tag{2.7}$$

where J is the diffusion flux, D is the diffusion coefficient, and C is the concentration.

In the discrete setting of the simulator, diffusion is treated as a reaction with a propensity proportional to the concentration difference between neighboring compartments [156]:

$$a_{\text{diff}}(x) = D\left|C_i - C_j\right| \tag{2.8}$$

where C_i and C_j are the concentrations in adjacent compartments [156].

Regarding MI, as it quantifies the amount of information obtained about one random variable through another. Thus, in the context of CalComSim the emphasis is specified in the MI between the transmitter and receiver states [156].

- Calculation of probabilities: Let X represent the state at the transmitter and Y represent the state at the receiver. The probabilities $P(X)$ and $P(Y)$ are calculated based on the states recorded during the simulation. The joint probabilities $P(X, Y)$ are also calculated [156].

- The MI $I(X; Y)$ is given by Eq. (2.9) [156].

$$I(X;Y) = \sum_{x \in X} \sum_{y \in E} P(x,y) \log\left(\frac{P(x,y)}{P(x)P(y)}\right) \tag{2.9}$$

where $P(x, y)$ is the joint probability of $X = x$ and $Y = y$, and $P(x)$ and $P(y)$ are the marginal probabilities [156].

The simulation code adheres to the mathematical frameworks delineated earlier. It commences by initializing the system, proceeds to calculate propensities, and methodically iterates through the SSA algorithm to model the temporal progression of the biochemical system. Diffusion is regarded as a specific instance of reaction dynamics, and MI is derived from the temporal sequences of transmitter and receiver states [156].

The use of CalComSim is wide in many calcium transmission investigations in MCs [157, 158]. For instance, in [155] is proposed an adaptive error control for cell signaling-based MC (i.e., that which encodes information based on the concentration of molecules) governed by reaction–diffusion processes mediated by gap junctions. Cell-signaling MC is found in cells such as astrocytes, epithelial cells, and smooth muscle cells (present in the heart, brain, kidneys, and muscles, among others). Networks based on cell signaling perform tasks directly in the tissue. Bio-devices coupled to natural tissues can monitor pathologies associated with failures in cellular communication that affect the performance of natural cellular processes.

An example of the mentioned biological analysis is utilized for MCs to detect tumors. Implantable devices with synthetic molecular receptors (i.e., developed through synthetic biology) have improved cancer detection efficiency and accuracy by identifying tumor-associated molecules in biological communication. Among the various pathologies that manifest in signaling cells and which these synthetic networks have the potential to ameliorate are breast cancer, attributable to genomic damage in epithelial cells, Alzheimer's disease linked to molecular dysregulations within astrocytes, and vascular pathologies in smooth muscle cells that may exhibit inherent anomalies as a consequence of monogenic disorders [155].

Another outstanding work is developed in [22] with a theoretical approach in line with existing models of cell-signaling MC. Advances in this research domain occur mainly through simulations that are used to identify and evaluate new communication solutions. This study focuses on the error control proposal for cell signaling-based MC channels. Thus, [22] uses astrocytes and inositol triphosphate (IP_3) and calcium (Ca^{2+}) molecules as reference models. Astrocytes are fundamental cells in the brain that control synaptic functions and present a high degree of molecular diversity. The astrocytes' most important information molecules are the IP_3 and Ca^{2+}, which are also important for other signaling cell types in the body (epithelial and smooth muscle cells, among others) [22].

The study in [22] examines a synthetic MC system involving bio-devices integrated inside a biological cellular milieu. The error control design presupposes a unidirectional single-hop MC system consisting of a hybrid transmitter, reception bio-devices (i.e., biological devices including electrical components), and a channel (biological cells situated between the transmitter and receiver bio-devices). Cell signaling is incapable of supporting bidirectional communication for several reasons, including [22]:

- The fact that transceivers have not yet been demonstrated for calcium–cell signaling-based communication (i.e., the transmitter always remains a transmitter and does not have a receiver mode and vice versa for the receivers) [22].
- Calcium flows unidirectionally depending on the source of the calcium signal [22].
- A bidirectional communication has not yet been demonstrated for cell signaling-based channels [22].

In the same investigation, the MC systems are composed by [22]:

- The transmitter (Tx) represents a bio-device with the capacity to either store or generate molecules for the purpose of encoding information, and it encompasses additional functionalities, such as (1) microfluidic channels whose dynamics are solely governed by diffusion reactions, which facilitate the controlled release of molecules based on their concentration and type. These channels may be fabricated by employing microfiltration methodologies and (2) a sensor adept at quantifying the concentration of molecules and converting the biochemical data into an electrical equivalent format of binary information (i.e., bit-0 or bit-1). A plausible approach is to deploy novel micro-/nanosensors that possess the capability to detect chemical and biological signals, for instance, utilizing principles of light emission and confocal microscopy. When specific molecules undergo energy release from chemical bonds, they are capable of emitting visible light through the phenomenon of bioluminescence. This mechanism can be harnessed to transduce biochemical signals into binary bits [22].

  The transmitter can encode data by concentrations of two types of information molecules. It follows Minimum-Shift Keying (MSK) modulation to transmit M1 molecules (Ca^{2+}) in symbol-1 periods (bit-1), whereas it transmits M2 molecules (IP_3) in symbol-0 periods (bit-0). The Tx node executes the retransmission procedures based on the adaptive error control that assumes an asynchronous single-hop MC system composed of a hybrid transmitter, receiver bio-devices, and channel. The synthetic communication (i.e., coding based on concentration and type of molecules released at the transmitter in a controlled manner) is propagated cell to cell (i.e., molecules pass for each biological cell until they arrive at the receiver) by a diffusion process. The biological cells follow the MC model presented in subsequent paragraphs and the two conditions for astrocytes; they are healthy and reactive [22].

  In the cited research [22], the authors assume unidirectional communication (i.e., the direction of encoded data flow only occurs from transmitter to receiver bio-device). This unidirectionally constrains error correction methodologies predicated upon two-way communications (for instance, facilitating the exchange of bidirectional messages among nodes upon the occurrence of an error and the receiver issuing a request for symbol retransmission). Consequently, the proposed error control mechanism assesses the necessity for retransmission in accordance with an algorithm that computes the bit error probability contingent upon channel noise. This methodology encompasses two distinct phases, namely, (1) molecular noise inference and (2) retransmission adaptation, which entails five sequential steps, as illustrated in Fig. 2.38.

Exploiting synthetic biology techniques has value in the designing of signal processing circuits. The artificial manipulation of cellular systems has facilitated the production of engineered biomolecules, which supports various strategies, including the advancement of synthetic cellular signaling pathways. Furthermore, the engineering of genetic circuits within biological cells contributes to the development of programmable synthetic bio-devices. The statistical analysis and information transfer among engineered cells, which is predicated upon molecular interactions and biochemical reactions, have been investigated for the purpose of noise measurement in error correction methodologies as indicated in [22].

- In [22], the receiver bio-nanodevice (Rx) possesses supplementary capabilities to quantify the concentration of various molecules and perform statistical analysis. Rx acquires the molecules and interprets the data they convey. Rx quantifies the concentration of molecules upon their interaction. Within the cellular environment, there exists a collection of IP_3 and calcium receptors that facilitate the distributed adhesion of molecules and the quantification of the concentration of the received molecules. In the receiver, symbol interference transpires when the transmitted signal is compromised by the overlay of noisy molecules within the originally designated spectrum (i.e., as a result of the continuous variability in molecular concentrations), leading to a distortion of symbols (i.e., bit-0 is erroneously recognized as bit-1 and vice versa) [22].
- The channel comprises the propagation of IP_3 and Ca^{2+} in [22]. It incorporates the phases of intracellular and intercellular signaling. Within the cellular environment (intracellular), various biochemical reactions modulate the concentration of these molecules. Intercellular communication transpires when gap junctions within adjacent cells are activated. Subsequently, molecules disseminate through the cytosol (i.e., the aqueous component that occupies the cell cytoplasm) between two proximate cells [22].

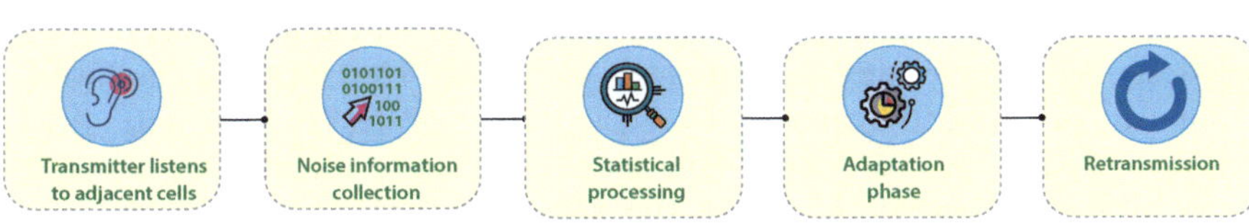

Fig. 2.38 Adaptive retransmission overview

> To achieve near-error-free transmission, [22] recommends employing the binary erasure channel idea. In the binary channel, the transmitter conveys a bit using one of the two symbols (i.e., bit-0 or bit-1). The Rx obtains either a symbol (bit-0 or bit-1) or an erasure symbol (indicating that no estimate for the transmitted signal is available). Consequently, if the Rx gets a symbol, it presumes that this was, indeed, the sent signal. If the receiver is unable to recognize the sign owing to interference, it discards it and awaits retransmission. The transmitter autonomously commences retransmission contingent upon the noise measurement at the source and its correlation with the bit error probability [22].

The authors in [22] posit that solely the nodes responsible for transmission and reception are bio-devices endowed with supplementary capabilities to encode and decode information in accordance with molecular concentration. These bio-devices leverage the intrinsic cellular components of the organism to facilitate the transmission of information (i.e., utilizing biological cells that are naturally occurring in the environment, devoid of genetic modification, and lacking additional capabilities). The employment of molecules that are inherently available within the organism as carriers of information, alongside the utilization of the biological milieu as a communication channel, constitutes the fundamental principle of MC frameworks for intrabody networks comprising implanted bio-devices, which notably emphasize considerations of biocompatibility [22].

The authors in [22] consider the cellular environment with an area composed of $i \times j \times k$ astrocytes (c) following a three-dimensional grid organization where $c_{i,j,k} = (i = 1, \ldots, I : j = 1, \ldots, J; k = 1, \ldots, K)$ indicates the position of an arbitrary cell that contains both intracellular and intercellular reactions and N distinguishable types of molecules $\{M_1, \ldots, M_N\}$. The cells are connected with a maximum of six neighboring cells through gap junctions. The model for connections between cells follows the study of topologies found in astrocytes. The network uses Ca^{2+} and IP_3 signaling in astrocytes, once these cells support the molecular propagation for long distances. The mesoscopic type of diffusion of Ca^{2+} and IP_3 molecules is mediated by gap junctions that connect the cytosol of two cells (Fig. 2.39). The communication model applies the Exact Stochastic Chemical Reaction–Diffusion ODE solution from the Gillespie algorithm. The ODE-based simulations produce accurate variability of the chemical reactions and serve to study noise effects caused by inherent stochastic behavior [22].

Since the neighboring cells of the communication system (between Tx and Rx) are part of the natural environment, they can receive and transmit molecules (i.e., as part of their regulatory process), which can cause interference in synthetic communication [22]. To simulate the behavior of the natural MC environment, this work follows a baseline model for calcium signaling based on the well-accepted ODEs that describe the molecules' oscillations in astrocytes. The algorithm employed let to simulate cellular MC channels to validate the use of Ca^{2+} and IP_3 molecules for multicarrier MC. The use of these molecules permits us to differentiate the symbols (i.e., Ca^{2+} for bit-1 and IP_3 for bit-0). The study found that the signal-to-noise ratio showed a low overlap between the concentrations of molecules, which reduces the

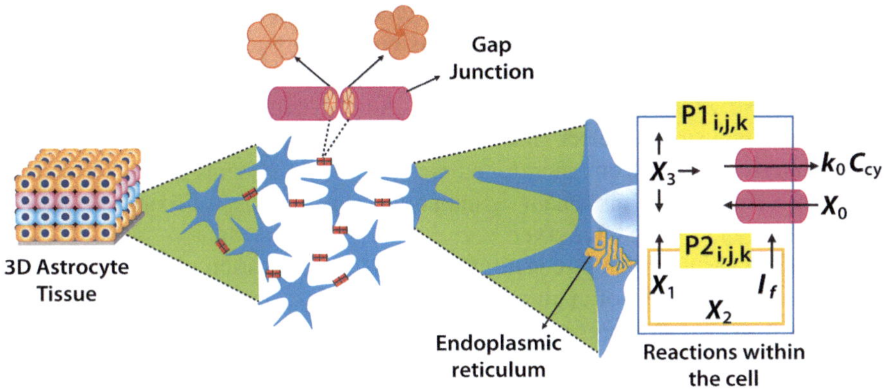

Fig. 2.39 Three-dimensional tissue with the intracellular and intercellular processes. P1 is the flow of IP_3 or Ca^{2+} from the cytosol into the extracellular space and P2 is the diffusion of Ca^{2+} from the ER to the cell cytosol

risk of interference between symbols. The model considers the molecules' storage areas (pools) for the variation of Ca^{2+} concentration in the cytosol (C_{cy}), Eq. (2.1); the variation of Ca^{2+} concentration in the endoplasmic reticulum (C_{er}), Eq. (2.2); and the variation of IP_3 concentration (IP_{3cy}) in the cytosol, Eq. (2.3) [22].

2.3.4 *COMSOL Multiphysics*

COMSOL Multiphysics Simulation Software[34] previously referred to as FEMLAB prior to 2005 serves as a sophisticated tool employed by engineers and researchers for the simulation of designs, devices, and processes across various domains of engineering, manufacturing, and scientific inquiry. This multiphysics software facilitates the simulation of real-world designs, devices, and processes within its comprehensive framework. The platform is engineered for the development of physics-based models and simulation applications. COMSOL Multiphysics encompasses the Model Builder, Application Builder, and Model Manager components. The Model Builder integrates all functionalities and operations necessary for constructing, solving, visualizing, and assessing models. The Application Builder provides users with the requisite tools to develop their own simulation applications. The Model Manager functions as a dedicated workspace for the organization and management of models and applications.

[34] All information related with COMSOL Multiphysics® and detailed in next paragraphs is taken from webpage at: COMSOL—Software for Multiphysics Simulation. The current software version is Version 6.2.0.415 (deployed in May 15, 2024). To use this feature, users need to have a COMSOL license with active subscription.

The COMSOL Multiphysics software offers a consistent user interface and experience that remains uniform across various engineering applications and physical phenomena. Supplementary modules furnish specialized capabilities for electromagnetics, structural mechanics, acoustics, fluid dynamics, thermal transfer, and chemical engineering. Select from a range of LiveLink™ products to establish direct interfaces with Computer-Aided Design (CAD) systems and other external software solutions. Implement simulation applications utilizing COMSOL Compiler™ and COMSOL Server™.

COMSOL Multiphysics® constitutes a versatile simulation platform employed across various domains of engineering, manufacturing, and scientific inquiry. This software provides an integrated environment for both fully coupled multiphysics and single-physics modeling, alongside comprehensive model management and intuitive tools designed for the development of simulation applications. Enhance the impact of simulation across users' design teams, manufacturing divisions, testing laboratories, clientele, and other stakeholders by disseminating users' applications via COMSOL Compiler™ and COMSOL Server™. Supplementary modules deliver targeted functionality for domains, such as electromagnetics, structural mechanics, acoustics, fluid dynamics, thermal transfer, and chemical engineering. Interfacing solutions are offered for CAD systems and additional third-party applications. All supplementary products within the suite integrate effortlessly with COMSOL Multiphysics, ensuring a consistent modeling workflow irrespective of the specific application being addressed.

To simulate real-world designs, devices, and processes, multiphysics software from COMSOL exhibits the following characteristics:

- General-purpose simulation software based on advanced numerical methods.
- Fully coupled multiphysics and single-physics modeling capabilities.
- Complete modeling workflow, from geometry to results evaluation.
- User-friendly tools for building and deploying simulation apps.

COMSOL Multiphysics represents a sophisticated simulation platform that offers comprehensive capabilities for both fully coupled multiphysics and single-physics modeling. The *Model Builder* encompasses all requisite stages in the modeling workflow—ranging from the definition of geometries, the specification of material properties, and the establishment of the physics that elucidate particular phenomena, to the execution of computations and the assessment of results. By adhering to a standardized modeling workflow with the Model Builder, all modeling endeavors conducted with COMSOL Multiphysics facilitate transitions between simulating electromagnetics, structural mechanics, acoustics, fluid dynamics, thermal transfer, and chemical reaction phenomena within a singular software environment. Furthermore, this software adeptly integrates physical phenomena from these diverse domains into a unified model. The Model Builder within COMSOL Multiphysics delivers an all-encompassing simulation environment and a coherent modeling workflow from inception to conclusion, irrespective of the specific design or process one seeks to analyze and develop.

The modeling workflow encompasses:

- Geometry and CAD
- Physics-based modeling
- Equation-based modeling
- Meshing
- Studies and optimization
- Solvers
- Visualization and results evaluation

The section *Model Builder Features and Functionality* allow users to administer models and simulations through the Model Manager. The Model Manager enhances and optimizes the processes of modeling and simulation by offering tools for the administration of models and applications. This encompasses model management as well as methodologies for collaboration and the centralized organization of models, incorporating version control that systematically records modifications and updates to the models. The Model Manager is directly accessible via the COMSOL Multiphysics user interface and provides the capability to utilize a local database or establish a connection to a remote server database.

There exist methodologies for the efficient preservation of models and applications (Fig. 2.40), encompassing solely pertinent information related to drafts and revisions, in addition to supplementary data types, such as CAD, mesh, and experimental datasets. The identification of models and applications within the Model Manager is facilitated through user-defined tags that assist in the systematic organization of files, as well as the capability to conduct searches for specific features or parameters. The regulation of access enables the management of models through the delineation of user groups and associated permissions.

When the designer has developed a model, the Application Builder is employed to convert it into a simulation application featuring a specialized user interface that can be accessed by collaborators and clients lacking expertise in simulation software. In order to facilitate the organization of the developed models and applications, the COMSOL Multiphysics platform encompasses the Model Manager, which serves as a tool for the management of modeling and simulation, offering version control and optimized storage solutions.

The Application Builder (Fig. 2.41) facilitates the development of user-friendly interfaces by simulation specialists for their computational models, resulting in bespoke applications that are ready for immediate use. Consequently, the application user can concentrate on the relevant input parameters and computational outcomes without necessitating prior knowledge of the fundamental model. The Application Builder is incorporated within the Windows version of COMSOL Multiphysics and is readily accessible through the COMSOL Desktop environment. While you need Windows to *build* apps, they can be run on macOS and the Linux® operating system as well. The Application Builder in COMSOL Multiphysics® provides all functionality needed to build and maintain comprehensive simulation apps. For example, users can:

Fig. 2.40 All add-on and interfacing products connect seamlessly via COMSOL Multiphysics®

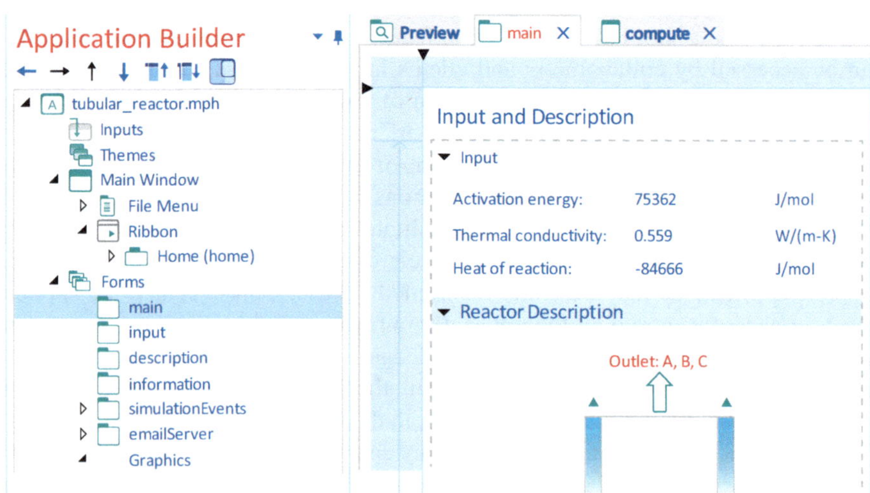

Fig. 2.41 Application Builder environment in COMSOL Multiphysics

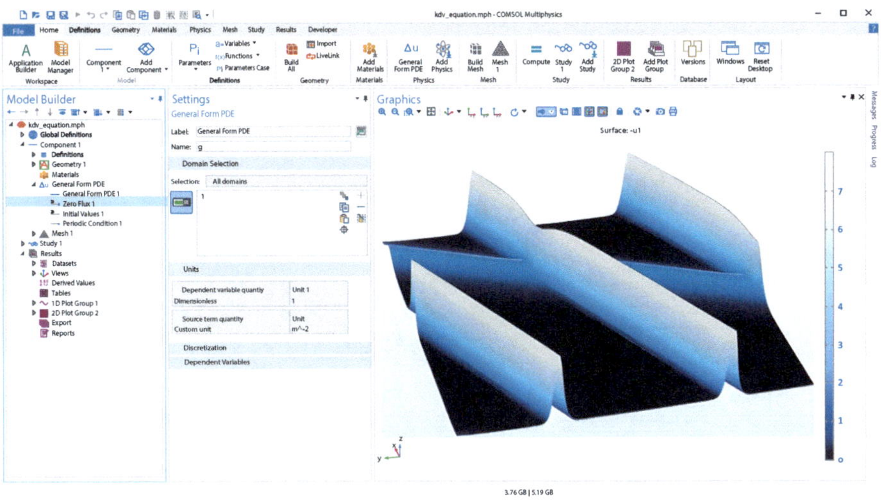

Fig. 2.42 Application environment for equation-based modeling COMSOL Multiphysics®

- Include user documentation, checks for "inputs within bounds," and predefined reports at the click of a button.
- Build the geometry around a parametric CAD model.
- Import text or binary files with experimental data.
- Configure the app to automatically send an email to a select set of recipients when a computation is finished.
- Generate reports in Microsoft® Word®, Microsoft® PowerPoint®, or HyperText Markup Language (HTML) formats. Beyond the built-in tools, users can write methods in the Java® programming language.

It is feasible to incorporate any assortment of supplementary products from the COMSOL product suite to the COMSOL Multiphysics software platform. This integration provides access to specialized functionalities tailored to meet the specific modeling requirements of the user within a user interface that maintains a consistent appearance, irrespective of the engineering discipline or physical phenomena involved. Furthermore, the aforementioned software presents a versatile environment applicable to any engineering domain. To achieve this, the COMSOL Multiphysics add-on modules seamlessly integrate into the software platform, enabling the modeling of both single-physics and multiphysics scenarios (Figs. 2.42 and 2.43).

The COMSOL Multiphysics software encompasses the Application Builder, which facilitates the construction and management of bespoke simulation applications predicated on the user's models. Through the deployment of user-developed applications, individuals are enabled to disseminate the capabilities of simulation among colleagues in different departments and clients external to the user's organization.

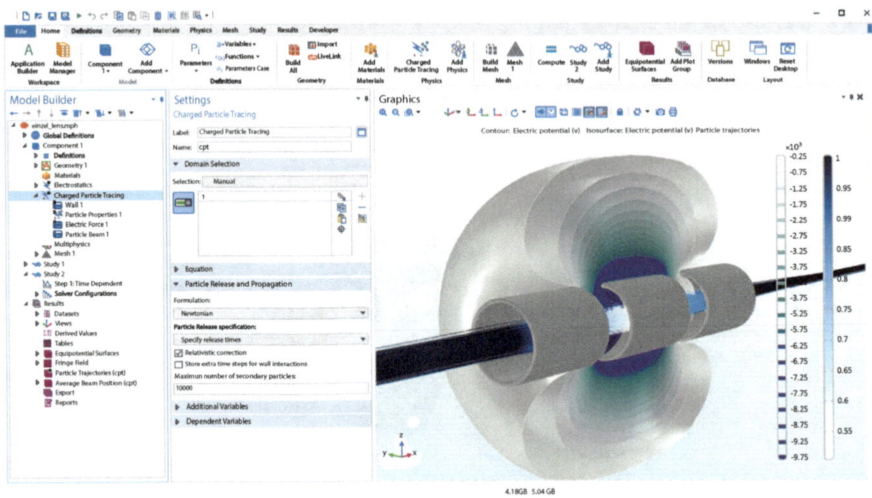

Fig. 2.43 Application environment for particle tracing COMSOL Multiphysics

To effectively manage the results obtained from simulations, Multiphysics modeling yields results of significant accuracy. The integration of modeling alongside experimental methodologies facilitates the optimization of processes and devices in a manner that is often more expedient and precise compared to the exclusive use of experimental techniques or prototype testing. Through the development of experimentally validated models tailored for user analyses, individuals can attain a more profound comprehension of the design or process, as it allows for a more convenient examination than is typically achievable within a laboratory setting.

Multiphysics is frequently essential for the precise representation of a user's design or process. As a participant in COMSOL Multiphysics, the user is liberated from the constraints typically associated with simulation software and possesses comprehensive authority over all dimensions of the user's model. The user is afforded the opportunity for creativity in a manner that is unattainable, or significantly more challenging, with conventional methodologies, owing to the capability to integrate an arbitrary number of physical phenomena. For additional customization, users can input bespoke physics descriptions, accompanied by relevant equations and expressions, directly within the user interface. Accurate multiphysics models take into account a broad spectrum of potential operating conditions and physical influences. This facilitates the application of models for the comprehension, design, and optimization of processes and devices under realistic operating scenarios.

Concerning the features and functionalities of the Model Manager, the COMSOL Multiphysics software encompasses a sophisticated tool designed for the efficient storage of databases and version control pertaining to models and associated files, including experimental data and CAD files. The Model Manager is seamlessly integrated within the COMSOL Multiphysics user interface. It facilitates organization

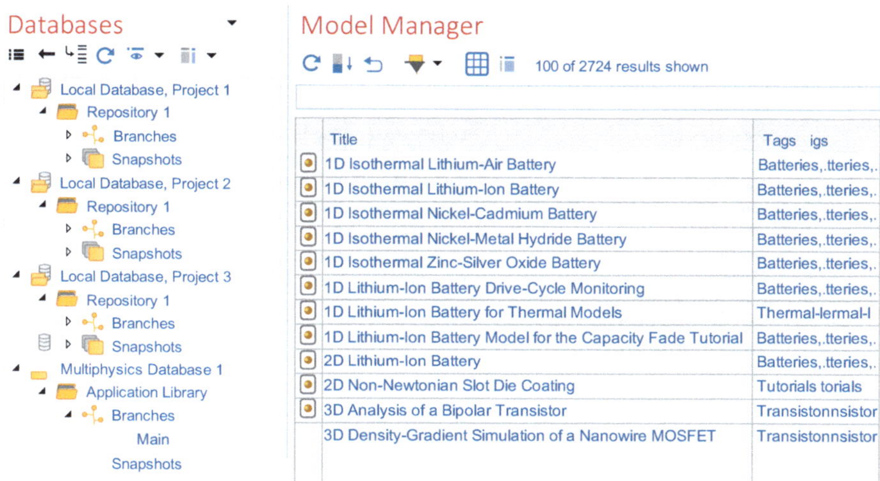

Fig. 2.44 Model Manager features and functionality in COMSOL Multiphysics

and provides advanced search capabilities, encompassing the ability to search for specific features within a model, as well as a comparison feature that elucidates the precise differences between two model versions. Model files are stored in the system with optimal efficiency, minimizing redundancy. In addition to enabling access to the user's model versions through the COMSOL Desktop, the Model Manager server incorporates a web interface for the administration of modeling and simulation projects, which includes user account management and asset oversight. The Model Manager serves to benefit both individuals and organizations through the provision of a structured workspace that fosters collaboration among colleagues, teams, and even external entities. It is incorporated within the COMSOL Multiphysics platform product as delineated below:

- Model and app management: Allow users to centrally organize models and apps.
- Version control: Racks changes and updates to user's models systematically.
- Efficient storage: Stores only relevant data of model drafts and revisions and stores auxiliary data: CAD, mesh, experimental data, etc.
- Searching for model contents: Quickly finds all models having a particular set of features or parameters and tags and organizes models.
- Access control: Allows multiple users to access managed models and data and ventrally controls user access.
- Connect to a local or remote database.

The Model Manager is equipped with an array of instruments designed to tackle these issues, all while remaining within the confines of the COMSOL Desktop modeling environment. The Model Manager encompasses a diverse range of robust tools dedicated to the management of modeling and simulation processes (Fig. 2.44).

From the COMSOL Desktop, an individual may establish a new local database on their personal computer to enhance organization and monitor their models and

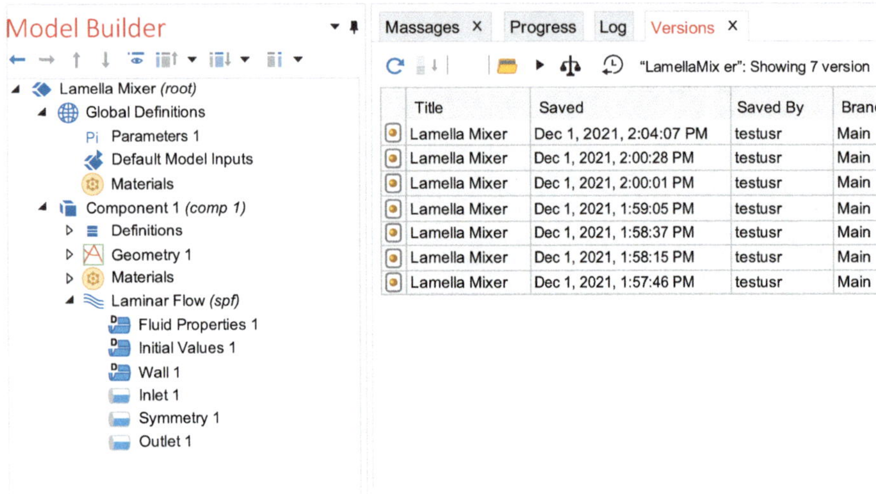

Fig. 2.45 Model Manager and Model Builder features in COMSOL Multiphysics

supplementary data files. Alternatively, individuals have the option to disseminate models and files by uploading them to a server database that is accessible through a Model Manager server (Fig. 2.45). The functionality to utilize a local Model Manager database is incorporated within all licenses, with the exception of the class kit license. An individual can connect to a Model Manager server using either floating network or single-user licenses. With a floating network license, individuals are also permitted to install and operate a central Model Manager server.

The Model Manager incorporates fundamental version control mechanisms, including the ability to review the version history of models along with auxiliary data files, the automatic identification of version conflicts during the saving process, and the capability to perform comparative analysis between different versions. For instance, users have the option to access a prior version to generate a wholly new model, complete with a distinct version history, or to examine all modifications executed on a model from one version to another. In the context of manipulating an existing model within a database, users can produce a draft of the model that possesses its own version control, thereby allowing experimentation with diverse simulation concepts without compromising the version history of the original model. Upon completion of the draft, users may opt to retain it as a new version of the original model or elect to discard it.

More sophisticated version control instruments, such as branching, merging, and reverting functionalities, are also provided. Branching permits users to engage in the development of an entire suite of models and supplementary data files in a manner that is isolated, while concurrently deferring the determination of whether the modifications made are worthy of preservation. The reverting function allows users to recover models and auxiliary data files that may have been inadvertently deleted.

In [3], a critical subject pertaining to the investigation of computational models is elucidated. Such systems are inherently multiphysics in character. In the foreseeable future, bioengineering modeling is anticipated to be routinely employed within clinical settings to customize a variety of individual therapies and treatment approaches predicated on patient-specific models. It is apparent that proficiency in computational modeling is a vital competency for biomedical engineers and should constitute an essential element of contemporary biomedical engineering curricula.

Then, [3] delineates the analysis of modeling organs, tissues, and devices, encompassing a wide array of topics in modeling, ranging from the electrical stimulation of tissues to diffusion, biomechanics, heat transfer, and fluid dynamics utilizing MATLAB and COMSOL Multiphysics, including numerical and analytical methods for resolving ordinary and PDEs. Furthermore, an overview of the various principles of multiphysics involved is provided, alongside practical implementations of such models within MATLAB and COMSOL in [3].

A further example study employing COMSOL Multiphysics is the one delineated in [159]. The authors note that creating dependable MCs systems for high-rate information transfer is difficult because of the intricate and dynamic characteristics of application settings, as well as the physical and resource constraints of micro/nanoscale transmitters and receivers. Consequently, microfluidics may address several practical issues by facilitating testbeds that accurately duplicate application media with meticulous control over flow parameters. Nonetheless, current microfluidic MC testbeds encounter considerable constraints in the creation of chemical signals with programmable waveforms, particularly regarding pulse width. A viable microfluidic MC transmitter design utilizing the hydrodynamic gating approach, a widely used method for chemical waveform production, has been developed to address this issue. [159] presents the experimental validation and characterization of this approach, assessing its accuracy regarding spatiotemporal control over the produced molecule concentration pulses.

The cited investigation details the fabrication of the transmitter, its working mechanism, and discusses its potential limitations based on empirical data. Therefore, [159] shows that the microfluidic transmitter is capable of providing precise, programmable, and reproducible molecular concentration pulses, which would facilitate the experimental research in MCs. [159] validated the accuracy of this model by comparing it with finite element simulations using COMSOL Multiphysics under various system settings.

Another study [160] evaluates the accuracy of its proposed modeling by comparing model-based analytical results to the numerical results obtained by solving the exact system model with COMSOL Multiphysics. There, the authors in [160] considered a microfluidic MCs system, where the concentration-encoded molecular messages are transported via fluid flow-induced convection and diffusion and detected by a surface-based MC receiver with ligand receptors placed at the bottom of the microfluidic channel. The overall system is a convection-diffusion-reaction system that can only be solved by numerical methods, e.g., finite element analysis. However, analytical models are key for the ICT, as they enable an optimization framework to develop advanced communication techniques, such as optimum detection methods and reliable transmission schemes [160].

Following the guidance of [160], they formulate an analytical model to estimate the anticipated temporal progression of bound receptor concentration, which represents the received signal utilized for decoding conveyed information. The model eliminates the necessity for computationally intensive numerical approaches by encapsulating the nonlinearities induced by laminar flow, which produces a parabolic velocity profile, and a finite quantity of ligand receptors that results in receptor saturation. The model additionally accounts for the impacts of the reactive surface depletion layer caused by mass transport restrictions and the shifting reaction boundary arising from the passage of a finite-duration molecule concentration pulse over the receiver surface. According to the suggested model, [160] formulates closed-form analytical equations that estimate the received pulse width, pulse latency, and pulse amplitude, which may be utilized to enhance the system from an ICT standpoint.

Conversely, the primary focus of [161] is to the construction of communication systems that can process and exchange information via molecular and chemical processes to facilitate computer and communication devices. [161] develops microfluidic devices based on chemical processes to implement Binary Concentration Shift Keying (BCSK) modulation and demodulation capabilities. A microfluidic MC transmitter design is proposed in [161], capable of producing constantly predetermined pulse-shaped molecular concentrations in response to rectangular triggering signals to fulfill the modulation function. The inquiry outlines the construction of a microfluidic MC receiver that can demodulate a received signal into a rectangular output signal through a thresholding reaction and an amplification reaction [161].

This chemical reaction-based MC system is replicable, and its parameters are subject to optimization. Moreover, it addresses the sluggishness, unreliability, and lack of scalability inherent in cellular biological processes. To elucidate design insights, [161] additionally develops the theoretical signal responses for its engineered microfluidic transmitter and receiver, which further enhance the optimization of the transmitter design. The theoretical findings in [161] are corroborated by simulations conducted using the COMSOL Multiphysics finite element solver; it illustrates the predetermined characteristics of the produced pulse and the demodulated rectangular signal, together with their reliance on design parameters.

Another example of the importance of COMSOL Multiphysics is found in [162]. The authors in that research discuss that macroscale MCs have recently garnered significant scholarly interest due to their prospective applications. Given that the majority of empirical investigations pertaining to microchannels predominantly concentrate on microscale scenarios, it becomes imperative to conduct and execute experiments aimed at examining the feasibility of the concept while concurrently validating the associated models and parameters.

A macroscale flow-based MCs platform utilizing a fluidic medium is constructed in a semi-cylindrical channel under laminar flow conditions in [162]. The transmission medium is water within the plexiglass pipe, where a transmitter emits hydrochloric acid molecules, and a chemical sensor functions as the receiver. [162]

provides a linear time invariant (LTI) model for the pH meter and derives its parameters from experimental findings. The subsequent section, [162], delineates the suitable encoding-decoding schemes and implements the optimal receiver for the verified model and acquired parameters. An ISI mitigation strategy with adaptive thresholds is employed to enhance the rate. A comprehensive theoretical model for this system is provided, corroborated by experimental and simulation findings utilizing the COMSOL software platform. The suggested model is validated in [162], and associated parameters are derived from both experimental and simulation data. The system's performance is evaluated based on bit error probability and communication rate.

Also, in [163] COMSOL Multiphysics is used to validate the accuracy of its model by comparing it with finite element simulations with the employment of the referred digital platform under various system settings. The proposal in [163] establishes a practical microfluidic transmitter architecture for MCs based on *hydrodynamic gating*, a widely utilized technique for generating chemical waveforms in microfluidic channels with high spatiotemporal resolution. The authors in [163] developed an approximate analytical model that can capture the fundamental characteristics of the generated molecular pulses, such as pulse width, pulse amplitude, and pulse delay, as functions of main system parameters, such as flow velocity and gating duration. This analytical model can enable the optimization of microfluidic transmitters for MCs applications in terms of minimizing Inter-Symbol Interference (ISI) and maximizing data transmission rate [163].

COMSOL Multiphysics is furthermore used in [164]. In that investigation is analyzed the blood-tissue barrier (BTB) inside the body forming the main communication pathway for molecular information exchange between the nanomachines as well as the intrabody nanonetwork and the bio-cyber interface on the Internet of Bio-Nano Things (IoBNT) network. But overcoming this barrier by the molecules is one of the main challenges for MCs in the body. Therefore, spatiotemporal modeling of MC across the BTB is of particular interest [164].

Therefore, in [164] is developed a mathematical model and stochastic particle-based simulator for MCs over high spatiotemporal resolution between mobile nanomachines in the blood capillary and the surrounding tissue. The transmitting bio-nanomachines used in this MCs system are modeled as a moving sphere with a continuous emission pattern over a specific duration. In this work, the blood capillary characteristics, including the BTB and blood flow, are modeled and their effect is examined on the molecular received signal. The authors in [164] modeled and solved this MCs system numerically using the COMSOL Multiphysics software, which is based on the finite element method. The numerical results are validated using this developed stochastic particle-based simulator over high spatial–temporal resolution. The mathematical model is numerically solved using the COMSOL Multiphysics software. A time-dependent solver is used to solve the advection-diffusion equations with the boundary conditions and the source term. This proposal is defined in Figs. 2.46 and 2.47.

In [165] is presented an application of COMSOL Multiphysics with a combination of MCs systems and wireless communications. Then, [165] mentions that

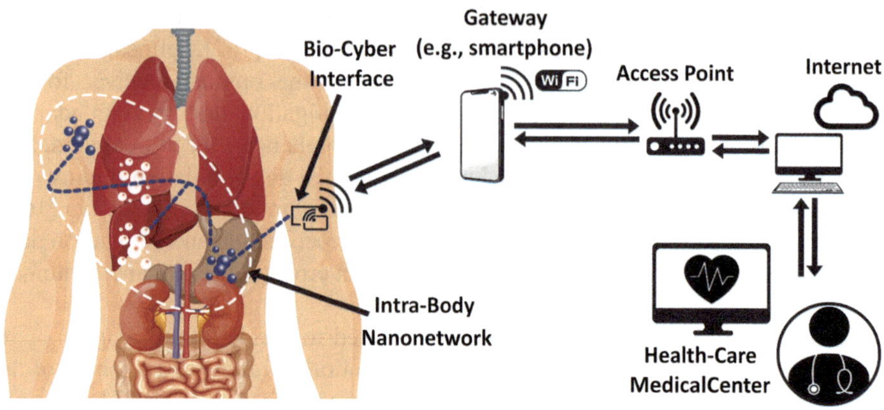

Fig. 2.46 Typical IoBNT architecture in [164]

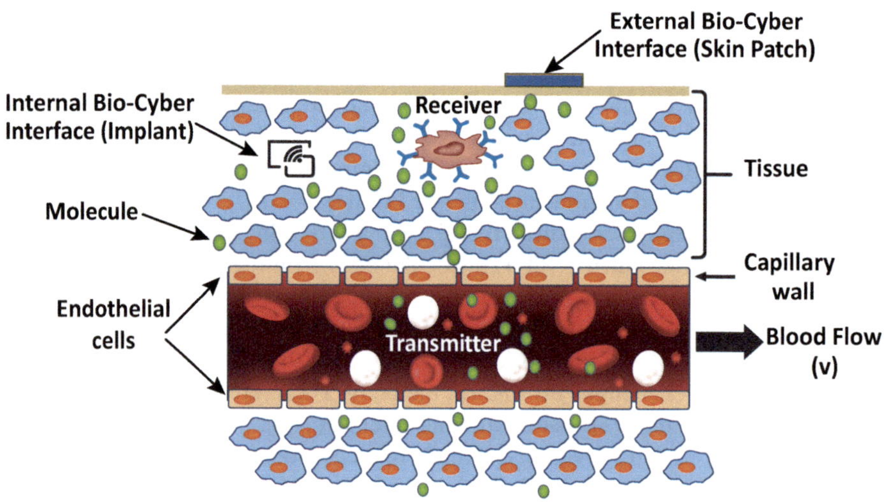

Fig. 2.47 Graphical representation of MCs between nanomachines in the blood capillary and the surrounding tissue across the BTB in [164]

Water Distribution Networks (WDNs) can suffer from a number of critical failures ranging from burst pipes, pressure loss, and contamination. While low-frequency ground-penetrating waves can provide intelligence, detailed data collection requires embedded sensors inside the WDN pipes. The key challenge is to extract data out from inside WDNs and other similar pipe networks (e.g., natural gas, oil, and sewage). WDNs tend to be extremely large (millions of junctions stretching over 100,000 km). The problem is illustrated in Fig. 4, where contamination occurs at a given node in the WDN. The contaminants quickly spread around the WDN, with varying levels and dynamic response signals. MC offers the opportunity to send artificial molecular signals from a number of sensors to a single data hub (relay

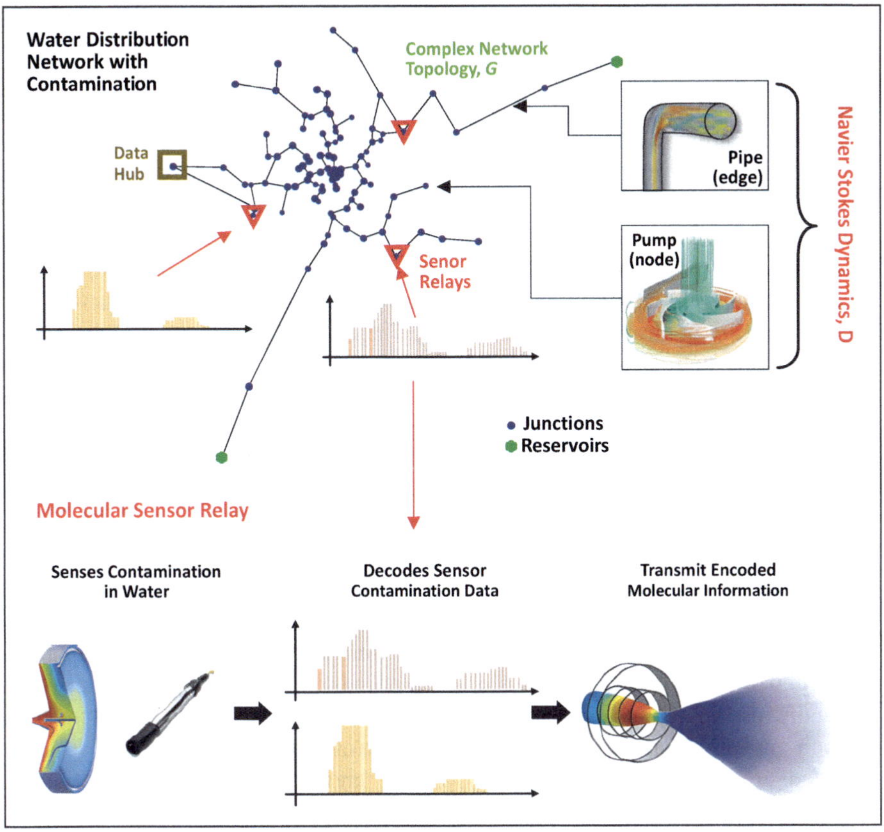

Fig. 2.48 Monitoring contamination in a WDN using molecular relay communication. Macroscale simulations using EPANET2 (EPANET is used to perform extended-period simulation of the hydraulic and water quality behavior within drinking water distribution systems.) and micro-simulations using COMSOL with computational fluid dynamics (CFD) and particle tracing modules

channel, Fig. 2.48). This can be formulated into a familiar relay channel, where (Fig. 2.48) [165]:

- The *WDN dynamics* is the *multipath channel*, where time-varying demands and multiple network paths give rise to multiple signals.
- The *contaminant* starts at the *source*.
- The *data hub* is the *destination* but cannot alone decode the attributes of the source pollutant due to its distance and the multipath loss through the WDN.

The *MC devices* are the *relays* distributed strategically in the WDN, which can sense the source of pollution and transmit an alternative harmless molecular signal to the data hub. The multiple MC relay devices together enable the destination data hub to understand the pollution dynamics in the WDN [165].

The use of COMSOL Multiphysics is exposed in [166], the discussion in that work specifies that *quorum sensing* as a special kind of communication in bacterial populations can be analyzed by means of methods and techniques of mathematical modeling and computer simulation. In [166] is studied, a modification of a deterministic mathematical model of bacterial quorum sensing is proposed, taking into account the law of multiphase population dynamics. The mathematical model is formalized by an initial-boundary value problem for a system of semilinear reaction-diffusion PDEs. The equations include generation terms in view of changes in the biomass density. The model describes space-time dynamics of concentrations of special substances (signaling agents and Lactonase enzymes) that characterize the quorum sensing in Gram-negative bacteria. The problem is solved by means of the finite element method using the COMSOL Multiphysics platform. Computational experiments are performed to estimate concentrations of key substances characterizing quorum sensing for *Pseudomonas putida* bacterial strains in an expanded range of population dynamics [166]

Moreover, experiments are performed in [167] using COMSOL Multiphysics, with a finite element multiphysics simulation software, and Python-based simulations to analyze the end-to-end communication model in terms of path loss, delay, and gain. This exploration in [167] arguments that as an alternative to ongoing efforts for vaccine development, scientists are exploring novel approaches to provide innovative therapeutics, such as nanoparticle-and stem cell-based treatments. Thus, understanding the transmission and propagation dynamics of coronavirus inside the respiratory system has attracted researchers' attention. In [167], the transmission and propagation dynamics of coronavirus within the respiratory tract are modeled, commencing from the nasal passages and extending to the alveoli, utilizing MC theory. The analytical findings elucidate the relationship between the characteristics of the transmission channel and the pathophysiological properties associated with coronavirus. For the initial 50% of the peak viral particle production rate, the path loss exhibits an increase exceeding 16 times that observed in the subsequent 50%. The immune system's delayed response, coupled with the heightened absorption of viral particles within the respiratory tract, contributes to a delay in the arrival of these particles at the alveoli. Moreover, the findings indicate that asthmatic patients exhibit a higher viral load in comparison to individuals without such conditions [167].

The digital platform described in this section is similarly utilized in [168] that proposes the designs of the microfluidic transmitter and receiver with Quadruple Concentration Shift Keying (QCSK) modulation and demodulation functionalities. To do so, [168] first presents an AND gate design and then applies it to the QCSK transmitter and receiver design. The transmitter is capable of modulating two input signals to four different concentration levels, and the QCSK receiver can demodulate a received signal to two outputs. More importantly, [168] also establishes a mathematical framework to theoretically characterize our proposed microfluidic circuits. Based on this, the authors in [168] first derive the output concentration distribution of their proposed AND gate design and provide the insight into the selection of design parameters to ensure an exhibition of desired behavior. They further derive the output concentration distributions of the QCSK transmitter and receiver. Simulation results obtained in COMSOL Multiphysics not only show the desired behavior of all the proposed microfluidic circuits but also demonstrate the accuracy of the proposed mathematical framework [168].

Besides, [169] is a book that contains vast recent applications of microscale and nanoscale modeling systems with finite element methods using COMSOL Multiphysics and MATLAB. Moreover, the book *Vascular Biomechanics: Concepts, Models and Applications* [170] establishes biological analysis with COMSOL Multiphysics simulations.

In [171], COMSOL serves as a computational tool for the examination of a natural modality of nanoscale communication, specifically neuro-spike communication, which motivates the integration of nanomachines within the human biological system for medical applications. Consequently, the elucidation of failure mechanisms in both healthy and pathological connections of nervous nanonetworks is of paramount importance. Thus, the article referenced in [171] examines the mechanisms of information transmission through a singular myelinated axonal segment. The authors of the aforementioned study proposed a sophisticated multi-compartmental model for a single myelinated segment by integrating the paranodal regions of the axon into the framework. Subsequently, they delineate the communication channel of the myelinated segment with respect to attenuation across a spectrum of frequencies. Building upon this foundation, they formulate the rate per channel use and establish an upper limit on information capacity. The performance assessments indicate that the proposed methodology yields significant enhancements concerning frequency response. The authors in [171] express that their results could have a significant effect on the characterization of demyelinated axons from the ICT perspective.

2.3.5 *Network Simulator-3 (NS-3)*

NS-3[35] is a discrete-event network simulator, targeted primarily for research and educational use. NS-3 constitutes an open-source software package, governed by the GNU GPLv2 license, primarily tailored for operating systems, such as Linux, OS X, and FreeBSD, and is readily accessible for scholarly pursuits and developmental endeavors. The NS-3 initiative is dedicated to establishing a robust simulation core that is thoroughly documented, user-friendly, and amenable to debugging, thereby addressing the comprehensive requirements of the simulation process, which encompasses simulation configuration through to the collection and analysis of traces. Additionally, the NS-3 software framework promotes the creation of simulation models that are sufficiently authentic, enabling NS-3 to function as a

[35] ns-3 a discrete-event network simulator for internet systems (nsnam.org). The information provided by these paragraphs is taken from this webpage.

real-time network emulator that interacts with actual networks, thereby facilitating the reuse of numerous existing real-world protocol implementations within the NS-3 environment. This simulator is equipped with a real-time scheduling capability that supports various "simulation-in-the-loop" scenarios for interfacing with actual systems. For example, users have the ability to transmit and receive packets generated by NS-3 on tangible network devices, and NS-3 can act as an interconnection platform to introduce link effects among virtual machines. A further focus of the simulator is on the integration of real application and kernel code. Current testing and evaluation efforts are underway for frameworks designed to execute unmodified applications or the complete Linux kernel networking stack within the NS-3 architecture.

NS-3[36] is a discrete-event simulator typically run from the command line. It is written directly in C++, not in a high-level modeling language; simulation events are simply C++ function calls, organized by a scheduler. An NS-3 user will obtain the NS-3 source code, compile it into shared (or static) libraries, and link the libraries to *main()* programs the developers. The *main()* program is where the specific simulation scenario configuration is performed and where the simulator is run and stopped. Several example programs are provided, which can be modified or copied to create new simulation scenarios. Users frequently modify the NS-3 library code (and subsequently reconstruct the libraries) to alter its functionality. NS-3 provides optional Python bindings for the development of scenario configuration programs utilizing Python (and adopting a Python-centric workflow); however, this introductory guide does not encompass those elements.

NS-3 has various optional extensions, but the main features just require a C++ compiler (*g++* or *clang++*), Python (version 3.8 or above), CMake, and a build-system (e.g., make, ninja, and Xcode). NS-3 is operational and compatible with systems that utilize a contemporary C++ compiler along with Python runtime capabilities. For Linux environments, one may employ either the g++ or clang++ compilers. In the context of macOS, it is advisable to utilize clang++ (which can be accessed through Xcode or the Xcode Command Line Tools). For Windows platforms, the Msys2 tools in conjunction with the MinGW64 toolchain may be employed (since the release of ns-3.37) for a majority of applications. For versions prior to ns-3.37, or in scenarios requiring emulation modes or Python bindings, it is strongly recommended to utilize a Linux virtual machine or the Windows Subsystem for Linux.

The NS-3 simulator represents a discrete-event network simulator primarily designed for academic research and pedagogical purposes. Initiated in 2006, the *ns-3 project* is an open-source endeavor associated with NS-3. The NS-3 framework has been meticulously developed to furnish an open and extensible platform for network simulation, catering to the realms of networking research and education. In essence, NS-3 offers comprehensive models that elucidate the operational dynamics

[36] Tutorial (nsnam.org). Supported by this URL, the information shown in next paragraphs are taken of it.

and performance metrics of packet data networks, alongside a simulation engine that enables users to execute simulation experiments. Some of the motivations for utilizing NS-3 encompass conducting studies that are challenging or infeasible with actual systems, analyzing system behavior within a rigorously controlled and reproducible setting, and enhancing understanding of network functionalities. Users may observe that the available model set within NS-3 predominantly emphasizes the modeling of Internet protocols and networks; however, it is noteworthy that NS-3 is not confined to Internet-centric systems, as a number of users are employing NS-3 to model systems that do not rely on Internet frameworks.

Many simulation tools exist for network simulation studies. Below are a few distinguishing features of NS-3 in contrast to other tools:

- NS-3 is conceptualized as an assemblage of libraries that can be synergistically integrated with one another as well as with additional external software libraries. In contrast to certain simulation platforms that furnish users with a singular, cohesive GUI environment for the execution of all tasks, NS-3 exhibits a more modular architecture in this aspect. A variety of external animation and data analysis as well as visualization tools can be utilized in conjunction with NS-3. Nevertheless, users should anticipate engaging with the command line interface and utilizing C++ and/or Python software development tools.
- NS-3 is predominantly employed on Linux or macOS operating systems, although compatibility is available for Berkeley Software Distribution (BSD) systems, as well as for Windows environments capable of compiling Linux code, such as Windows Subsystem for Linux or Cygwin. Currently, native Windows Visual Studio is not supported; however, a developer is actively engaged in the prospect of future support. Windows users may also opt to utilize a Linux virtual machine.
- NS-3 is not an officially supported software product of any company. Support for *NS-3* is done on a best-effort basis on the NS-3-users forum[37].

There are several important resources of which any *NS-3* user must be aware. The main website is located at https://www.nsnam.org and provides access to basic information about the *NS-3* system. Detailed documentation is available through the main website at https://www.nsnam.org/documentation/. It is possible also to find documents relating to the system architecture from this page. There is a Wiki that complements the main *NS-3* website which to find at https://www.nsnam.org/wiki/. Besides, user and developer FAQs can be found there, as well as troubleshooting guides, third-party contributed code, papers, etc. The source code may be found and browsed at GitLab.com: https://gitlab.com/nsnam/. There researchers will find the current development tree in the repository named ns-3-dev. Past releases and experimental repositories of the core developers may also be found at the project's old Mercurial site at http://code.nsnam.org.

[37] ns-3-users@googlegroups.com

NS-3 is architected as a cohesive suite of software libraries that function synergistically. User-defined programs can be developed that interface with (or draw upon) these libraries. Such programs are formulated utilizing either the C++ or Python programming languages. NS-3 is disseminated in the form of source code, thereby necessitating that the target system possesses a software development environment to initially compile the libraries, followed by the compilation of the user program. In principle, NS-3 could be disseminated as pre-compiled libraries for select systems, and there exists the potential for such distribution in the future; however, at present, a considerable number of users engage in modifications to NS-3 itself, rendering the availability of the source code for library reconstruction particularly advantageous. Should any individual wish to undertake the development of pre-compiled libraries and packages for operating systems, they are encouraged to reach out to the ns-developers mailing list. Seasoned Linux users may at this juncture question the rationale behind NS-3 not being provided in a manner analogous to most libraries that utilize a package management system. Although some binary packages are available for various Linux distributions (such as Debian), the majority of users ultimately find themselves modifying and recompiling the NS-3 libraries independently, thus making the presence of the source code considerably more pragmatic.

In NS-3, each event is meticulously linked to its corresponding execution time, and the simulation advances by executing events in accordance with the chronological sequence of simulation time. Upon the processing of an event, it has the potential to produce zero, one, or multiple subsequent events. Throughout the course of the simulation, events are systematically consumed; however, the generation of additional events may occur (or may not). The simulation is programmed to terminate automatically when no additional events remain in the event queue, or upon the identification of a designated "Stop" event. This framework is specifically designed for research purposes, taking into account the requirements of the scholarly community and promotes a collaborative model that encourages community engagement both for the development of novel modules and for the execution of validation or peer review endeavors across various research teams. NS-3 operates under an open-source framework and adheres to a free software philosophy. The simulation environment is structured to facilitate its application as a real-time emulator, thereby integrating simulated components alongside physical components within a unified network [172].

This degree of integration is facilitated by a real-time scheduling mechanism, in addition to the provision of simulation support, thereby enabling simulation-in-the-loop methodologies. The simulator is designed for utilization through its API within programmatic representations of user-defined scenarios and configurations, which can be developed in accordance with conventional workflows and methodologies. The API serves to generate outputs in the form of traces that are congruent with those that can be acquired by monitoring traffic in pragmatic network environments using tools, such as Wireshark[38] or TCPdump[39], thus allowing for analysis through the same instruments that are already employed for experimental configurations or diagnostics. The developers of NS-3 explicitly articulate their intention to enable the repurposing of actual implementations of protocols within NS-3, thereby minimizing the effort required for the design and creation of highly detailed simulation components to the greatest extent feasible. The overarching objective is to furnish a highly realistic simulation framework for a diverse array of network infrastructures, communication and routing protocols, technologies, and layers, thereby permitting network-enabled software to operate atop the simulation with minimal exertion (e.g., virtual machines) [172].

Various simulation tools have adopted diverse methodologies for modeling, encompassing the implementation of specialized modeling languages, code-generation utilities, and component-oriented programming paradigms. Although high-level modeling languages and simulation-centric programming paradigms possess certain benefits, the accurate simulation of actual implementations is not generally regarded as one of their proficiencies. The elevated level of abstraction may lead to considerable discrepancies between simulation outcomes and empirical results, thereby necessitating a focus on realism. For instance, NS-3 selected C++ as its programming language due to its facilitation of the incorporation of C-based implementation code; notably, this choice also proved advantageous for integration with Alchemist. Furthermore, the architecture of NS-3 bears resemblance to that of the Linux operating system, featuring internal interfaces (network to device driver) and application interfaces (sockets) that align effectively with contemporary computer architecture. Additionally, considerable attention has been directed toward enhancing debugging ease, improving alignment with modern programming languages, and optimizing performance. The designers of NS-3 prioritized C++-based models to achieve superior performance and facilitate debugging processes. Figure 2.49 succinctly depicts the structural composition of NS-3 [173].

The primary characteristic of this architecture resides in the reality that simulations are, in fact, C++ programs themselves, rather than being constructed in a scripting language or within a configuration file. While this aspect may not inherently confer an advantage to the simulation itself, the design of the simulator's interface, which is tailored for external accessibility, has proven advantageous for the integration with alternative simulators. Furthermore, this suggests that a simulation within NS-3 possesses access to both its inherent functionalities and those of external sources: in addition to its reliance on NS-3, it operates as a fully developed C++ program. NS-3 systematically categorizes its functionalities into distinct modules. A module represents a segment of the simulator that addresses a specific facet of its operation. The following are the most significant modules: Core: This module encompasses all fundamental facilities utilized by the remaining modules. These facilities comprise smart pointers, callback handlers, an attributes manager, logging

[38] https://www.wireshark.org/

[39] https://www.tcpdump.org/

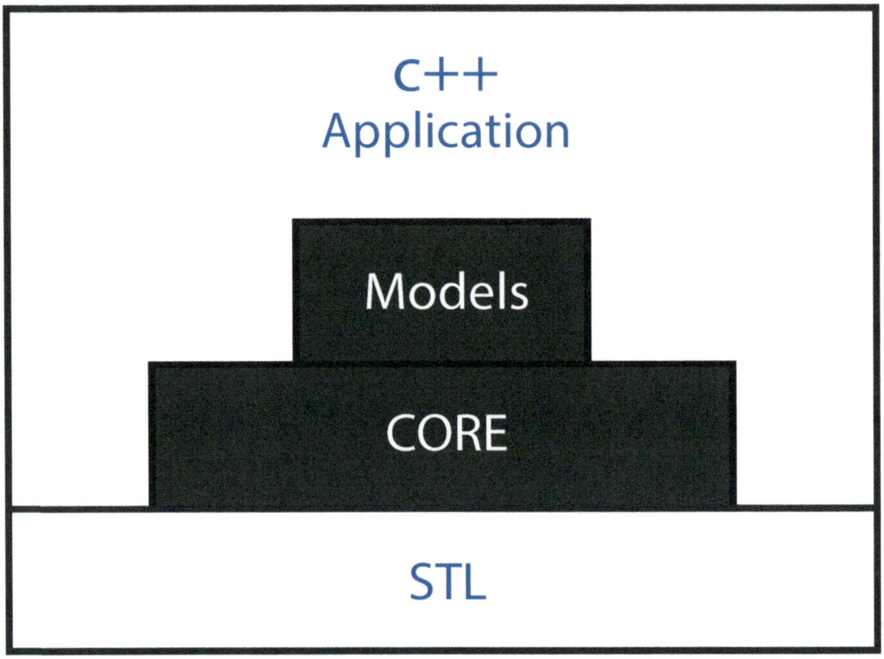

Fig. 2.49 Graphical representation of the NS-3 architecture. The entire software is based on the C++ Standard Template Library. The core module provides the basic facilities used by all the models. The simulation is a C++ application that has access to all these functionalities

utilities, and random variables. Network: This module contains the models of all the basic components of a network. It is worth citing some of these [173]:

- Node: A node is a communication endpoint. An example of a node is a computer in a LAN (local area network) [173].
- NetDevice: A NetDevice is the means used by a node to interact with the network, like an Ethernet adapter inside a computer. The network module also contains a model of the corresponding Media Access Control (MAC) address [173].
- Socket: A socket, just like in operating systems, is a software abstraction which enables an application to communicate across the network [173].
- Application: An Application is the equivalent of a program running on a computer. An Application runs on a specific Node. It can use sockets to communicate with other Application running on other Nodes. NS-3 contains various predefined Application (such as an Application which periodically sends a User Datagram Protocol (UDP) packet) in an ad hoc module [173].
- Internet: The Internet module contains the model for all the protocols normally used on the Internet, such as Internet Protocol version 4 (IP v4), Internet Protocol version 6 (IPv6), Internet Control Message Protocol (ICMP), TCP (with its various versions), UDP, etc. [173].

- Devices modules: There are many modules which model various kinds of devices, such as Carrier Sense Multiple Access (CSMA) devices or Wi-Fi devices. Figure 2.50 illustrates how ns-3 models a computer network [173].

The architecture of a computer network is closely mirrored in this context; indeed, programming an application within NS-3 bears a strong resemblance to programming within an authentic environment. This facilitates the process of simulation development. A notable advantage of NS-3 lies in its precise emulation of actual devices and protocols in conjunction with its performance metrics. Moreover, NS-3 ranks among the most rapid network simulators currently accessible, wherein identical simulations—comprising numerous nodes interconnected within a wireless network and disseminating messages—are executed across various network simulators [173].

The procedure of code creation in NS-3 is shown in Fig. 2.51 [174]. The detailed source code showing use of different classes in NS-3 is shown in Fig. 2.52 [174].

NS-3 serves as a discrete-event network simulator tailored for Internet systems, although it was not initially conceived for the simulation of MCs. Nonetheless, its adaptable architecture has facilitated the incorporation of fundamental components pertinent to MCs. Specifically, two simulation tools have been instantiated within NS-3: one adhering to the guidelines established by the IEEE (Institute of Electrical and Electronics Engineers) standards 1906.1 and 1906.1.1 (discussed in Chap. 3, Sect. 3.3), and another dedicated implementation for bacterial communication

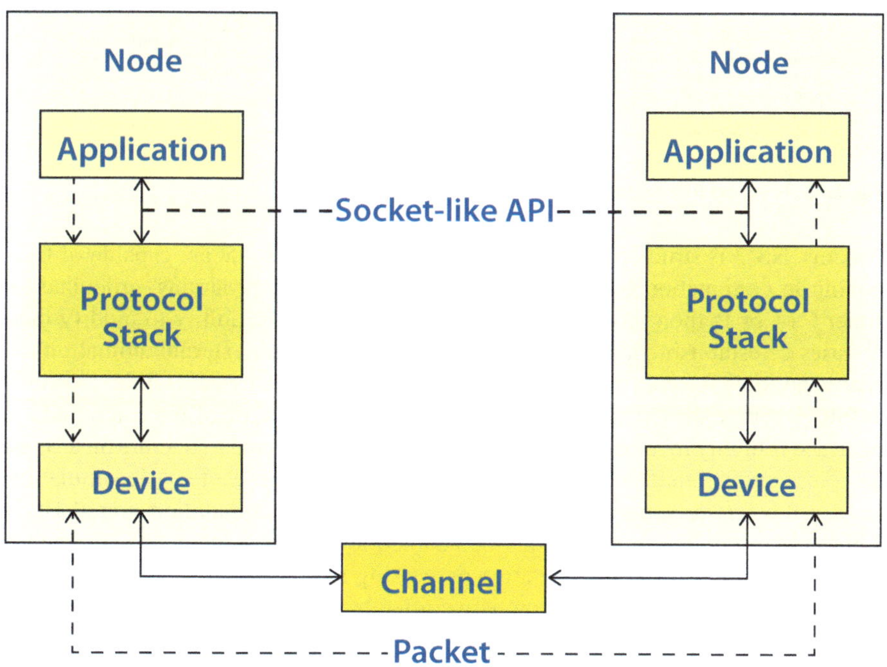

Fig. 2.50 Graphical representation of the NS-3 network stack

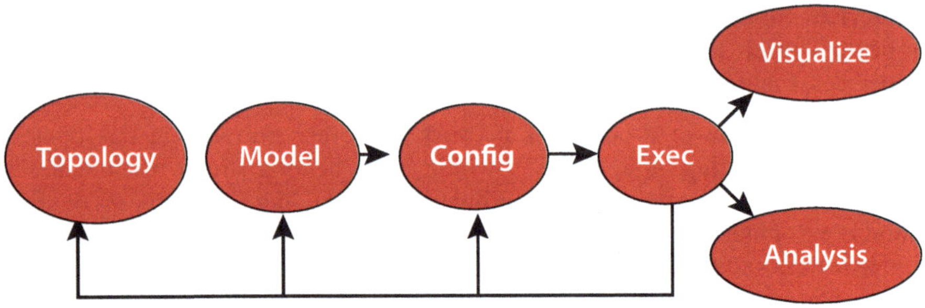

Fig. 2.51 NS-3 source code

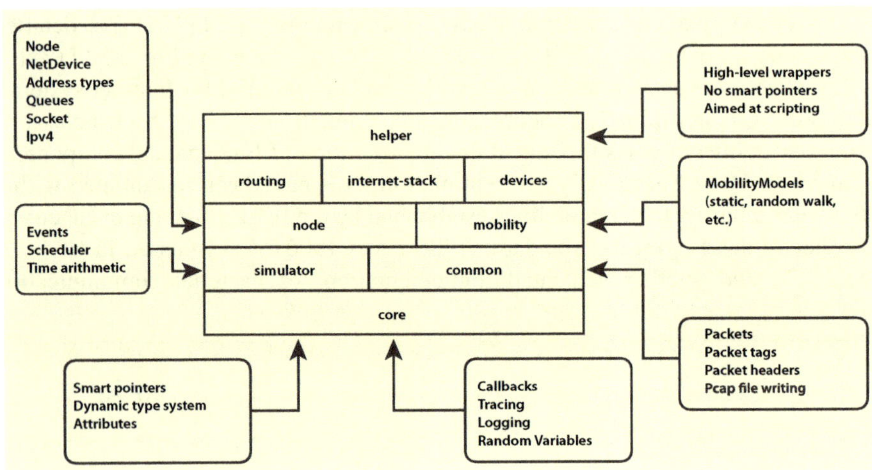

Fig. 2.52 NS-3 detailed source code

systems. NS-3 is structured into various software libraries that are capable of functioning in conjunction with one another. User-generated programs, articulated in either C++ or Python programming languages, possess the ability to modify these libraries or establish connections with them. Furthermore, external animation and analytical tools for data analysis and visualization are readily accessible. However, to fully leverage the extensive capabilities of NS-3, utilization of the command line interface is highly recommended. The simulator is architected to operate on a singular computational machine. The minimal requirements to execute basic simulations are a *gcc* or *clang* compiler and Python interpreter. The simulation granularity is defined by users. Event design can be adapted to any granularity and is synchronized in the simulator implementation processing loop [175].

The simulator has been meticulously crafted by adhering to a hierarchical architecture. Its foundational models, components, and entities articulated within the communication framework delineate essential parameters and functionalities that

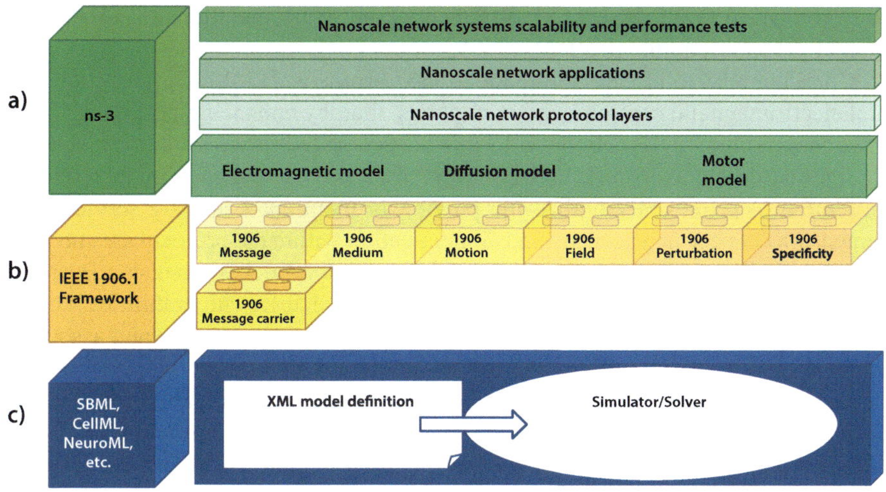

Fig. 2.53 Simulator integration use case where (**a**) is a network simulator, such as network simulator version 3 (NS-3). (**b**) is the IEEE 1906.1 standard. (**c**) is a biological simulation tool, such as SBML

are ubiquitous across all communication schemes applicable at the nanoscale, thereby facilitating the interaction of components and entities during the transmission of a specified message. The fundamental structure of the simulator necessitates enhancement to accommodate the implementation of distinct communication paradigms. As an initial measure in this regard, prototypes of both EM and MC schemes have been formulated [176].

The overarching objective of the IEEE Standard 1906 series is to function as a cohesive integrative framework that facilitates the interoperability of simulation tools across various scales and diverse academic disciplines. Figure 2.53a illustrates a network simulator that employs a variety of distinct small-scale biological tools, such as SBML. Figure 2.53b delineates the components as articulated within the aforementioned IEEE standards—Fig. 2.53c. The IEEE 1906.1.1 Yet Another Next Generation (YANG) module is instrumental in enabling the interoperation of these tools across different scales [177].

The nanoscale characteristics inherent in IoNT technology present numerous security challenges regarding secure data transmission, attributable to the substantial influence of IoNT instruments within sectors, such as healthcare, intelligent residential environments, and national defense. Consequently, authentication emerges as a critical prerequisite for ensuring secure data transmission, particularly within the IoNT framework. This technological advancement has opened up new possibilities for enhancing the efficiency and effectiveness of these areas and has the potential to revolutionize their approach [178]. The study presented in [178] consists of a novel elliptic curve cryptography-based authentication protocol for IoNT in this context. Since the nanoscale devices have less computational capabilities, this protocol leverages a secure hash function and XOR operations to ensure lightweight yet robust authentication.

Consequently, to effectuate the mentioned protocol, [178] integrates molecular and electromagnetic communication methods, thereby enhancing both security and efficiency within IoNT networks. In this particular hybrid methodology, the configuration of parameters and the facilitation of message exchanges are meticulously structured to proficiently realize the previously mentioned security services. The dimensions concerning security and authentication within the protocol are subjected to a thorough examination utilizing the real-or-random model alongside Burrows–Abadi–Needham logic, while its robustness against security vulnerabilities is evaluated through the automated validation of Internet security protocols and associated applications. In [178] is also analyzed communication costs, computational overhead, and real-world feasibility. Finally, by employing the Nano-sim (integrated with NS-3 simulation framework) and N3Sim tools, [178] simulates and demonstrates the IoNT environment (i.e., expressed in terms of the packet loss ratio and the average amount of time required for propagation of authentication messages through MC). This research stands as a pivotal contribution, fortifying the foundations of IoNT through heightened security and enhanced reliability.

Research in [178] indicates that the security of Unmanned Aerial Systems (UASs) is a significant component of the combat capability of modern armed forces. Specifically, these systems that handle private information are vulnerable to security breaches. An Intrusion Detection System (IDS) has been developed in the design presented in [179] to safeguard against security issues utilizing the Human Immune System (HIS). Intrusion Detection Systems (IDSs) are employed to identify and react to efforts to breach the target system. The operation of UASs in the real world presents challenges in the testing and validation of these systems utilizing various sensors. This design draws inspiration from HIS. In the mapping, unsecure signals are analogous to antigens that are identified by antibody-based training patterns and eliminated from the operational cycle. Among the primary applications of the suggested design. In [179] are the rapid identification of invasive signals and the isolation of their activity. Furthermore, the SUASHIS approach is assessed using comprehensive simulations conducted in the NS-3 environment. The simulation findings demonstrate enhancements in the performance parameters of the UAS network, specifically regarding the false positive rate, false negative rate, detection rate, and packet delivery rate [179].

HIS constitutes the fundamental protective apparatus of humanity, enabling survival against environmental hazards and various diseases. Furthermore, drawing parallels between the internet and human systems in multiple dimensions suggests that an immune system analog could be devised for the Internet, grounded in the principles of HIS. The immunity system encompasses all biological mechanisms tasked with safeguarding the organism from harmful environmental agents, including microorganisms and their byproducts, chemicals, pharmaceutical substances, and pollen grains. HIS is organized into three collaborative defensive tiers. The initial layer comprises the skin, mucosal membranes, and integumentary secretions. In relation to the second layer, it encompasses phagocytic leukocytes, antimicrobial proteins, and inflammatory responses as its components. Ultimately, the third layer, which embodies specific defensive strategies, incorporates lymphocytes and antibodies. Antibodies engage with particular microorganisms, abnormal somatic cells, toxins, and other entities identified by foreign molecules in distinctly tailored manners. HIS is classified into two categories: innate immunity and acquired immunity. Figure 2.54 delineates the lymphoid organs alongside their primary functions [179].

HIS represents a highly intricate mechanism, proficient in safeguarding the organism against an extensive array of extraneous pathogens. The underlying mechanism upon which the HIS operates is remarkably effective, particularly in the differentiation between self and non-self-antigens. A non-self-antigen refers to any external entity capable of triggering an immune response, such as bacteria or viral attacks. Conversely, self-antigens are fundamentally distinct from non-self-antigens. Self-antigens comprise the intrinsic cells of the organism itself. The main theories regarding HIS algorithms are affinity, clonal selection, and negative selection. The referred aspects are the support to design an algorithm in [179] with simulations in NS-3.

The research in [180] focuses on the MCs between nanodevices in a human body environment which is a very promising interdisciplinary technique employing the mechanisms from different research areas. Biological models inspire and encourage the utilization of nanotechnologies in order to interact with organs and tissues in human body. Some of the envisaged applications due to the specified investigation are [180]:

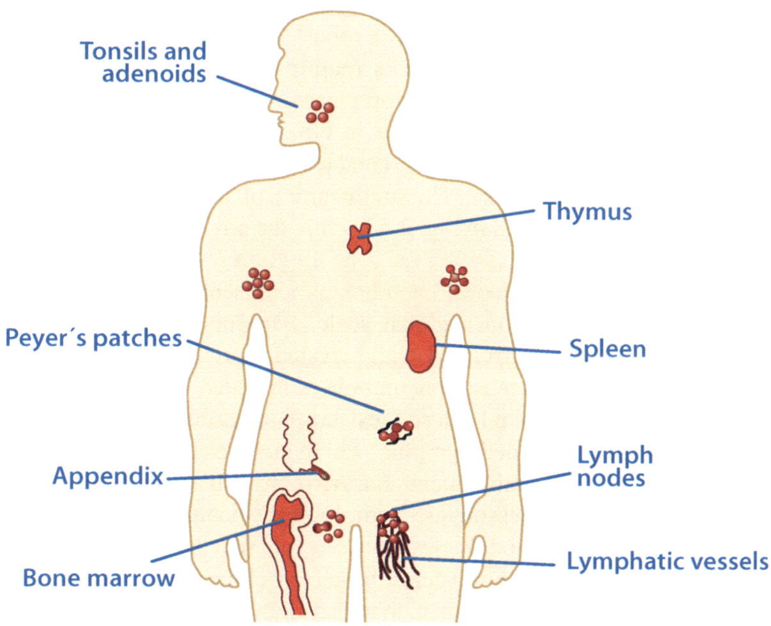

Fig. 2.54 HIS structures

- Immune system support: The immune system which protects the organism against diseases could be potentially based on nanodevices (nano, micro, and macro systems). These nanodevices can be used, e.g., for the detection and elimination of malicious agents and cells [180].
- Biohybrid implants: Biohybrid implants focus on supporting or even replacing organs. This is made possible by creating a "friendly interface" between the implant and human environment [180].
- Drug delivery systems:. Drug delivery systems could include specific types of implants like systems for compensating for metabolism diseases (diabetes). Drug delivery systems can also help mitigate neurodegenerative diseases by delivering neurotransmitters [180].
- Health monitoring: In-body nanodevices can be used for health monitoring, such as cholesterol level, hormonal disorders, or early diagnosis [180].
- Genetic engineering: Refer under the term genetic engineering [180].

The authors in [180] manifest that manipulation and modification of nanostructures in human body have potential application areas for nanodevices. MCs can be realized through several approaches which depend on the way the molecules are propagated in the human body and what their communication range is. [180] focuses on the MCs in short and medium range where the information is encoded by the nanotransmitter into the molecule and received by a nanoreceiver which is placed at hundreds of micrometers from the nanotransmitter. As the signal propagation approaches, the calcium signaling is used.

[180] employs the NS-3 simulation tool for the examination of MC, which is highly esteemed for its efficacy in modeling both wired and particularly wireless network technologies and is supported by a robust and engaged community. Already referring, the modeling of nanonetworks requires completely different approach and techniques in comparison to common macroscale communications; therefore, standard models and functions available in NS-3 must be adopted. An existing extension, referred to as nanonetworks, enhances NS-3 by incorporating the capability to simulate nanonetworks. The architecture of the intrabody nanonetwork under consideration is depicted in Fig. 2.55, while the physical transmission process is represented in Fig. 2.56.

Health care-associated bacterial pneumonias represent a significant contributor to morbidity and mortality on a global scale. Particularly, pneumonia acquired within hospital settings is correlated with elevated mortality rates. At present, the effective management of bacterial pneumonia is impeded by two primary factors: first, a substantial escalation in bacterial resistance to antibiotic agents, and second, a modification in the patient demographic, which includes elderly, immunocompromised, or chronically ill individuals. Furthermore, all diagnostic modalities—encompassing clinical assessments, radiological evaluations, microbiological analyses, and molecular methodologies—possess inherent limitations. Consequently, antibiotic therapy is often administered based on clinical acumen, without prior identification of the specific pathogen. Whether delivered orally or via intravenous routes, these therapeutic interventions exert systemic effects, thereby influencing

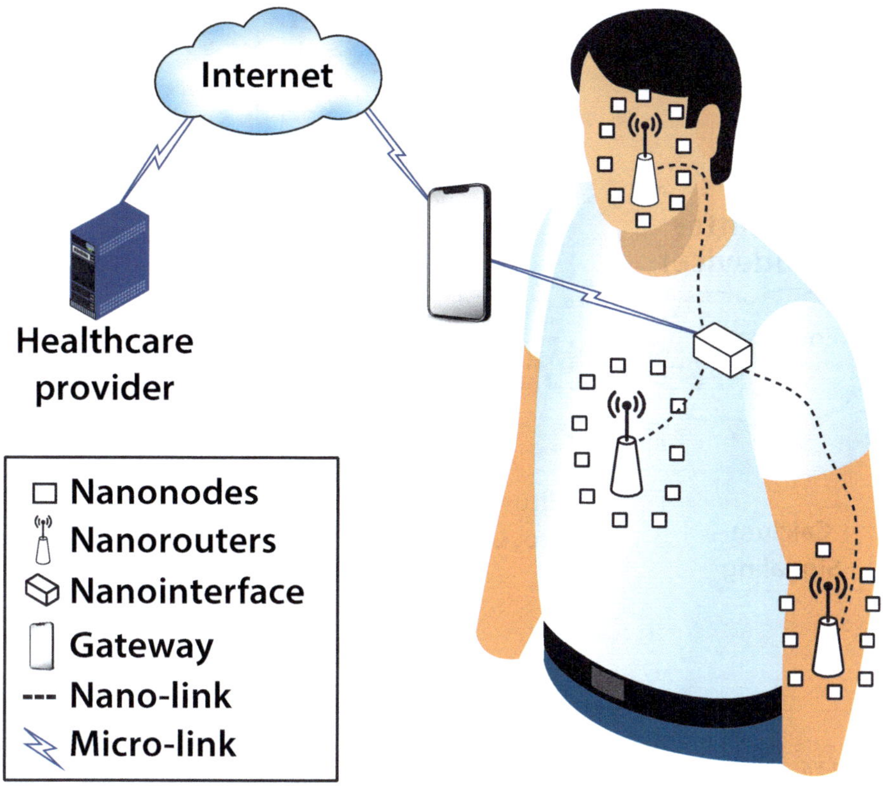

Fig. 2.55 Proposed intrabody nanonetwork's scheme (The developed model of the environment with changeable concentration of nanodevices was created. The initial nanonetwork's topology (simplified version) is displayed in Fig. 2.56. In [180] used Random Walk 2D Mobility Model as a propagation model where the communication area bounds were set to 1.5 mm × 1.5 mm.)

the entire organism. The systemic administration of antibiotics in the absence of a clearly defined pathogen exacerbates and accelerates the challenge posed by the rising prevalence of resistant bacterial strains. Thus, the advancement of novel diagnostic and therapeutic strategies holds significant clinical importance.

Then, in [181] is envisioned an MC application to nanomedicine. Then, in that investigation a system is designed to detect infections early, locally, and specifically, estimate their relevance, and check their severity. This design consists in a biological communication that after consulting with an *analysis and control station* as shown in Fig. 2.57, the proposed system responds to an infection in a suitable manner. This system comprises a (yet to be specified) number of nanomachines, which act from within the human body as sensors or actuators.

When endowed with communication functionalities, these entities establish an intrabody network that facilitates the reception, transmission, and relay of information. Furthermore, implants or wearable microsystems constitute a body area network (BAN), acting as intermediaries between the nanoscale devices and a control

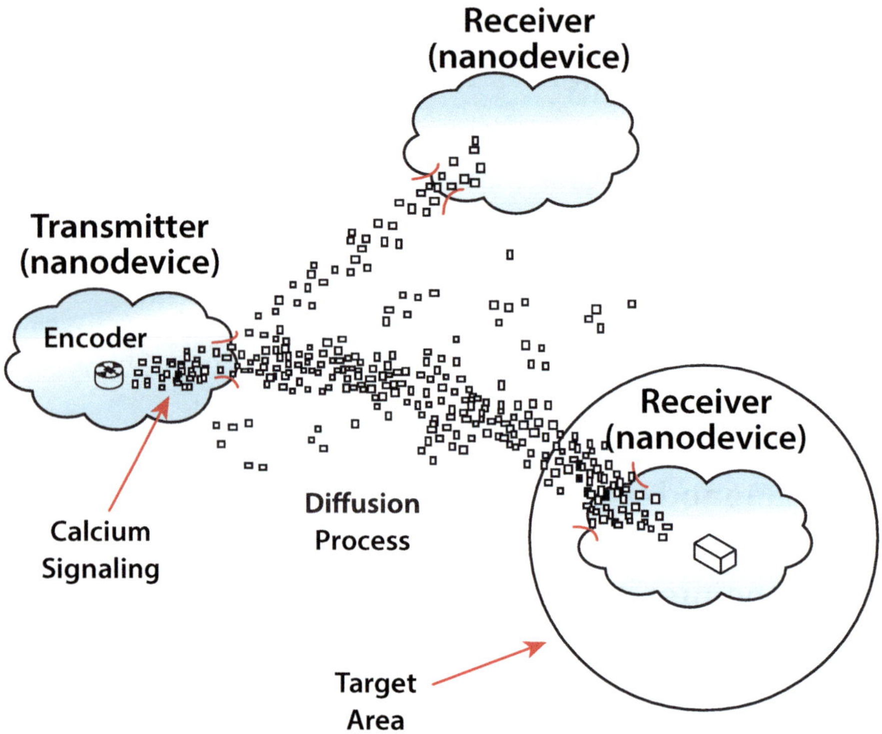

Fig. 2.56 Signal propagation in calcium signaling communication system

hub. The control hub (for instance, a smartphone) is situated externally to the human organism, where it furnishes requisite computational resources and memory to archive and analyze the data acquired by the nanoscale devices, subsequently relaying this information to the patient or a healthcare professional. A control hub with Internet connectivity further enables the inclusion of external participants in the procedural framework, such as medical databases or healthcare practitioners for enhanced analysis, oversight, or consultation. The whole system forms an IoNT [181].

In the delineated scenario, one or more nanoscale devices may facilitate the identification of a bacterial pulmonary infection directly at the alveolar interface, for instance, by quantifying proinflammatory cytokines liberated by the effector cells of the immune system, or by ascertaining specific bacterial virulence determinants, such as lipopolysaccharide, as an indicator of the presence of Gram-negative bacteria. The devices relay the quantified concentrations to the control station to ascertain whether the data constitutes a sufficient basis for a significant and pathogenic infection. If deemed necessary, the nanoscale devices may subsequently be instructed to selectively administer antibiotics and/or anti-inflammatory compounds at the site of infection. Additionally, in [181] is presented the requirements for a nanonetworking

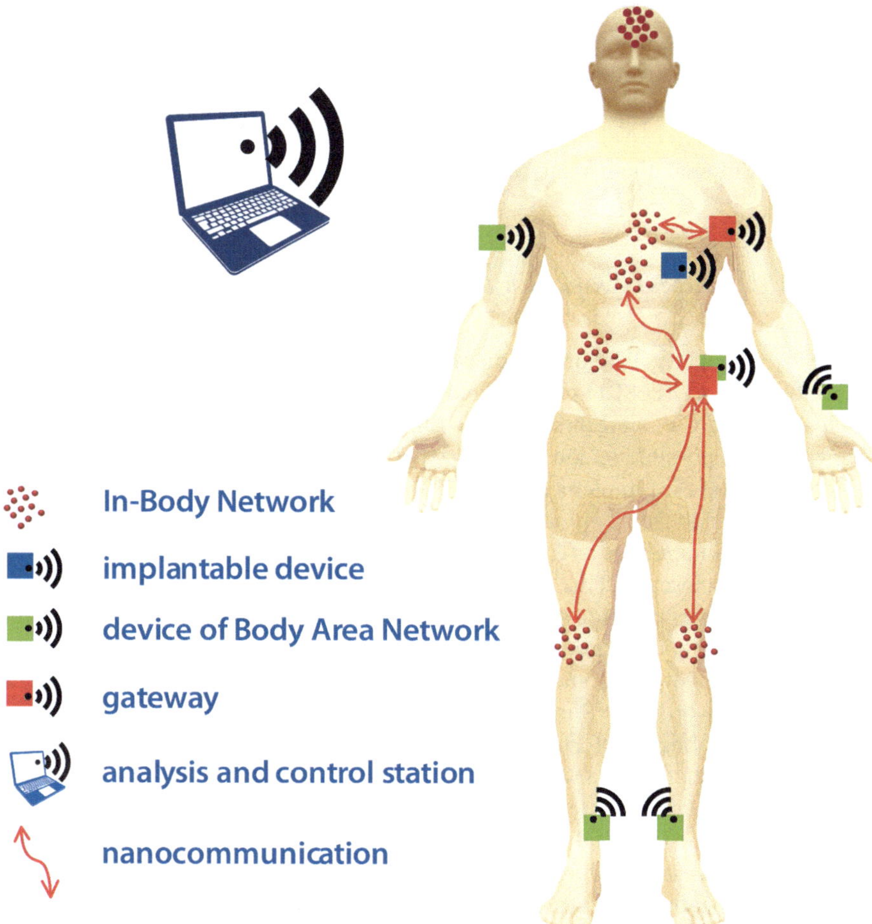

Fig. 2.57 In-body and BAN Architecture

simulator and proposed elements of a suitable architecture based on the NS-3 simulator and its module Nano-Sim.

In [182], the analysis and implementation of a diffusive propagation model for MC are performed in NS-3. The work is based on the IEEE 1906.1-2015 standard recommendation, which seeks to create a reference framework for MCs. The standard provides a simulation module in NS-3, which contains only the components of the general structure of MC and their interaction between them. The components mentioned are *Message Carrier, Motion, Field, Perturbation,* and *Specificity*. The transmitter uses CSK (Concentration Shift Keying) modulation. In the medium, Brownian motion (BM) with and without drift is used for the motion of the molecules, and ISI is considered. In the receiver, amplitude detection is used. The whole process is applied in four scenarios: Free BM, BM with drift, free BM bounded by the medium, and BM with drift bounded by the medium are considered. As a result, the pulse train of the mean concentration of molecules as a function of time at the receiver is obtained. In addition, the results are compared with an investigation performed in N3Sim to validate the results. Finally, it is validated that the mean concentration at the receiver using the diffusive propagation model implemented complies with the mathematical model established by Fick's second law [182].

2.3.6 *Bacteria Network Simulator (BNSim)*

BNSim[40] (or properly, BNSim2) is an open-source bacteria-based network. BNSim is developed by CMU SLD Group[41] and released under GPLv2 license. BNSim is developed using C++ and thread under Linux and macOS. The simulator is fully extensible; it is possible to add regulatory networks to the library and different cell types to the zoo [183] implementations [184].

BNSim simulator conceptualizes bacterial-based networks that are constructed utilizing either indigenous or engineered strains of bacteria that facilitate communication at the nanoscale. This delineation encompasses the microscale molecular transport systems employing chemotactic bacteria for the precise delivery of cargo, in addition to genetic circuits that facilitate intercellular interactions, such as quorum sensing or photonic communication. To accurately characterize the dynamics of bacterial networks, BNSim serves as an open-source, parallel, stochastic, and multiscale modeling platform that amalgamates various simulation algorithms alongside genetic circuits and chemotactic pathway models within a sophisticated three-dimensional environment. Furthermore, the BNSim platform is applicable for modeling synthetic bacterial consortia that execute a XOR function and aggregate adjacent bacteria via light-based communication. As a result, the findings illustrate how BNSim is capable of forecasting diverse properties of realistic bacterial networks and offering insights for their corresponding experimental implementations in wet laboratories [184].

Regarding the physical characteristics of the molecular channel implementations, [184] establishes that bacteria with high motility in viscous fluids have been considered as possible carriers for long-distance directed transportation. Indeed, owing to their capacity to detect a multitude of chemical signals released from specific target sites, these microorganisms have been deemed as optimal candidates for precision cargo delivery mechanisms, particularly those designed for diagnostic purposes and pharmaceutical administration. A scenario is illustrated in Fig. 2.58, wherein cancer cells (acting as network receivers) exude various chemical substances within the three-dimensional milieu that function as chemical signals for the

[40] BNSim: bacteria network simulator—System Level Design (utexas.edu), GitHub - weiguopeng/BNSim2: BNSim2

[41] http://www.ece.cmu.edu/~sld/

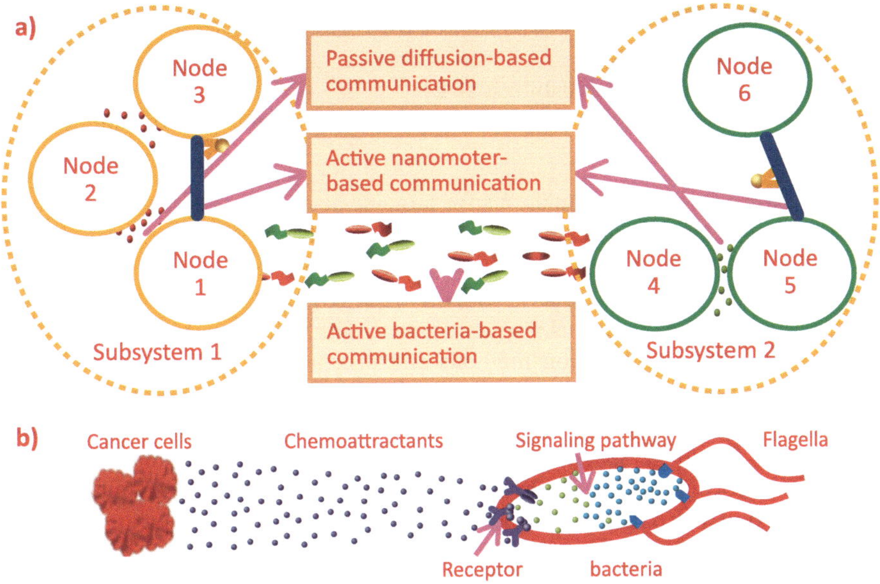

Fig. 2.58 (**a**) An illustration of a biological nanonetwork [Mechanisms of passive transport (e.g., diffusion-based processes) and active transport (e.g., nanomotor-driven systems) are adept at facilitating information exchange within diminutive networks. Conversely, for extensive nanonetworks, communication strategies that employ self-propelling bacteria equipped with sensors and multiple flagellar motors are regarded as one of the most promising methodologies.] (**b**) A targeted drug delivery system wherein malignant cells excrete various chemical substances into the surrounding environment, which subsequently function as chemoattractants for motile bacteria

motile bacteria. Upon recognizing these chemical signals through chemoreceptors, the motile bacteria are capable of navigating toward the designated areas and either delivering or synthesizing the therapeutic agents upon contact with the targets [184].

Conversely, the progression of synthetic biology uncovers numerous innovative applications that utilize genetically modified bacteria, encompassing functionalities such as temporal regulation, quantification, temporal measurement, logical operations, pattern recognition, intercellular signaling, and resilient multicellular computation, many of which entail populations of bacteria interconnected via quorum sensing molecules that function as imperceptible "chemical conduits." Within these networks of bacteria, the genetic circuitry responsible for communication, environmental sensing, and protein synthesis constitutes the fundamental framework of collaborative bacterial communities. Nevertheless, owing to the intricate physical and biological complexities (e.g., diverse intracellular dynamics, intercellular signaling, and physicochemical environmental variables), bacteria-based nanonetworks have yet to be accurately characterized at both the cellular and population levels; consequently, to ameliorate this situation, BNSim has been developed [184].

BNSim integrates three simulation methods [184]:

- Gillespie's exact SSA
- Stochastic differential equations (SDE)
- A new hybrid simulation algorithm based on the hierarchy and dynamics of biochemical systems

Besides, to speed up the simulations for large-scale chemical system, approximate methods based on chemical Langevin equation, or other SDE, have been developed. However, unlike the exact SSA, these algorithms have difficulties in capturing the nonlinear noise generated by complex protein structures or large fluctuations due to the small quantities of various reactant species. The hybrid methods integrate both exact SSA and approximation methods and take advantage of the stiffness that exists inherently in most chemical systems. More precisely, the presence of stiffness indicates that the evolution of a chemical system involves several (widely varying) time scales. Chemical species with relatively small quantities are modeled as discrete stochastic processes, while species that are present in larger quantities are modeled using approximate methods [184].

To support the simulator creation, the authors defined that at this time hierarchical structure of biological systems has not been considered as an effective means to speed up computations. Therefore, simulator creation is focused on finding the key subsystem whose noise characteristics have the greatest impact on the overall system behavior, then simulate the subsystem using the exact SSA to preserve the noise characteristics. In the case of bacterial chemotaxis, the methylation dynamics have been identified as the key subsystem due to the complex chemoreceptor structure, large fluctuations of the protein *CheR* and *CheB*, and the limited methyl group binding sites. Its noise characteristics have been shown to be the origin of noise-induced Levy walks of *E. coli*. Therefore, the chemoreceptor dynamics are simulated using the exact SSA, while all other subsystems are simulated using approximate methods under the partial equilibrium assumption. Therefore, the entire system is then synchronized at every reaction event of the key subsystem (Fig. 2.59) [184].

The simulator performance was done using the new hybrid multiscale stochastic method. Thus, in [184] the comprehensive bacterial chemotaxis pathway was simulated employing a detailed individual receptor model elucidated in the subsequent sections, resulting in a remarkable enhancement in computational speed, achieving an increase of 5–7 orders of magnitude relative to the traditional exact stochastic simulation algorithm (see Fig. 2.59c). Furthermore, to facilitate the parallel simulation of a substantial quantity of interacting bacteria, BNSim implements a sophisticated thread-scheduling mechanism that optimally distributes the computational load associated with simulating intracellular dynamics as well as the interactions between bacteria and their environment across all available threads executing on a multicore architecture (refer to Figs. 2.59c and 2.60c) for the results obtained from running BNSim with 8 threads [184].

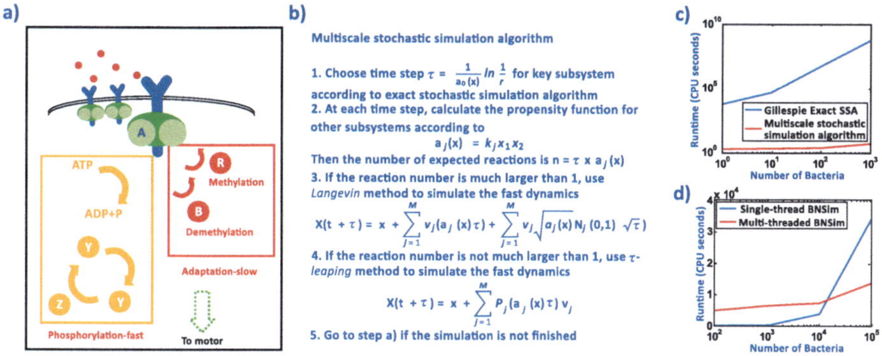

Fig. 2.59 Partitioning of the bacterial chemotaxis pathway and the introduction of a novel multiscale stochastic simulation algorithm. (**a**) A schematic representation of the bacterial chemotaxis pathway structure is presented [The fast subsystem, identified as the phosphorylation relay, is delineated within the left box, while the slow subsystem, characterized by the adaptation process (i.e., methylation kinetics), is encapsulated within the right box.] (**b**) The innovative multiscale stochastic simulation algorithm proposed in Reference [184] is outlined [The principal subsystem is modeled utilizing an exact Stochastic Simulation Algorithm (SSA), whereas other subsystems are approximated through methods contingent upon the anticipated reactions within the time step established by the exact stochastic simulation algorithm for the slow key subsystems.] (**c**) The computational runtime (measured in CPU seconds) of the exact stochastic simulation is compared to that of the multiscale stochastic simulation as a function of bacterial population size (As anticipated, the multiscale stochastic simulation framework significantly enhances the temporal progression of the chemical system by several orders of magnitude.) (**d**) A comparative analysis of multithreaded BNSim against single-threaded BNSim is conducted (The multithreading capability of BNSim facilitates the efficient simulation of a considerable bacterial population, whereas the performance of single-threaded simulations experiences a substantial decline once the bacterial count exceeds 10,000. It is noteworthy that multithreaded BNSim is not optimized for simulating a minimal bacterial population due to the associated synchronization overhead.)

[184] specifies briefly the biological models integrated (i.e., chemoreceptor modeling, phosphorylation relay, and flagella motor actuation) and presents a validation of its models against experimental data[42] as:

- Biological Models for *Escherichia coli*: The authors in [184] establish the following characteristics for BNSim.
  - Fine-grained Chemoreceptor Model simulated with SSA: To simulate the chemoreceptors, [184] uses the Monod-Wyman-Changeux model in which the receptor homodimers assemble into fully cooperative signaling teams that switch rapidly between active (*on*) and inactive (*off*) states. In order to capture the essential methylation and demethylation noise, the authors in [184] model 370 receptor clusters individually. For wild-type bacteria, each receptor

[42] The complete details regarding the modeling and parameter values used in BNSim models are available in the online supplementary material at: Microsoft Word—supplementary-v1.docx (cmu.edu).

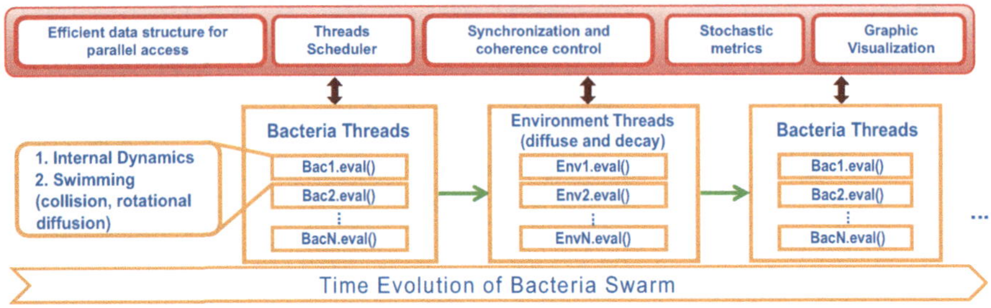

Fig. 2.60 Schematic view of BNSim structure for multithreaded simulation (As shown in the figure, the time evolution of the stochastic simulation follows a tick-tock fashion. During "tick," bacteria sense the environment, update their internal dynamics, and swim in the environment simultaneously. This is done by exploiting the multithreading feature of BNSim. The 3D space is partitioned into cubes larger than the size of bacteria. A typical cube size is 10 μm × 10 μm × 10 μm, while *E. coli* is a sphere with a radius of 1 μm and its swimming speed is 20 μm/s. The neighborhood information of each cube can be accessed conveniently and safely using mutex. The cubes of the environment are updated in a similar manner during "tocks." All threads are synchronized using two cyclic barriers. Therefore, it is possible to ensure that chemical and physical interactions between bacteria and environment can be efficiently simulated)

cluster has 13 Tsr receptors (for serine) and 6 Tar receptors (for aspartate). The activity of each individual receptor cluster depends on its methylation level and ligand occupancy with the using of the "assistance neighborhoods" model [184]:

$$A = \frac{e^{F^{-on}}}{e^{F^{-on}} + e^{F^{-off}}} = \frac{1}{1 + e^{F}} \tag{2.10}$$

where A represents the average activity of the receptors in the cluster, and F is the sum of the individual free-energy differences between the receptor *on* and *off* states. Since the entire receptor cluster is considered to be either in an *on* or *off* state, the offset energy of the receptor at different methylation levels follows the original "assistance neighborhoods" model. The methylation level of each receptor has a range from 0 to 8. At each simulation step, one receptor inside one of the 370 receptor clusters is selected to be methylated or demethylated by the protein *CheR* and *CheB*, respectively. Also, in [184] is used a Wiener process to describe the fluctuation of the protein *CheR* as:

$$dR(t) = \beta x d\eta \tag{2.11}$$

where $R(t)$ is the concentration of *CheR*, β is the noise parameter, and $\eta(t)$ describes the Wiener process. Note that this adaptation process of the chemoreceptor complex (i.e., methylation and demethylation dynamics) is the key subsystem of the bacterial chemotaxis pathway (step 1 in Fig. 2.59 b)).

– Coarse-grained Chemoreceptor Model Simulated with SDE: The free-energy difference F in units of k_BT between the active and inactive states of the receptor complex can be simplified to [184]:

$$F = \epsilon_0 + \epsilon_1 m + \sum_r \sum_C \log\left(\frac{1 + C / K_r^{off}}{1 + C / K_r^{on}}\right) n_r \tag{2.12}$$

where m is the methylation level of the receptor complex, r indicates the receptor species (aspartate or serine receptor), and C represents the chemoattractant concentration. K_r^{off} and K_r^{on} are the binding constants between a certain ligand and a receptor complex r in either "*on*" or "*off*" state, and ϵ_0 and ϵ_1 are two parameters tuned to fit the experimental data. The methylation kinetics responsible for bacteria adaptation to the environment are modeled by the Ornstein-Uhlenbeck process [184]:

$$\frac{dm}{dt} = 1 - \frac{1}{\tau}\left[m(t) - \bar{m}(C)\right] + \sigma_m \sqrt{\frac{2}{\tau}}\,\Gamma(t) \tag{2.13}$$

where τ is the relaxation time, σ_m is the variance of fluctuations in the methylation dynamics, $\Gamma(t)$ is a normally distributed homogeneous random process with zero mean and unit variance, and $\bar{m}$ is the average methylation level of a receptor cluster in its adapted state given C [184].

– Phosphorylation relay and flagella motor: Active receptors enhance autophosphorylation activity of the receptor-associated kinase *CheA*, which transmits the signal to the flagella motors by phosphorylating a diffusive response regulator *CheY*. Since the phosphorylation relay is not the key subsystem and the number of reactant proteins involved are large (i.e., thousands), there must be used an approximate method to model the signal relay (steps 2 and 3 in Fig. 2.61b) Fig. 3b). Thermal fluctuations and upstream signaling cause the flagella motor to spontaneously change between alternating rotational states. More precisely, in [184] simulates the flagella motors with a two-state model with each state, describing clockwise (CW) and counterclockwise (CCW) rotation of the flagella, sitting in a potential well. The energy barriers for CCW to CW transition and CW to CCW transition are $G_0(Y_p(t))$ and $-G_0(Y_p(t))$, with transition rates k^- and k^+, respectively. The definition of $G_0(Y_p(t))$, k^- and k^+ are specified in Eqs. (2.14), (2.15), and (2.16), respectively [184]:

$$G_0\left(Y_p(t)\right) = \frac{g_0}{4} - \frac{g_1}{2}\left(\frac{Y_p(t)}{K_D + Y_p(t)}\right) \tag{2.14}$$

$$k^+ = w_0 \exp\left(G_0\left(Y_p(t)\right)\right) \tag{2.15}$$

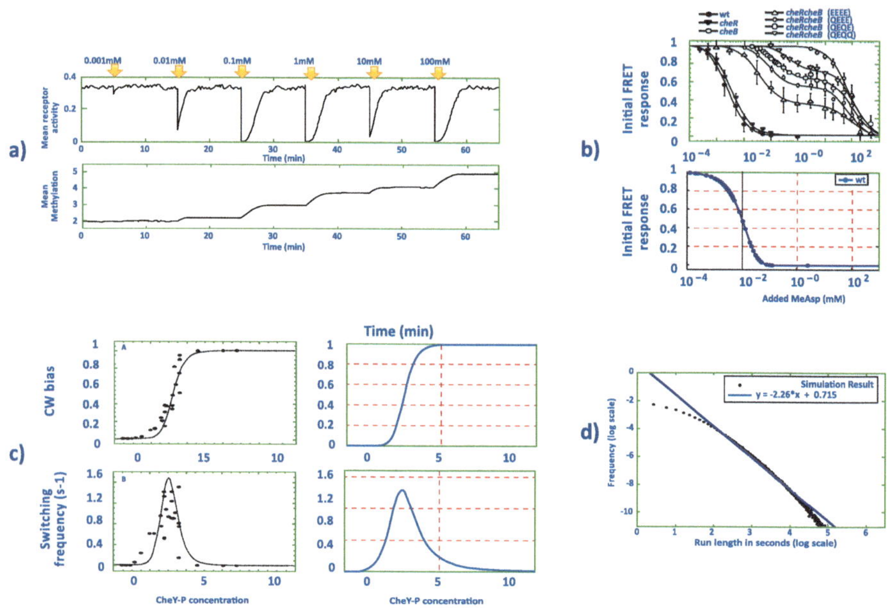

Fig. 2.61 Single-cell model calibration. (**a**) Precise adaptation of bacteria (The bacterial chemotaxis network exhibits a remarkable sensitivity to minute alterations in chemical concentrations across an extensive spectrum of ambient concentrations. The figure is derived from the average outputs of BNSim simulation results encompassing 370 receptor clusters). (**b**) Normalized initial receptor activity (The upper figure presents the experimental findings, whereas the lower figure illustrates the BNSim simulation results pertaining to wild-type bacteria. The precise receptor activity curve guarantees that bacteria accurately respond to external stimuli.) (**c**) The CW bias and switching frequency of a flagellar motor as a function of *CheY-P* concentration (The experimental data is displayed on the left, while the BNSim simulation results are provided on the right). (**d**) BNSim simulation findings indicate that bacterial run intervals conform to a power-law distribution in environments devoid of nutrients

$$k^{-} = w_0 \exp\left(-G_0\left(Y_p\left(t\right)\right)\right) \tag{2.16}$$

where Y_p is the concentration of the phosphorylated protein *CheY*, and parameters w_0, g_0, g_1, and K_D are chosen to fit the experimental data. Similarly, the flagella motor is simulated by stochastic ODESs and updated in response to each reaction event of the key subsystem [184].

– Molecule diffusion and swimming dynamics: For the molecule diffusion in the environment, in [184] is assumed that there is no chemical interaction between the molecules, so a reaction-diffusion equation based on Fick's second law to model the diffusion process must be used:

$$\frac{\partial C\left(x,y,z,t\right)}{\partial t} = D\frac{\partial^2 C\left(x,y\,z,t\right)}{\partial x^2} + D\frac{\partial^2 C\left(x,y\,z,t\right)}{\partial y^2} + D\frac{\partial^2 C\left(x,y\,z,t\right)}{\partial z^2} \tag{2.17}$$

where $C(x, y, z, t)$ is the chemical concentration at location (x, y, z) and time t, and D is the diffusion coefficient. Bacteria use runs and tumbles to effectively swim in liquids. When bacteria tumble, a new direction is chosen from a Gamma distribution which fits the distribution of tumbling angles [184].

- Single-cell validation: The validation of the fine-grained chemoreceptor models against experimental results is effectuated here. Figure 2.61a shows the simulation results of bacteria chemosensitivity and its precise adaptation. Bacteria adaptation is a key feature of the chemotaxis system, which shows its ability to return to the same level of chemoreceptor activity over a wide range of chemical gradients. Figure 2.61b shows the dose-response curve reported in experiments and simulated with BNSim. The dose-response curve for wild-type *E. coli* is obtained by adding a certain level of stimuli to an environment containing only a single bacterium [184].

  The response curve shows how much a chemoreceptor is suppressed when a given amount of stimuli is added to a previously nutrient-free environment. The flagella motor of bacteria either turns CW or CCW according to the *CheY-P* level. The CW bias is the fraction of time a flagella motor is in the CW state for a certain time window [184].

  As shown in Fig. 2.61c, a low *CheY-P* concentration increases the probability that a motor is in the CCW state, while a high *CheY-P* concentration makes a motor more likely to be in the CW state. When a motor turns CCW, its filaments rotate as a bundle and push the bacterium body steadily forward; the bacterium is then said to "run"; otherwise, the bacterium tumbles and chooses another direction with a bias toward its previous direction. However, in a nutrient-free environment, the fluctuations of *CheR* propagate to the motor regulatory protein *CheY* and result in a power-law distribution of run intervals and an exponential distribution of the tumble intervals. Figure 2.61d indicates that the power-law exponent obtained from BNSim matches the experimental data very well. The extended runs allow bacteria to perform Levy walks with extended runs; this turns out to be a good strategy (compared to the pure Brownian movement) when food is distributed randomly and sparsely throughout the environment [184].

• Bacterial transport mechanisms: Bacteria exhibit remarkable motility within a fluid medium characterized by low Reynolds number. Their capacity to detect chemical stimuli and navigate toward favorable environments has been harnessed for extended directed transport. The inherent stochasticity of individual bacterial entities, combined with rotational diffusion during locomotion and the gradients of chemoattractants established by receivers (i.e., targets) within the environment, collectively results in bacterial movement adhering to a biased random walk paradigm. To explore the design trade-offs associated with this phenomenon, the study referenced in [184] simulated a population of bacteria under different conditions and then collected statistics at population level, with the subsequent features.

  - Simulation setup
  - Impact of bacteria number on communication delay and mean square displacement
  - Impact of chemoattractant emission rate on communication delay and mean square displacement
  - Communication delay as a function of distance

- Bacterial cell-to-cell interactions based on chemotaxis [184].
- Bacterial cell-to-cell interactions based on quorum sensing and light communication. Particularly, the characteristics adopted by BNSim are as follows [184]:

  - Bacterial Consortia perform XOR Function.
  - Bacteria phototaxis.
  - Simulation results.

BNSim can be characterized as a biochemically precise, yet adaptable and effective simulation framework for nanonetworks based on bacteria, which incorporates intracellular biochemical reactions, intercellular communications, and environmental dynamics. Furthermore, the BNSim framework facilitates the examination of various modalities of communication, including chemotaxis, quorum sensing, and light-mediated intercellular signaling. BNSim produces a high degree of accuracy and performance within a parallel modeling platform for bacteria-based nanonetworks, in comparison to numerous experimental investigations, and has elucidated several potential trade-offs that may be realized under particular design constraints [183, 184].

BNSim amalgamates a diverse array of simulation algorithms along with genetic circuit frameworks and chemotactic pathway models within a sophisticated three-dimensional environment. The proposed framework adeptly encapsulates intracellular biochemical reactions while concurrently managing the emergent physicochemical interactions present within the environment. In an endeavor to enhance communication efficacy, the authors consider pivotal subsystems as integral components of the hierarchical architecture inherent in biological systems. For example, the authors incorporate methylation dynamics as a constituent of a multi-faceted chemoreceptor structure. Within the particular investigation, the chemical signaling pathway within each bacterium, as well as the chemical gradients established by the receiver nodes, was meticulously modeled [185].

BNSim systematically models both the intracellular biochemical processes and the chemophysical interactions within intricate three-dimensional environments, thereby enabling the exploration of the dynamics inherent to large-scale microbial communities with remarkable precision. Furthermore, BNSim incorporates an efficient scheduling mechanism, which enhances its execution speed and scalability across many-core computational platforms. Within the BNSim framework, the progression of the stochastic simulation adheres to a tick-tock methodology. During the "tick" phase, bacterial entities perceive their surroundings in a three-dimensional space and exhibit locomotion contingent upon the conditions of their environment. The comprehensive three-dimensional simulation space is partitioned into cubic segments of sufficient size to accommodate numerous bacterial clusters. Computational tasks associated with the entirety of the three-dimensional simulation space are subsequently allocated randomly among multiple threads to promote equitable workload distribution. Throughout the execution of the tick phase, data regarding the movement of bacteria is continuously exchanged among various cubes (threads). In the "tock" phase, environmental data is disseminated among cubes, and the status flags are updated to facilitate the efficient simulation of the chemical and physical interactions between bacterial entities and their environment [186].

Consequently, the execution of BNSim necessitates the exchange of data among the various cubes throughout both the tick and tock phases. Given that the total workload is allocated randomly across numerous threads, the inter-cube data exchange induces irregular inter-core traffic patterns [186].

Simulations using BNSim are realized in [187]. There is presented a modeling and experimental framework to characterize the chemotaxis of *Serratia marcescens* relying on two-dimensional and three-dimensional tracking of individual bacteria. [187] focuses on single-cell tracking and measuring the chemotactic drift velocity from the biased tumble rate of individual bacteria on exposure to a concentration gradient of L-aspartate. The chemotactic response of *Serratia marcescens* is quantified over a range of concentration gradients (10^{-3} to 5 mM/mm) and average concentrations (0.5×10^{-3} to 2.5 mM). Through the analysis of a large number of bacterial swimming trajectories, the tumbling rate is found to have a significant bias with respect to the swimming direction [187].

[187] additionally verifies the relative gradient sensing mechanism in the chemotaxis of *Serratia marcescens* by measuring the change of drift velocity with the average concentration and the gradient. The applied full pathway model with fitted parameters matches the experimental data. Measurements based on individual bacteria lead to the determination of the motility coefficient μ ($7.25 \times 10^{-6}\ \frac{\text{cm}^2}{\text{s}}$) of a population. The experimental characterization and simulation results for the chemotaxis of this bacterial species contribute toward using *S. marcescens* in chemically controlled biohybrid systems [187].

Also, in [188] BNSim is utilized as a digital platform. The topic analyzed in [188] deals with quorum sensing that is a chemical communication process bacteria use to sense population density and regulate their collective behavior. By using quorum sensing inhibitors that degrade quorum sensing molecules and inactivate their receptors, one can inhibit bacterial pathogenesis, such as the production of virulence and biofilm development. To keep the level of quorum sensing molecules below the activation threshold, [188] proposed a biological controller that can generate different concentration levels of inhibitors under different environment conditions. The authors in [188] provide a detailed analysis of our proposed controller that includes system response, stability, and sensitivity analysis. Additionally, in [188] is discussed the autonomous controller design under specified environment constraints and validate our results via simulation. [188] represents a first step toward a paradigm change in reducing bacterial pathogenesis via controlling the dynamics of bacterial cell-to-cell communication network.

Another case which uses BNSim is found in [186]. The theme in [186] (consisting in understanding the role of competition and cooperation among multiple interacting species of microorganisms that constitute the microbiome and deciphering how they enforce homeostasis or trigger diseases) requires the development of multiscale computational models capable of capturing both intra-cell processing (i.e., gene-to-protein interactions) and inter-cell interactions. The multi-tiered interdependence that orchestrates the interactions ranging from genetic elements to protein structures within a cellular context and from molecular signalers to cellular entities and microbial assemblages in environmental settings presents a plethora of computational and communicational challenges. Internal cellular processing cannot be accurately replicated without a comprehensive understanding of the extracellular milieu. Likewise, intercellular communication cannot be entirely conceptualized without consideration of the intricacies of internal processing and the diffusion dynamics of molecular signalers [186].

Therefore, to address the compute-and communication-intensive nature of modeling microbial communities, [186] proposed a novel reconfigurable Network-On-Chip (NoC)-based many-core architecture capable of simulating a large-scale microbial community. The reconfiguration of the NoC topology is achieved through the fractal analysis of NoC traffic and use of the on-chip wireless interfaces. More precisely, in [186] the computational and communication workloads are analyzed and the authors exploit the observed fractal characteristics for proposing a mathematical strategy for NoC reconfiguration. Experimental results demonstrate that the proposed NoC architecture achieves 56.6 and 62.8 percent improvement in energy delay product over the conventional wireline mesh and flatten butterfly based high radix NoC architectures, respectively [186].

Similarly, in [189] the simulation tool is BNSim; the theme is focused on bacterial swimming dynamics, specifically fractal swimming dynamics of populations of *Serratia marcescens* bacteria both in vitro and in silico, while accounting for realistic conditions like volume exclusion, chemical interactions, obstacles, and the distribution of chemoattractant in the environment. While previous research has shown that bacterial motion is non-ergodic, in [189] is demonstrated that, besides the non-ergodicity, the bacterial swimming dynamics are multi-fractal in nature. In [189] is demonstrated also that the multi-fractal characteristic of bacterial dynamics is strongly affected by bacterial density and chemoattractant concentration [189].

On the other side, population bottlenecks and founder events reduce genetic diversity through stochastic processes associated with the sampling of alleles at the time of the bottleneck and the recombination of alleles that are identical by descent. At the same time, bottlenecks and founder events can structure populations through the stochastic distortion of allele frequencies [190]. There, the authors undertake an empirical assessment of the impact of two independent bottlenecks of known size from a known source and consider inference about evolutionary process in the context of simulations and theoretical expectations. In [190], a comparable degree of diminished variation is observed in the concurrent bottleneck events, wherein the population initially comprising a smaller cohort of females experienced a more pronounced effect. The residual level of genetic diversity aligned with theoretical model forecasts, contingent upon the assumption that population resurgence occurred predominantly in an exponential manner and that polygenic influences were negligible during the initial phases. A significant degree of differentiation was evident when compared to the source population and among the two populations that underwent bottlenecking, illustrating the stochastic perturbation of allele frequencies [190].

Consequently, in [190] empirical evidence was presented which corroborates the theoretical predictions that a substantial degree of genetic diversity may persist subsequent to a significant bottleneck event, provided that there is a swift demographic recovery, and that populations originating from a common source can rapidly undergo differentiation. These processes may hold substantial significance in the evolutionary dynamics of population genetic structure for species experiencing rapid alterations in their available habitat.

In [191], it is articulated that contemporary advancements in synthetic biology have significantly augmented humanity's capacity to manipulate biological mechanisms at the cellular level. Specifically, a substantial corpus of recent research concentrates on the treatment of human pathogens with an emphasis on countering escalating antibiotic resistance. In pursuit of this objective, an innovative approach has been proposed to interfere with the pathogen communication network (i.e., quorum sensing), thereby impeding the pathogens' ability to synchronize the secretion of virulence factors. Hence, in [191] is proposed a new molecular model that captures the effects of quorum sensing inhibitors on the quorum sensing network of *P.* aeruginosa and formulates a constrained optimization problem to find the optimal combination of quorum sensing inhibitors while accounting for the selective pressure of developing drug resistance. Optimization results in [191] show significant reduction of selective pressure on pathogens, while preserving the efficacy of the multidrug quorum quenching therapy; these results were obtained with BNSim.

2.3.7 High-Level Architecture (HLA)

Given that MCs simulation models entail simulating a substantial quantity of nanoscale entities, the simulation tool's scalability is a significant consideration. In [192], a simulator architecture is presented that addresses critical design concerns of the MC model, with an emphasis on scalability. The HLA, specified by IEEE 1516, is utilized for the design and development of a distributed simulation tool for MCs. The findings indicate that several scalability solutions can be employed to leverage increased processing capacity, hence reducing execution time. This facilitates the modeling of extensive systems, which may otherwise be unfeasible. Should a consensus be reached across many academic organizations and industry on a common architecture that facilitates software component reuse and high scalability, a burgeoning library of simulation components may be established, enabling the

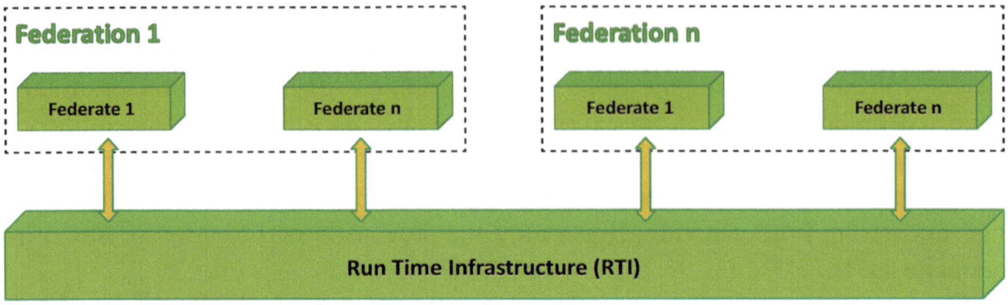

Fig. 2.62 Generic HLA architecture

modeling of intricate situations in a distributed environment [192]. This target is similar to the one established by the Circle project (with MolComML, Chap. 3, Sect. 3.2) and also with IEEE 1906 standards (Chap. 3, Sect. 3.3).

Consequently, the study in [192] endeavors to delineate an architecture grounded in the standardized HLA, which is extensively utilized in large-scale simulations. As HLA offers the fundamental framework for networked simulations, the referenced simulator embodies a specialized design for MCs, emphasizing interoperability, reusability, and scalability. The architecture and parameter analysis influence parallelism and the load distribution capacity in a distributed MC simulation context. An HLA-based distributed simulator facilitates comprehensive simulation scenarios even the interaction with systems that include living organisms. A distributed simulation is a collaborative system in which each simulation unit functions on an independent computing unit and communicates to recreate a situation inside a shared logical temporal framework [192].

The primary objective of HLA is to delineate the component models and their interconnections. These elements (named federates) are developed to facilitate software reuse and abstraction. A Run Time Infrastructure (RTI) handles the interaction among the components utilizing common protocols to create a simulation model known as a federation. Every federation must possess a Federation Object Model (FOM) developed in compliance with a standardized Object Model Template (OMT). This facilitates autonomous design and development of components, as well as decentralized execution of the simulation. Figure 2.62 illustrates a generic HLA design. The RTI serves as the foundation of the federation, facilitating synchronization, communication, and data sharing among the federates. Each federate may represent an individual event or time-driven simulation, real-time simulations involving human interaction, or actual systems and equipment. The HLA does not limit the modeling within a federate; it delineates the interactions between them. RTI offers six categories of services [192]:

- Management of the federation: Essential capabilities necessary for establishing and implementing a federation.
- Management of declarations: Administration of data interchange among federates, utilizing the data supplied by federates.

- Objects' management: Object-level services include creation, deletion, and identification.
- Management of ownership: The dynamic transfer of ownership of objects/attributes during execution.
- Temporal organization: Coordination of runtime simulation data interchange.
- Management of data dispersion: Data routing among federates during federation operation.

The method of accessing these services is delineated by the definition of the HLA interface; however, the implementation of the services falls beyond the scope of the specification. Multiple RTI systems have been created that adhere to the specification suitable for developing an HLA-based simulation environment. The HLA interface definition delineates the method of accessing these services; however, the implementation of the services falls beyond the scope of the specification. Numerous RTI systems have been created that adhere to the criteria applicable for developing HLA-based simulation environments [192]:

2.3.7.1 Concept of Simulation

At the conceptual level, HLA delineates autonomous entities known as federates and the interactions between them to establish a federation. The flexible component design enables the clear abstraction of interactions with living entities, other simulation systems, or passive observers overseeing the simulation process. If the communication parameters of a live creature or lab-on-a-chip are abstracted and communicated in real time, they may be integrated into a more complex model that includes both experimental systems and simulation components [192].

Decoupling and standardizing these linkages enable the creation and implementation of hybrid scenarios. Moreover, decoupling via HLA standards ensures that components may be deployed to independent processing units, with the RTI overseeing all communication among the federates. This notion is depicted in Fig. 2.63 [192].

2.3.7.2 Design of the Simulation

HLA possesses scalability, reusability, and interoperability as its main design criteria starting from the initial phase of its design. These criteria lead us to an architecture based on HLA, which addresses these concerns. Using HLA, it is possible for modules to be executed on different platforms to communicate. Common software libraries can be developed and used to create large-scale simulations, and it is possible to run simulations in distributed computer systems. There are four primary principal benefits of executing a simulation program across multiple computers that are mentioned in subsequent lines [192].

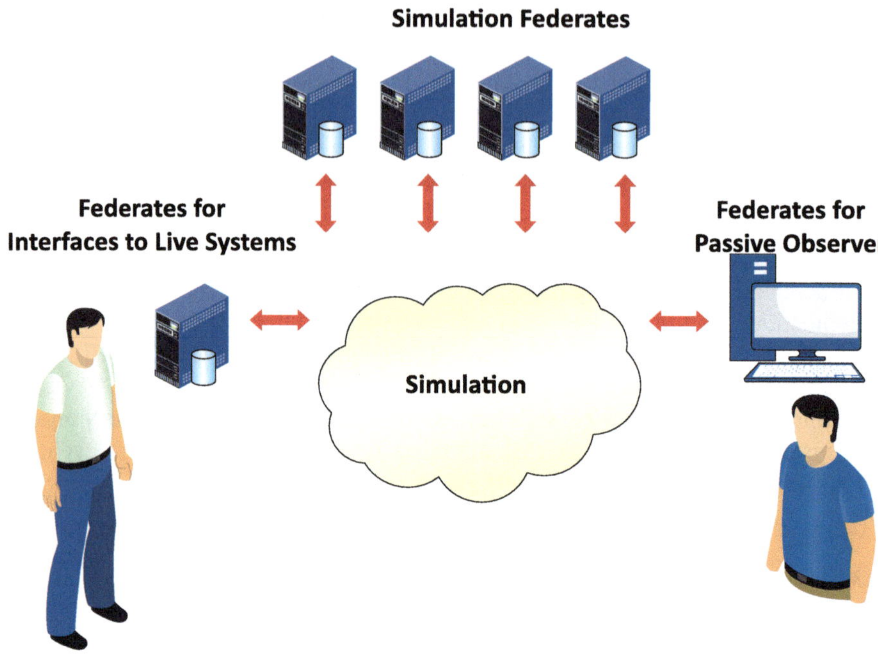

Fig. 2.63 Simulation deployment

- A shorter execution time: Execution time might be decreased by breaking up the large simulation calculation into smaller sub-computations and distributing these tasks across different processor units [192].
- The distribution of space: By running the simulation on geographically separated computers, it is possible to communicate with individuals or living things located in many locations [192].
- Including simulators that run on equipment made by several vendors. Contact with real systems may require the use of certain platforms, and different research teams may aim to develop components on different platforms. Interaction across many platforms is facilitated by a distributed technique [192].
- Tolerance for faults: By using several computing units, fault tolerance may be implemented, allowing another unit to take over a failing unit's tasks in the case of failure [192].

2.3.7.3 Components of the Simulation

Figure 2.64 illustrates the proposed architecture. The simulation is defined in an MC Federation. Separate federates are defined for molecules, nanomachines, and the medium. The federates communicate with each other over RTI based on a common FOM. Portico, an open-source, cross-platform HLA RTI implementation is used to

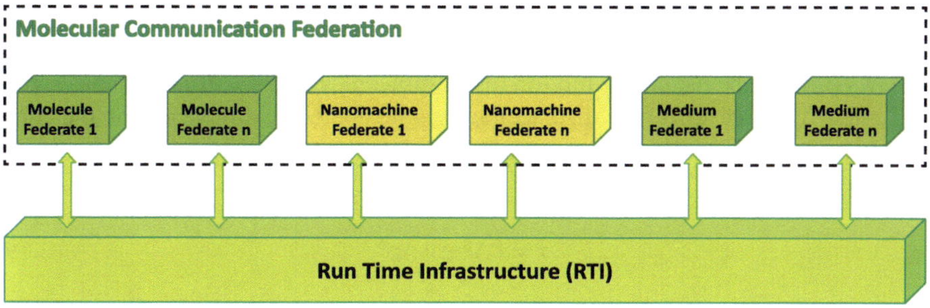

Fig. 2.64 Simulation architecture

develop the simulation tool. Portico uses another open-source project, JGroups, for communication among federates. Communication among federates includes attribute updates and interactions. It is mainly dominated by the number of molecules. The number of messages sent at each simulation time step can be formulated as indicated in Eq. (2.18) [192].

$$M = m \cdot \left(1 + P_t + P_r + P_d + P_{rb} + P_s\right) \tag{2.18}$$

where

m: Number of molecules
P_t: Probability that the molecule is transmitted by a nanomachine
P_r: Probability that the molecule is received by a nanomachine
P_d: Probability that the molecule is diffused out of the simulation boundaries and destroyed
P_{rb}: Probability that the molecule is rolled back due to a collision
P_s: Probability that the molecule moves to another medium slice since an additional message is transmitted for each one of these events

Since the sum of P_t, P_r, P_d, P_{rb}, and P_s is equal to one, the upper bound for the number of messages is $2 \cdot m$. Therefore, it is concluded that the value of M is between m and $2m$ [192].

By assigning tasks connected to molecules to Molecule Federates, tasks related to nanomachines to Nanomachine Federates, and dividing the 3D area into slices, each of which is overseen by a different medium federate, scalability is achieved. Because of Portico's cross-platform flexibility, each federate may be written using either Java or C++ and run on Windows or Unix-based computers. Each federate operates independently. Unix is used for simulation execution, and Java is used for HLA implementations [192].

- Molecule Federate: The simulation's molecules are included within the Molecule Federate. It makes molecular movements inside the medium easier. It complies with the Medium Federate characteristics, meaning that the RTI communicates any changes to these attributes to the Molecule Federate. The Molecule Federate calculates the molecules' next places and adjusts the properties appropriately. Although there is now an implementation for Brownian motion, other diffusion laws can also be used. The motion of the molecules in three dimensions is represented by three different displacements, one for each dimension of the three dimensions. The following equations calculate the total displacement $\vec{r}$ in one-time step [192].

$$\vec{r} = (\Delta x, \Delta y, \Delta z) \tag{2.19}$$

The displacement of a molecule in one-time step is a normally distributed random variable in each dimension of the 3D space [192].

$$\Delta x \sim \mathcal{N}(\mu, \sigma^2), \Delta y \sim \mathcal{N}(\mu, \sigma^2), \Delta z \sim \mathcal{N}(\mu, \sigma^2) \tag{2.20}$$

where μ has a value of 0, and σ is computed as Eq. (2.21) manifests.

$$\sigma = \begin{cases} \sqrt{\dfrac{k_b \cdot T}{2 \cdot \pi \cdot \eta \cdot r_s} \Delta t}, & \text{if } S_{\text{molecule}} \gg S_{\text{fluid}} \\ \sqrt{\dfrac{k_b \cdot T}{3 \cdot \pi \cdot \eta \cdot r_s} \Delta t}, & \text{if } S_{\text{molecule}} \approx S_{\text{fluid}} \end{cases} \tag{2.21}$$

where k_b: Boltzmann constant, T: Temperature of the environment, η: Viscosity of the fluid, r_s: The Stokes' radius of the propagating molecule

The standard deviation depends on the comparative size of the propagating molecule $S_{molecule}$ and the size of fluid molecules S_{fluid}. The environmental coefficients are established as components of medium slices, and upon a molecule entering a slice, it begins to travel in accordance with the new coefficients. The attributes and interactions of the molecular object in FOM, as published and subscribed, are explicitly enumerated in [192]; these qualities and interactions are conveyed to other federates that subscribe to them. The pseudo-code for the Molecule Federate is presented in Algorithm 1, located at [192].

- Nanomachine Federate: It simulates the nanomachines and is accountable for the execution of molecular transmission and reception procedures. Besides, it is feasible to delineate many categories of transmitter and receiver nanomachines exhibiting distinct characteristics. Diverse implementations of a transmitter nanomachine can discharge molecules either instantaneously or sequentially, from a singular spot on its surface or from multiple locations. A receiver nanomachine can be designed to capture molecules throughout its entire surface, or alternatively, it can be configured to receive solely through receptors located on its surface [192].

  The transmitter procedure outlined in [192] releases molecules instantaneously at the onset of the symbol duration from a singular location on its surface. The receiver nanomachine is designed to capture molecules immediately upon their collision with its surface, or, if receptors are specified, it captures molecules solely upon their collision with one of the receptor surfaces. The attributes and interactions of the nanomachine object in FOM, as published and subscribed, are enumerated in [192]. These attributes and interactions are conveyed to other federates that subscribe to them. Algorithm 2 in [192] presents the pseudo-code for the Nanomachine Federate.

- Medium Federate: It abstracts the medium slices. It is possible to define medium slices to construct a larger 3D space for simulation. It also performs collision handling for molecules. The implementation of this simulator includes basic collision management, in which the molecule movements are evaluated in random order by the Medium Federate. If the target location is occupied by another molecule, a rollback is performed at an intermediate time step [192].

  It is also possible to extend this implementation to implement more complex collision management by considering underlying physical and chemical laws during a collision process. The simulation scalability can be achieved by assigning different medium slices to different medium federates. The published and subscribed attributes and interactions defined for the medium object in FOM are listed below. The published attributes and interactions are communicated to other federates which subscribe to these attributes or interactions. The pseudo-code for the Medium Federate is given in Algorithm named 3 and can be found in [192].

2.3.7.4 Simulation Experiments

- Simulation scenario: It is illustrated in Fig. 2.65; this scenario involves eight transmitter nanomachines and eight receiver nanomachines positioned within a three-dimensional environment. The simulation parameters are detailed in Table 2.1. The transmitter nanomachines release molecules at the beginning of the symbol durations, each emitting one hundred molecules to signify binary one and zero to denote binary zero. The randomly created input sequences are identical for all transmitter nanomachines. Upon transmission, the molecules diffuse through the medium, and if they encounter a receiver nanomachine, they are captured by it. The molecules diffuse from one medium layer to another, and if a molecule exits the simulation borders, it is eliminated from the simulation [192].
- Performance evaluation: For the performance evaluation of the distributed simulation, it is defined the speedup (S) as follows [192]:

$$S = \frac{T_s}{T_m} \tag{2.22}$$

where T_s is the execution time with a single federate and T_m is the execution time with multiple federates. The simulation tool is implemented on a cluster of Linux machines that communicate via a LAN. The analysis evaluates the scalability of Medium Federates, Nanomachine Federates, and Molecule Federates, utilizing

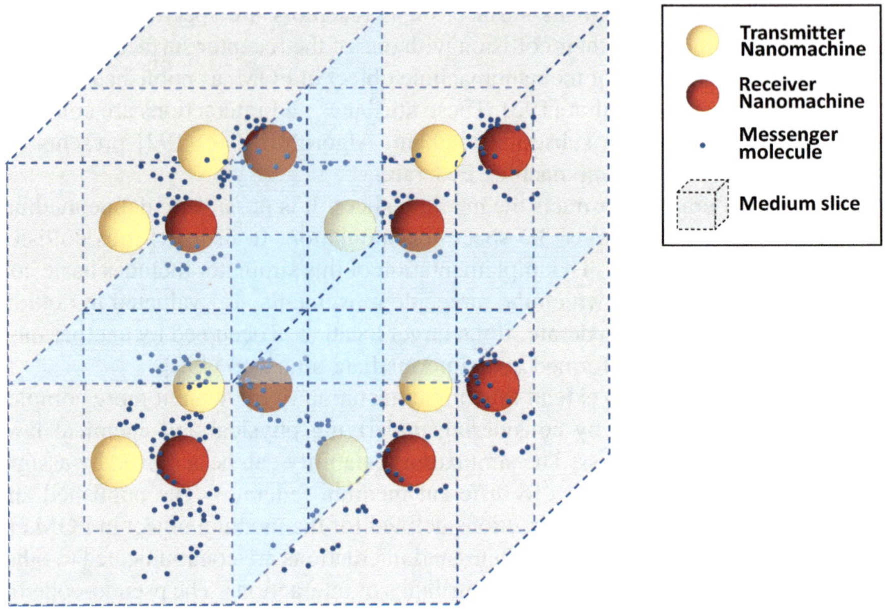

Fig. 2.65 Simulation scenario

Table 2.1 Simulation parameters

Parameter	Value	Unit
Temperature	310	K
Fluid viscosity	0.001	kg/s m
Medium dimensions (x/y/z)	200/200/200	μm
Transmitters' radius of the transmitters	10	μm
Receivers' radius	10	μm
Molecule's radius	2.86	nm
Symbol duration	30	ms
Distance between the transmitter and the receiver	1	μm

one, two, four, and eight federates for each category. Each Federate operates on an independent Linux node. A solitary Medium Federate oversees the entirety of the simulation space [192].

The simulation space is split into two segments when there are two federates present, and each federate is in charge of one section. The same justification is used for Medium Federates four and eight. The movement of a certain percentage of molecules throughout the medium is supervised by each Molecule Federate. In a similar manner, each Nanomachine Federate is in charge of the nanomachine objects assigned to it [192].

To initially evaluate the simulator's seamless functionality, a fundamental analytical model for random walk is utilized. Examine a spherical shell source

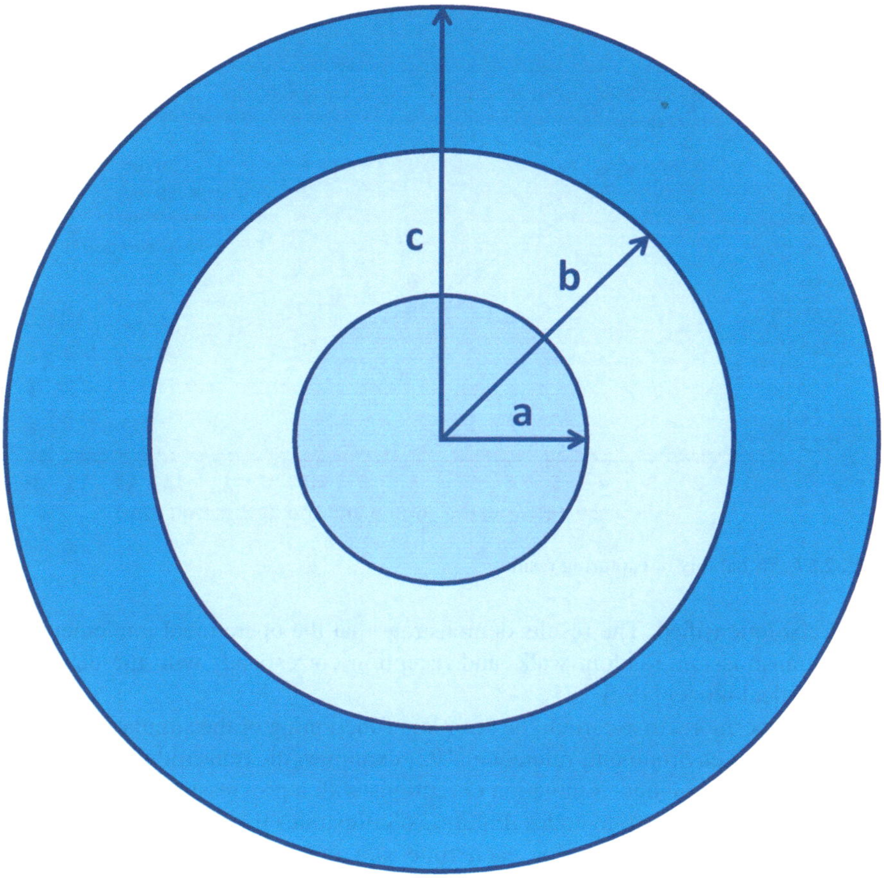

Fig. 2.66 Probability of capture

with radius b situated between a spherical absorber of radius a and a spherical shell absorber of radius c, as illustrated in Fig. 2.66. The likelihood of a particle discharged at $r = b$ being adsorbed at $r = a$ is described by Eq. (2.23) [192].

$$S = \frac{a(c-b)}{b(c-a)} \tag{2.23}$$

The probability a/b holds at the limit where $c \to \infty$. The authors in [192] use the same scenario and approximate the model by selecting c as $100 \cdot a$. Figure 2.67 depicts the probability of a hit as a function of the distance between the source and the destination over different simulated time values, with the results of the analytical model. Since simulated time increases, the likelihood of a strike approaches the analytical value; moreover, since the meantime of arrival for greater distances rises, the influence of simulated time on the probability of a hit

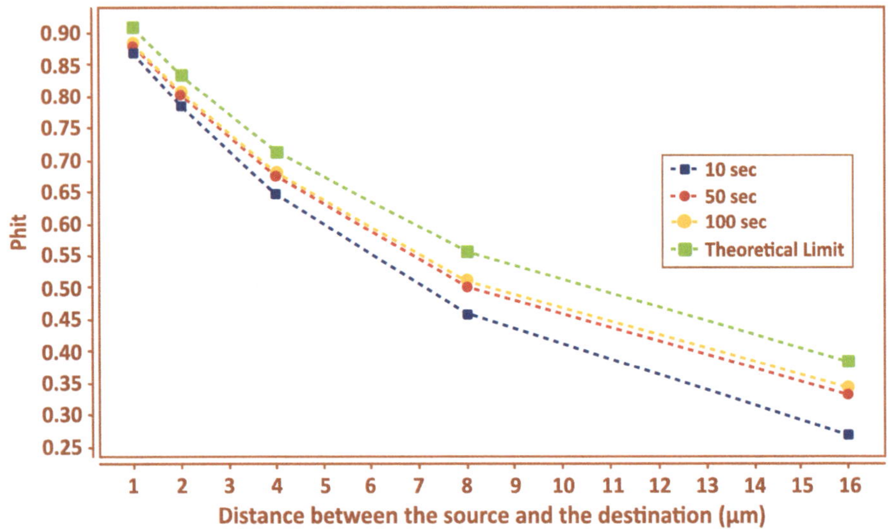

Fig. 2.67 Probability of capturing results

is also intensified. The results demonstrate that the operational implementation of transmission, random walk, and reception corresponds with the designated analytical model [192].

Furthermore, to ascertain the seamless functioning of the simulation within a distributed environment, reference [192] examines the transmitted and received molecules throughout simulation executions with a progressively growing number of medium federate nodes. Figure 2.68 illustrates the quantity of transmitted and received molecules for each time slot considering the input sequence "1101101011." The quantity of transmitted and received molecules during each time slot is illustrated within the symbol duration. A comparable trend is noted with an escalating quantity of nodes. The trend shows that as the degree of parallelism increases, comparable output is produced, and the fluctuation is caused by the stochastic nature of Brownian motion [192].

- Implementation considering collision and collision-free scenarios. [192] introduces the simulation scenario for a collision-free medium, wherein molecules can occupy the same position, alongside the collision control method provided in [192]. The collision management is executed in a fundamental manner, wherein molecular movements are assessed in a random sequence; if the designated place is occupied by another molecule, the movement is aborted. This implementation demonstrates the impact of a collision mechanism on speedup [192].

  The findings are depicted in Fig. 2.69a. Increasing the number of medium federates does not increase speed in the collision-free scenario, but introducing collisions does. This result shows that the computational bottleneck does not exist in collision evaluation in a collision-free environment, and the communica-

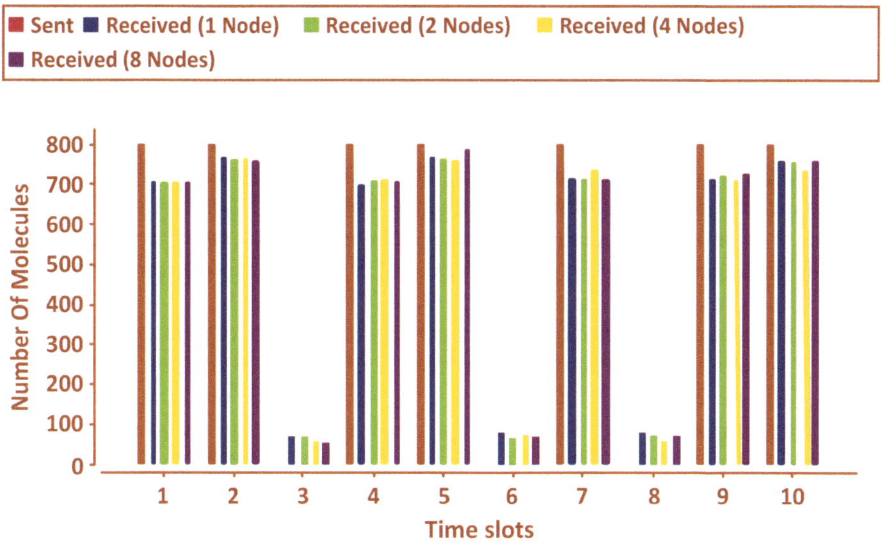

Fig. 2.68 Number of transmitted and received molecules

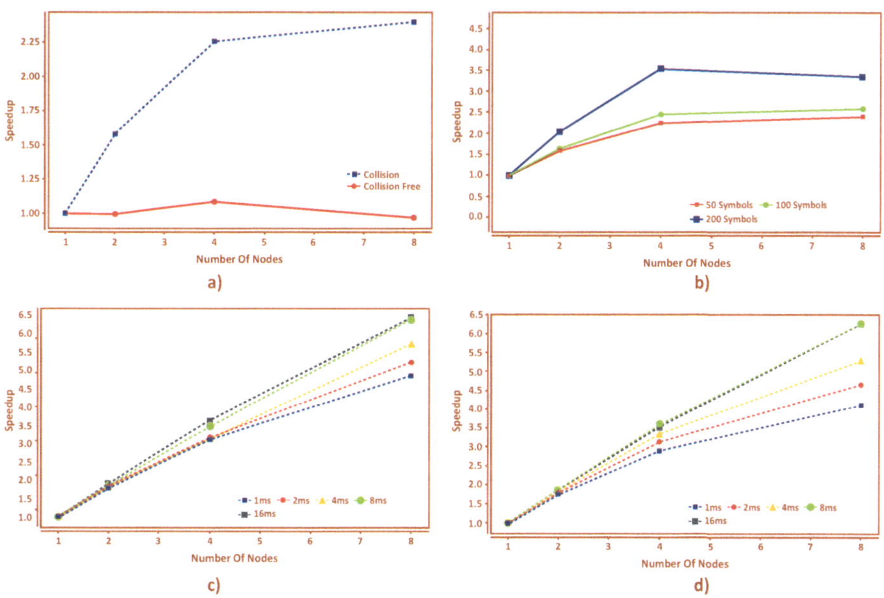

Fig. 2.69 (**a**) Collision and collision free. (**b**) Simulation duration effect. (**c**) Algorithm complexity effect. (**d**) Confined space effect

tion cost between several nodes reduces their benefit. This suggests that the strain on the Molecule Federate and the Nanomachine Federate must be divided across several nodes in order to employ the distributed architecture in such circumstances [192].

The next sections show the results for various circumstances. Because each medium slice must evaluate motions and run the collision algorithm, collision management bears the computational load of collision implementation. Execution time is decreased by distributing this demanding computing activity among many nodes. After four nodes, parallelism loses efficacy as communicational overhead takes control, as seen in Fig. 2.69a [192].

- Effect of simulation duration: To comprehend the impact of simulation duration, the simulation scenario including collisions is conducted for several lengths of time. Figure 2.69b illustrates that an increase in simulation duration correlates with an increase in speedup. This phenomenon can be attributed to the heightened computational burden associated with assessing collisions resulting from the accumulation of molecules within the medium. It is noted that when simulation time increases, alongside the enhancement in speedup, the impact of communication overhead is also encountered sooner. Consequently, there is virtually no advantage for eight nodes in comparison to four nodes during an extended simulation period [192].
- Effect of collision algorithm complexity: According to the results shown in the previous paragraphs, using a computationally intensive collision management technique results in higher speedups when the simulation is distributed across more nodes. Techniques that need 1 ms, 2 ms, 4 ms, 8 ms, and 16 ms of computing time for each molecule movement have been developed to improve the study of this effect. Figure 2.69c shows that system performance increases up to 8 ms when the medium federate processing time increases. After that point, communication cost prevents improvement from moving closer to the theoretical linear speedup limit. In situations where managing collisions requires significant processing power, the system achieves almost linear speedup [192].
- Effect of confined space simulation: The results presented in the preceding section pertain to an unconfined environment, wherein a molecule is eliminated from the simulation upon reaching the simulation limits. A limited space simulation is also implemented, wherein molecules collide with the simulation borders and remain within the simulation. Figure 2.69d illustrates the outcomes of this scenario. When molecules remain in the simulation, the quantity of molecules requiring management escalates. A reduction in algorithmic complexity leads to improved exploitation of system resources, hence enhancing speedup. In cases of elevated algorithm complexities, communication overhead prevails, resulting in no observable increase in speedup [192].
- Nanomachine Federate scalability: When collision management is not the bottleneck, different parallelism options need to be considered. This may be done by using several Nanomachine Federates to manage a group of nanomachines. The scalability of the simulation will rely on the complexity of the algorithms utilized for transmission and reception operations [192].

Receptor implementation effects are first discussed in Fig. 2.70a, where the effects of distributed execution on the basic reception mechanism (a molecule is captured by the nanomachine when it collides with the receptor nanomachine surface) and the effect of several randomly distributed receptors on the receptor nanomachine surface are shown. For a given signaling chemical, the number of cell surface receptors in nature might vary from 500 to more than 100,000 per cell. The results for 1,000, 10,000, and 100,000 receptors are shown in Fig. 2.70 a). Adding more Nanomachine Federates does not shorten execution time in a simple reception when the full surface is used since networking overhead renders the additional processing power useless. As the number of nodes increases, the networking overhead becomes more substantial. The more receptors there are, the more processing power is required to evaluate receipt, which increases the speedup as well [192].

The results for dummy algorithms that need 10 ms, 20 ms, 40 ms, 80 ms, and 160 ms for receiving operations are shown in Fig. 2.70b. System acceleration increases in tandem with algorithm complexity when more complicated algorithms are used. This suggests that using many Nanomachine Federates is necessary to achieve system scalability when replicating intricate transmission or reception models [192].

- Molecule Federate scalability: Brownian motion and more complex dummy algorithms that need specific processing durations to evaluate simulation possibilities, such as complex molecular movement models, are compared in [192]. In

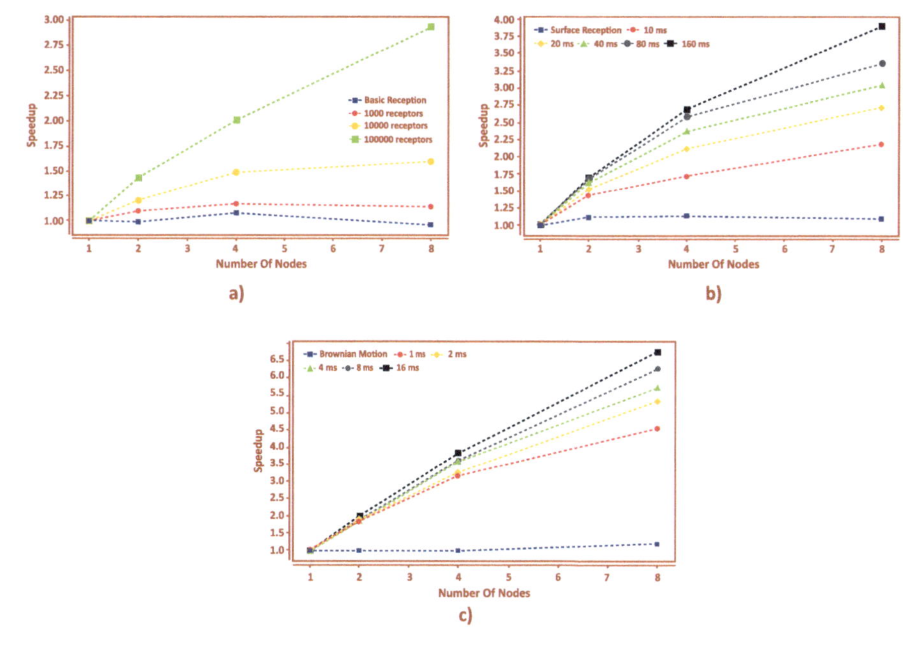

Fig. 2.70 (**a**) Receptor of confined space. (**b**) Generic Nanomachine Federate scalability. (**c**) Molecule Federate scalability

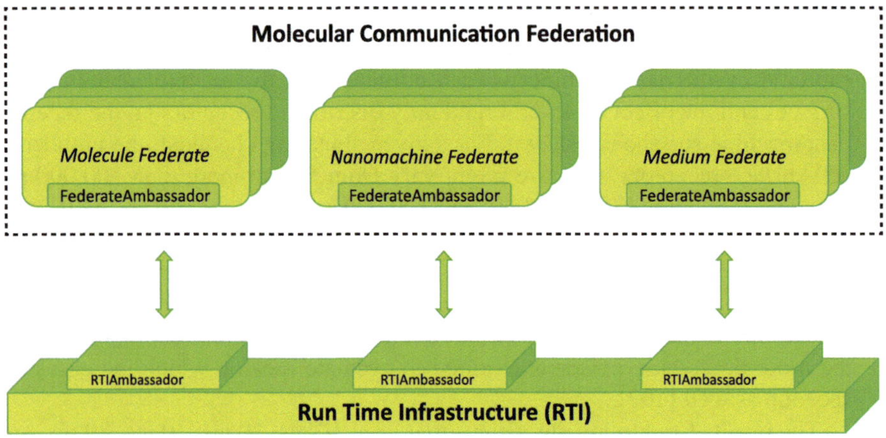

Fig. 2.71 Simulation architecture

the case of Brownian motion, as shown in Fig. 2.70c, the network overhead wins out and there is no gain in execution time. System acceleration is improved by using more complex algorithms that need 1 ms, 2 ms, 4 ms, 8 ms, and 16 ms for the assessment of molecular mobility. According to the results, Molecule Federates' scalability is crucial in simulation scenarios where the analysis of computationally intensive movement models is necessary.

Simulations utilizing HLA to examine MC systems are detailed in [193], which introduces the Distributed MC Simulator (dMCS). Consequently, [193] delineates an architecture and evaluates the performance of said architecture for a use case, including several transmitter and receiver nanomachines. The architecture created in [193] is illustrated in Fig. 2.71. The simulation is characterized as an MC Federation. Distinct federates are established for molecules, nanomachines, and the medium. The federates interact with one another via RTI utilizing a shared FOM. Scalability is attained by allocating molecule-related work to Molecule Federates, nanomachine-related duties to Nanomachine Federates, and partitioning the 3D area, with each slice managed by a designated medium federate [193].

The interfaces of the federate to other federates are shown in Fig. 2.72 [193].

The proposed architecture makes it easier to build reusable and interoperable simulation components that can run in a distributed setting. The architectural elements are described, and a simulated scenario is used to demonstrate the effectiveness of the simulation tool. The benefits of a distributed architecture are shown by analyzing computationally expensive collision algorithms, transmission or reception techniques, and molecular movement algorithms. Their results reveal that their proposed architecture enables the use of many scalability options to greatly reduce execution time. This makes it easier to model large and complex system simulations. Diffusion-based MC processes in in-body networks of modified cells or artificially built nanomachines may be simulated using this simulation technique [193].

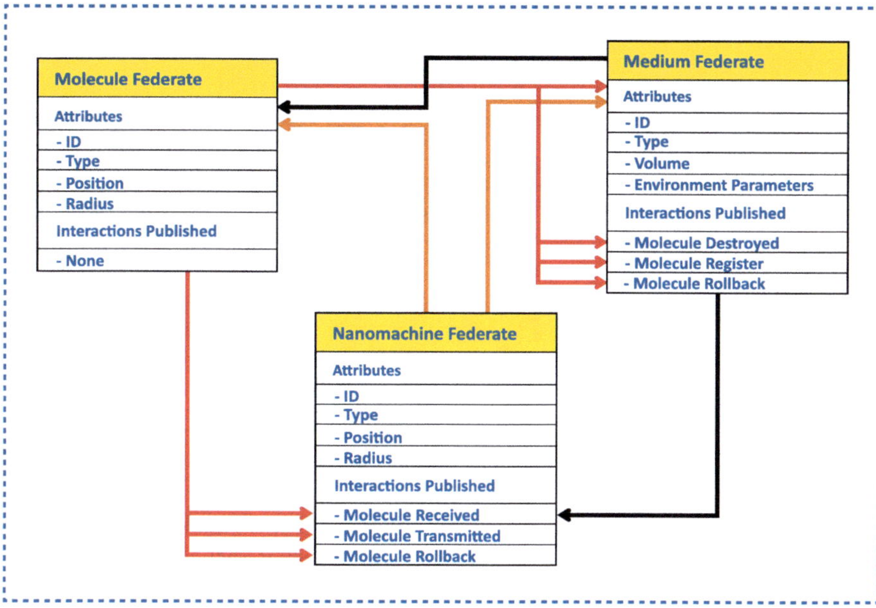

Fig. 2.72 Federations used in dMCS

Another example of the use of HLA simulator is introduced in [194]. The research issue in the cited article is based on communication via diffusion which according to [194] is a key topic since it dominates the movement at the nanoscale. [194] seeks to elucidate the dynamics of diffusion from a communication engineer's viewpoint while offering valuable insights for effective simulation design by addressing essential complexities in the development of MC simulations. The nuances involve determining the necessary number of replications regarding movement and bit sequence to achieve rapid yet significant outcomes. The authors in [194] examine the creation of intricate communication through diffusion systems, including a tunnel-based methodology that employs destroyer molecules and a distributed simulator architecture grounded in HLA. The implementation of advanced communication using diffusion systems markedly enhances data rates and overall communication, effectively connecting human-scale and nanoscale systems and facilitating the feasibility of nanonetworking technology.

2.3.8 *MolecUlar CommunicatIoN (MUCIN)*

A MolecUlar CommunicatIoN (MUCIN) simulator that is both modular and end-to-end is introduced in [195]. This MATLAB-designed simulator can send successive signals with any symbol interval and may be extended for other modulation techniques. MUCIN gives the number of molecules received at each sample period and

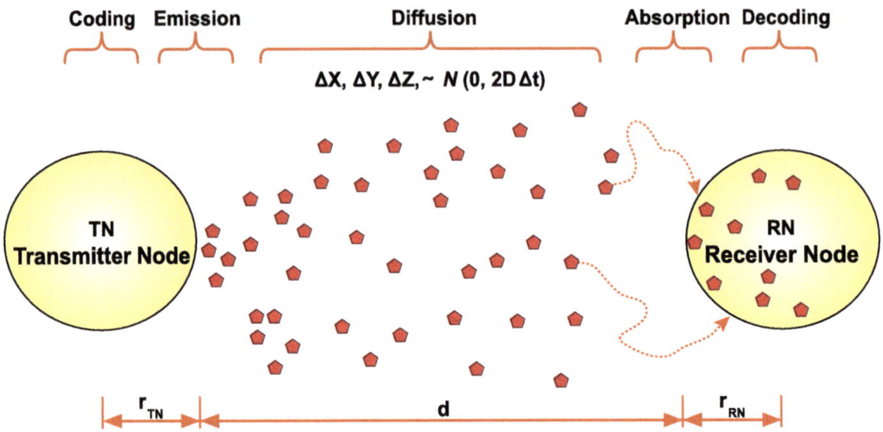

Fig. 2.73 MCvD system model in a 3D environment to present MUCIN simulator

the final demodulated symbols for signal and symbol-level analysis. The MUCIN simulator evaluates systems that employ MC via Diffusion (MCvD). This simulator is openly accessible under a BSD license[43]. The following is a summary of the MUCIN simulator's primary features [195]:

- One-dimensional, two-dimensional, and three-dimensional settings
- Transmitting successive symbols
- Probability of receptor response to suboptimal signal reception
- Point and spherical (reflective) molecular emission source
- Modulations based on concentration
- Modulations based on frequency
- ISI mitigation function at the receiver end

The MUCIN simulator facilitates the creation of a multicomponent system comprising a fluid environment and two devices, each referred to as a nanonetworking-enabled node. In MCvD, information is conveyed between the transmitter and receiver through the diffusion of certain molecules. These molecules can consist of several types of chemical compounds, including DNA fragments, proteins, peptides, or specifically designed/engineered molecules. The MCvD system for distance d is depicted in Fig. 2.73; the radii of transmitter and receiver node are denoted by r_{TN} and r_{RN}, respectively [195]. The five main processes of the MCvD system are [195]:

- Emission
- Absorption
- Propagation
- Encoding and decoding

[43] https://www.mathworks.com/matlabcentral/fileexchange/46066-molecular-communication-mucin-simulator

Before modulating the symbols during transmission, certain channel-coding techniques could be applied at the encoding stage. The diffusion channel's characteristics have a big impact on how the released messenger molecules spread. Once at the receptor, certain messenger molecules engage chemically with the receptors on the receptor's surface. The properties of these received molecules (such as concentration and type) combine to generate the received signal once they are released from the channel upon striking the receiver. If channel coding is used, decoding comes after the demodulation function, which takes place during the absorption phase [195].

In [195], the synchronization of the communication pairs in an MCvD system is assumed. As a result, time is divided into equal segments where a single sign is sent. These spans of time are represented by t-S and are known as symbol intervals. Numerous physical attributes of the messenger molecules, such as their amount, sort, or other attributes of the arriving molecules, are encoded with information [195].

In the subsequent paragraphs, MUCIN simulator is explained according to the analysis of the four processes mentioned previously [195].

- Emission process: The particle emission process task is to modulate the particle concentration $N_{Tx}(t)$ at the transmitter according to the input symbol $S_{Tx}(t)$, modulation type, and the waveform of the signal. The number of released molecules may be spiked in one t_s duration, like at the start of the slot or molecule emission may be spread over the slot [195].

  This behavior is contingent upon the transmitter node's capabilities, modulation scheme, and input signal. The transmitter node boundary serves as a reflective barrier for the released messenger molecules due to the absence of equivalent receptor structures. Simultaneously releasing all molecules for a symbol may be more advantageous than distributing them across the symbol slot for symbol demodulation. However, the transmitter node's capabilities may require its distribution over the symbol duration. The Peak-to-Average Molecule Ratio (PAMR) at the transmitter node is defined as [195]:

$$PAMR_{Tx} = \frac{\max N_{Tx}(t)}{avgN_{Tx}(t)} \tag{2.24}$$

  It will be necessary to dispatch the generated messenger molecules before the storage spaces fill up if the transmitter node has limited storage space. Larger storage capacity and a higher number of molecules in the storage facilities are needed to release the molecules exactly at the beginning of the symbol durations. The transmitter node may need less storage space if it spreads the emission out across the symbol period. Consequently, the transmitter node's capabilities determine whether it is feasible to send a peak at the start of the symbol duration. Because of its capabilities, a high PAMR value can violate the transmitter node's limitations [195].

  Two strategies for sending a symbol can be considered [195]: (1) emitting all the molecules at the start of the symbol duration and (2) emitting half of the molecules at the start and half at the middle of the symbol duration. The PAMR value of the former strategy is twice the value of the latter [195].

- Propagation process: The messenger molecules serve as the informational particles for an MCvD system. At this size, the stochastic movement and diffusion of particles inside a fluid is represented by Brownian motion. The motion is dictated by the aggregate forces exerted on the messenger molecule by the liquid molecules as a result of thermal energy. Brownian motion is characterized by the Wiener process, a continuous-time stochastic process. The Wiener process, denoted as Wt, is defined by the following characteristics [195]: (a) $W_0 = 0$. (b) W_t is almost surely continuous, and W_t has independent increments. (c) $W_t - W_s \sim \mathcal{N}(0, t-s)$ for $0 \leq s \leq t$.

  where $\mathcal{N}(\mu, \sigma^2)$ denotes the Gaussian distribution with mean μ and variance σ^2. For simulating the Brownian motion in n-dimensional space, time is divided into small intervals and at each consecutive time steps random movement is applied to each dimension as [195]:

$$\vec{r}_{(t+\Delta t)} = \vec{r}_t + \Delta\vec{r} \tag{2.25}$$

  The total displacement of a molecule, $\Delta\vec{r}$ in one step, Δt, can be found as [195]:

$$\Delta\vec{r} = (\Delta x_1, \ldots, \Delta x_n) \tag{2.26}$$

  where Δx_i is the displacement of a molecule at the ith dimension. The movement at each dimension in one-time step follows a Gaussian distribution and is modeled independently [195]:

$$\Delta x_1 \sim \mathcal{N}(0, 2D\Delta t) \ \forall i \in \{1, \ldots, n\} \tag{2.27}$$

  where Δt is the time step. These dynamics govern how molecules spread throughout their surroundings. For simplicity, collisions between the message molecules are disregarded in the model, which treats the transmitter, receiver, and messenger molecules as spherical spheres [195].

- Reception process: In nature, whenever a messenger molecule hits the body of the receiver, the molecule is received and removed from the environment. If the receptors do not remove molecules from the environment, they have other mechanisms to guarantee that each molecule contributes to the signal only once (such as acetylcholinesterase does in neuromuscular junctions). Therefore, after that point, the hitting molecule cannot move further and constitutes the signal just once. This process is referred to as first passage or the hitting process. The focus in reception process according to [195] is the probability that a diffusing particle first reaches a specified site at a specified time.

  In a 1D environment, [195] defines a closed-form solution for the first hitting probability function. In 1D and 2D environments, diffusing particles hit the receiver in the long run with the probability of one (recurrent process). When

3-dimensional environments are considered, however, there is a positive probability for a diffusing particle to miss the receiver. In a 1D environment, the first hitting probability for a point source is [195]:

$$f_h(d,t) = \frac{d}{\sqrt{4\pi Dt^3}} e^{-d^2/4Dt} \tag{2.28}$$

where d is the distance between the transmitter and the absorber. The first hitting probability in a 1D environment is inversely proportional to $t^{3/2}$ [195].
The first hitting probability for a point source in a 3D environment is [195]:

$$f_h(d,t) = \left(\frac{r_{RN}}{r_{RN} + d} \right) \frac{d}{\sqrt{4\pi Dt^3}} e^{-d^2/4Dt} \tag{2.29}$$

where r_{RN} is the radius of the receiver node. First, hitting probability in a 3D environment exhibits the same behavior as a normalization factor. The authors in [195] proposed and validated this formulation in the nanonetworking domain to enable more realistic and analytical works by removing the unrealistic reception assumptions such as the receiver being transparent to molecules [195].

In [195], simulations have been conducted to assess the first hitting process in more realistic situations by manipulating messenger molecules in accordance with [196] and either reflecting or eliminating particles that contact the transmitter or receiver. When a molecule contacts the receiver, two outcomes are possible: one is the absorption of the molecule with a probability of P_{react}, which models imperfect reception, and the other is reflection by the receiver node. They are beyond the scope of this paper.; however, one can derive P_{react} for different densities of receptor deployments and utilize the MUCIN simulator with that P_{react} value [195]. In [195], it considers the absorption process for the signal reception in a 3D environment.

- In MUCIN simulator, all information symbols are transmitted sequentially without concentrating on the length of a single symbol. Consequently, all the residual molecules contribute to the ISI for the subsequent symbols. The ISI effect in a 3D context is explored in [195]; hence, to achieve more realistic findings, symbols are transmitted consecutively without assumptions regarding the ISI. Consequently, possessing a comprehensive simulator allows us to fully observe the impact of ISI inside a three-dimensional environment. The entities and flow diagram of the MUCIN simulator are depicted in Fig. 2.74.

  The message molecules, the transmitter and receiver nodes, and the molecular diffusion channel make up the simulator entities. The modulator module receives the N_{sym} symbols that are randomly generated for transmission by the transmitter node, which has a message source [195].

  Depending on the current symbol and the modulation scheme selected, the modulator module calculates how many molecules will emit $N_{Tx}(t)$ at a specific moment. Emitted molecules move according to Brownian motion; some of them

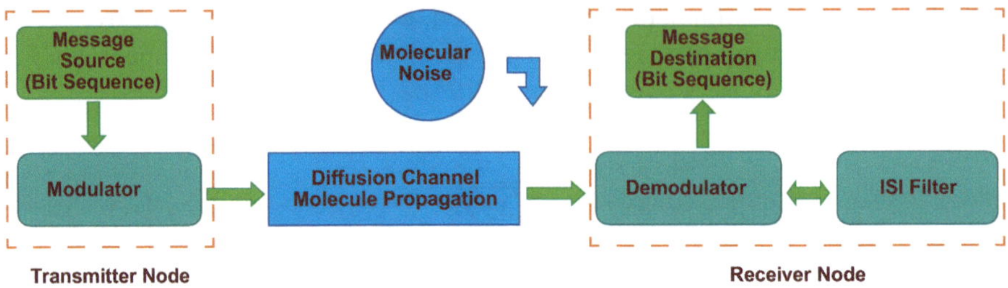

Fig. 2.74 End-to-end block diagram of MUCIN simulator

strike the receiver before being taken up by the receptors and removed from the surroundings. These processes are repeatedly carried out for each symbol length, and molecular noise is added from sources other than diffusion noise for each symbol duration [195].

Two signals are sent to the RN node from the channel: one with molecular noise and the other with only diffusion noise and no molecular noise. We may investigate diffusion noise alone or diffusion noise in combination with molecular noise by providing two channel outputs. If both noises are considered, the output of the diffusion channel at the receiver side, $N'_{Rx}\left(t,t+t_S\right)$, is composed of the diffusion channel output and the molecular noise term [195]:

$$N'_{Rx}\left(t,t+t_S\right)=N_{Rx}\left(t,t+t_S\right)+w\left(t,t+t_S\right) \tag{2.30}$$

where $w(t,\, t + t_S)$ follows Gaussian distribution $\mathcal{N}\left(0,\ \sigma_w^2\right)$ for each component independently. This molecular noise must not be conflated with the diffusion channel noise. The noise is not attributable to the randomness of particle movement, but rather to other causes, such as counting and decomposition noise. The MUCIN simulator intrinsically incorporates diffusion channel noise by monitoring the location of each particle. The demodulator module gets N $N'_{Rx}\left(t,t+t_S\right)$ and tries to decode the intended symbol. It also has connection with the ISI mitigation module and removes the ISI effect of the previous symbol by filtering the received signal (Fig. 2.75) [195].

Modulation techniques in the MUCIN simulator enable the encoding of symbols using various messenger-molecule emission properties at the transmitter, including molecule type, frequency, concentration, and phase, in order to produce a signal. Carrier waves are used in certain modulation systems but not in others. Two main modulation techniques are used by this simulator [195]: (1) Amplitude-based modulations: The concentration of the transmitted/received messenger molecules serves as the signal amplitude in CSK. Thus, it conveys information through fluctuations in the amplitude of a carrier wave or pulse. CSK is comparable to Amplitude Shift Keying (ASK) in traditional communication systems. Representing b bits necessitates 2b distinct amplitudes. If there are two different concentration levels to indicate binary "1" and "0," respectively, it is called BCSK. BCSK symbols may be modulated on a square wave or a single

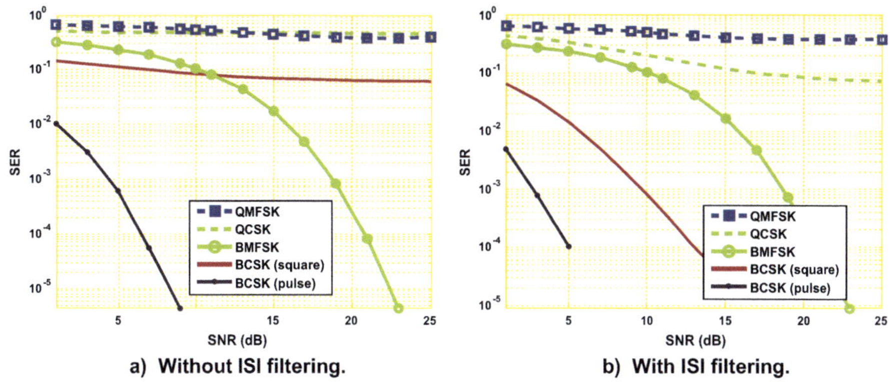

Fig. 2.75 ISI factors in MCs according to MUCIN simulator

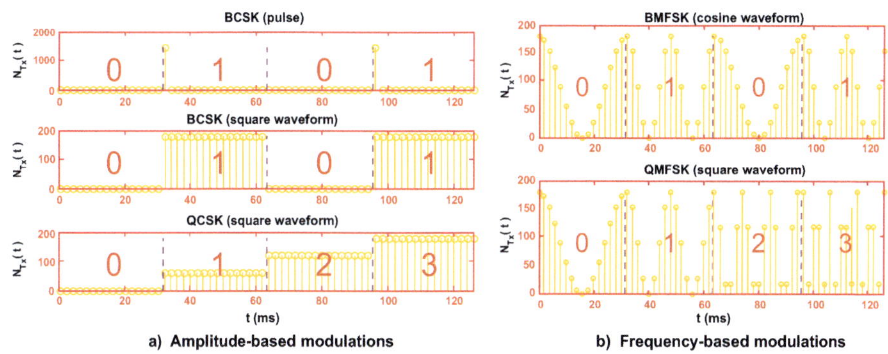

Fig. 2.76 $N_{Tx}(t)$ for different modulations

point pulse. Molecular Concentration On–Off Keying (MC-OOK) consist of BCSK symbols modulated on a single point pulse. If there are four amplitude levels for two-bit symbols, then it is called QCSK [195]. These modulations are shown in Fig. 2.76a, where the emission of zero molecules to reduce energy consumption serves as one of the symbols. In BCSK (pulse), the transmitter node releases a substantially larger number of molecules at the beginning of the symbol slot since the average energy consumption is the same for all modulations [195]. (2) Frequency-based modulations: Molecular Frequency Shift Keying (MFSK) refers to the representation of data by discrete frequency variations of a carrier wave, achieved by emitting varying quantities of messenger molecules within a symbol time. In this modulation technique, the quantity of molecules sent at emission intervals (defined by the sampling period) is variable throughout the symbol duration and fluctuates in accordance with a cosine wave. As transmitting a negative quantity of molecules is unfeasible, the simulator must adjust the amplitude higher along the y-axis [195]. When the carrier wave is a cosine wave and two frequencies are present, it is referred to as Binary MFSK (BMFSK).

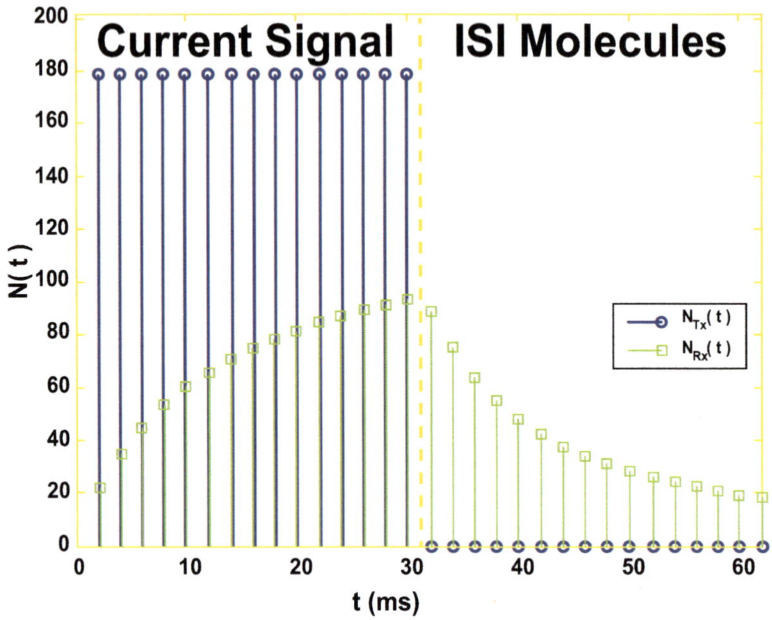

Fig. 2.77 ISI example for a CSK-based symbol

If there are four frequencies for two-bit symbols, it is referred to as Quadrature MFSK (QMFSK). These modulations are seen in Fig. 2.76 [195].
It is assumed in MUCIN that a single symbol is sent at each symbol period. The message molecules, which are derived from the stochastic nature of Brownian motion, travel erratically and might not always reach the receiving node. After being released at the current symbol location, each messenger molecule has a chance of interacting with the recipient within a specific timeframe, t_s, ISI, which is brought on by leftover molecules from earlier transmissions, is one example of how the received molecules affect the received signal. The heavy-tailed distribution of messenger molecule arrival in MCvD causes ISI to greatly hinder the symbol demodulation process. Instead of increasing the symbol interval, MUCIN suggests reducing ISI and using simulations to assess the performance improvement brought about by the ISI filter (Fig. 2.77) [195].

Figure 2.77 depicts an $N_{Tx}(t, t + t_s)$ and $N_{Rx}(t)$ example at the first and the second symbol duration. In the second symbol duration, as noted in [195], there are only surplus molecules remaining from the previous symbol. Circle and square top signals belong to the sent and received signals, respectively. In [195], these signal signatures are obtained by comprehensive Monte Carlo simulations, allowing for the elimination of the prior symbol's effect by subtracting the received signal vector corresponding to that symbol [195].

The ISI mitigation module processes the received signal and produces an ISI-cleaned signal for the symbol-detection module. The present signal is predominantly influenced by the preceding symbol. Therefore, a vector-based 2-tap Decision Feedback Equalizer (DFE) filter must be utilized at the receiver, taking into account only the prior symbol detection decision and mitigating its impact on the present signal. The output signal is represented as [195]:

$$N_{Rx}\left(t,t+t_s\right)=H_0N_{Tx}\left(t,t+t_s\right)+H_1N_{Tx}\left(t-t_s,t\right) \tag{2.31}$$

where H_k denote the weighting diagonal matrices, $N_{Rx}(t, t + t_s)$ are the received signal vector, and $N_{Tx}(t, t + t_s)$ is the transmitted signal vector during a symbol duration starting from t. The weighing coefficients are square matrices of dimension t_s/t_{ss} due to the presence of signal vectors. $N_{Rx}(t, t + t_s)$ can be seen as a sum of the two received signal vectors; one is due to the current sent symbol and the other is due to the previous symbol. These two components of the received signal vector are estimated via extensive Monte Carlo simulations depending on the distance (d), messenger molecule, transmitter and receiver node radii (r_m, r_{TN}, r_{RN}), diffusion coefficient (D), symbol duration t_s, and sampling duration t_{ss} [195]. In MUCIN, the particles' collisions are not considered since in this scale inertial forces are negligible and reduce the runtime. Ignoring collisions does not change the dynamics of the movement in diffusion since the governing force is the anti-gradient of concentration. In diffusion theory, even if the particles have an initial velocity, due to the governing force, it diminishes very quickly and starts pure diffusion [195].

In [197], MUCIN was employed as a simulator for an MC scenario. The subject of the referenced study indicates that synchronization is a pivotal concern in MCs. The absence of a suitable time-slotted framework for MCs systems impedes comprehensive investigation of desynchronization. Consequently, in [197], both difficulties are addressed. The referenced research introduces a time-slotted architecture for MC systems, drawing inspiration from natural oscillators, where the oscillation time instances define the limits of the time slots. Biological oscillators effectively meet biocompatibility standards. In [197], the authors refer to this system as a Time-Slotted MC-based (TS-MOC) system. A TS-MOC system will be advantageous for numerous MCs applications, particularly when multiple nanomachines must concurrently transmit data to a control or sink node or share the channel via time-division multiplexing [197].

Second, oscillation disturbances cause desynchronization due to the misalignment of timeframes. Desynchronization, coupled with significant propagation delay, leads to a temporal discrepancy between the signal's arrival and the commencement of a time slot. This occurrence is referred to as time-slot error, which can impair a system's performance. Consequently, the primary objective is to alleviate time-slot discrepancies. According to [197], two synchronization strategies are offered based on the initiation type: sender-initiated time-slot alignment and receiver-initiated time-slot alignment. A model for analyzing time-slot error is also developed. This analysis illustrates that the proposed approaches are both resilient and energy-efficient, achieving comparatively low error rates that signify robustness and reduced synthesis energy costs that indicate energy efficiency. The research indicates a strong concordance between the simulations and the analytical model [197].

MUCIN is also utilized in [198] for the analysis of an MC system. The identified problem indicates that microfluidic technologies provide distinctive functionalities for fluid manipulation at the nanoscale, facilitating exact regulation of molecular interactions. Potential uses of microfluidic MCs in IoBNT include healthcare, environmental monitoring, and biotechnology. The MultiPhase Particle-In-Cell (MPPIC) solver in OpenFOAM was initially designed for applications involving sedimentation, separation, and granular flows. Conversely, the approaches outlined in [198] examine the viability of employing the MPPIC solver in OpenFOAM for simulating microfluidic MCs within the framework of the IoBNT. These methods thus establish a modeling framework for examining the complex interaction between microfluidics and MCs. In [198], OpenFOAM simulations are defined in relation to an analytical model and a particle-based simulation technique. The simulation findings indicate that, under conditions of uniform particle release on the transmitter side and a flow-dominated regime, the Channel Impulse Response (CIR) derived from both the analytical and particle-based approaches aligns closely with the results from OpenFOAM simulations. As the shape extends in the direction of fluid flow, the OpenFOAM simulations align with the analytical model, particularly near the tail of the CIR. In [198], the OpenFOAM simulation results are compared with measurements obtained from a fluidic MC testbed at the millimeter scale. The OpenFOAM simulation results closely align with the experimental measurements for the flow-dominated domain, exhibiting a root-mean-square error (RMSE) of less than 10% [198].

In [199], MUCIN serves as the simulator for a scenario involving the transportation of messenger molecules in MCvD, which adheres to a Gaussian distribution function. One of the primary issues is to mitigate ISI caused by the diffusion process in a Gaussian distribution channel, owing to the random movement of information molecules (or messenger molecules). Consequently, in [199], a new Poisson distribution channel model for MCs is developed and assessed as an innovative way to enhance system reliability and mitigate the effects of ISI. In this methodology, the randomly disseminating messenger molecules adhere to a Poisson distribution function. [199] examines and contrasts these two estimates of the binomial distribution for propagating molecules in a three-dimensional MCvD system. The study examines whether the Poisson or Gaussian channel model is superior regarding ISI, Symbol Error Rate (SER), and transmission delay through numerical simulations [199].

In MCs, chemical signals (molecules) transmit information within the human body or among nanomachines. This mode of communication is notably intriguing because of its simplicity and minimal energy expenditure compared to alternative communication technologies. In [200], the impulse response estimated for an MC diffusive channel between a point transmitter and a fully absorbing receiver in three-dimensional space is analyzed, and the confidence interval for the distance from the transmitter to the receiver is estimated based on the peak value of the channel impulse response. The research employs the established Clopper-Pearson approach for confidence interval assessment, with simulations conducted in MUCIN [200].

Reference [201] examines the utilization of MCs to improve the efficacy of sequential drug administration in combination therapy for multidrug treatment protocols. To attain optimal efficacy, it is crucial to sustain an interval between successive drug doses. Consequently, [201] suggests a coordination approach that facilitates the regulation of release timings for a network of drug-laden nanomachines to guarantee the preservation of the delivery time interval. Specifically, MC is utilized to create a centralized network, in which the release timings of the pharmaceuticals from the drug-delivering nanomachines are regulated by a controller nanomachine. This nanomachine is designated as the internal controller nanomachine, as it is situated within the human body. The efficacy of the suggested strategy is assessed based on the error in the delivery time interval. Additionally, the analytical formulation of the mistake is established, and its accuracy is confirmed using simulations. The efficacy of the proposed strategies was assessed by the development of a network simulator utilizing the MUCIN simulator [201].

Reference [202] addresses a medication release timing concern in MC-based targeted drug delivery systems. A trigger source often emits a signal to many nanomachines, instructing them to concurrently release the medications. In a practical scenario where nanomachines are positioned at varying distances from the source, and the propagation delay of an MC channel is proportional to the cube of the distance, the trigger signal may arrive at disparate times, resulting in the nonsimultaneous release of drugs by the nanomachines. This results in release-time errors. Consequently, in [202], a concurrent drug-release protocol is presented to ascertain the exact timing for emitting the trigger signal, considering propagation delay data to reduce release-time inaccuracies. The study employs both analytical and simulation methods to illustrate the efficacy of the proposed technique in minimizing release-time mistakes, emphasizing its resilience to significant propagation delays. Furthermore, [202] demonstrates that the analytical model aligns well with the simulation in MUCIN.

Additionally, the MUCIN simulator program is employed to investigate the properties of the MCvD channel. This simulator employs the Binary Concentration Shift Keying (BCSK) approach for modulating binary information symbols, supports a one-dimensional environment, and transmits signals sequentially. The primary concern of the MCvD system is the ISI that occurs when molecules associated with the preceding symbol interfere with the current symbol. The conventional MCvD method displays an extended tail in the received molecular histogram, resulting in increased ISI. In this [203], the displacement of a messenger molecule is augmented to diminish the quantity of extraneous molecules in the MCvD channel. The proposed technique demonstrates the first striking time distribution to ascertain the maximum reception of information-carrying molecules by the receiver. The performance of the suggested technique is examined in [203] for various step lengths for ISI, symbol detection, and transmission delay. The findings in [203] demonstrate that the use of the proposed technique markedly enhances the performance of the MCvD system.

2.3.9 Actor-Based Communication via Reaction-Diffusion (AcCoRD)

AcCoRD is a general reaction-diffusion solver built in C, intended for communications analysis. It is an actively developed open-source project on GitHub;[44] it has the following primary features [204, 205]:

- AcCoRD is a hybrid solution that combines two simulation models to delineate 3D worlds with adaptable local precision. Every local region is categorized as either microscopic or mesoscopic. Microscopic areas delineate each molecule uniquely and progress through distinct time intervals. Mesoscopic areas quantify the number of molecules in non-overlapping virtual bins (termed subvolumes) and progress through time steps with continuous granularity. This simulator enhances scalability by allowing the arrangement of neighboring mesoscopic areas with varying subvolume sizes. This feature is derived from an extension of the two-dimensional system [204].
- Actors can be disseminated across the environment as active molecular sources (i.e., transmitters) or passive observers (i.e., receivers). Transmitters emit molecules based on the modulation of a stochastic binary data sequence. The exact quantity of molecules emitted and the timing of their release within a certain symbol interval may be either deterministic or stochastic [204].
- Receivers may document the quantity of molecules and, optionally, their locations at any given period. Future development will integrate these two actor classes to facilitate transceivers that respond based on their observations [204].
- Zeroth-, first-, and second-order chemical reactions can be specified either locally or globally, including the full propagation environment or specific regions. This framework can encompass events like molecular breakdown, enzyme kinetics, reversible or irreversible surface binding, ligand-receptor interactions, transitions across boundary membranes, and simplified molecular crowding. Surface binding reactions encompass absorption, which refers to consumption; adsorption, denoting adherence; and desorption, indicating release from a surface. In this simulator, the phrase generally denotes adsorption as reversible absorption [204].
- AcCoRD executes some tiny behaviors in continuous time. The emission of molecules by active agents, zeroth-order processes, and the majority of first-order reactions can transpire at any moment. Consequently, a molecule may participate in several reactions during a single microscopic time interval, and the precision of these events is not contingent upon the selected time step [204].
- Independent simulations can be executed an unlimited number of times (and on several machines) and subsequently combined to ascertain the average behavior and channel statistics [204].

[44] GitHub—adamjgnoel/AcCoRD: Actor-based Communication via Reaction-Diffusion. A molecular communication simulator. Simulates stochastic reaction-diffusion systems.

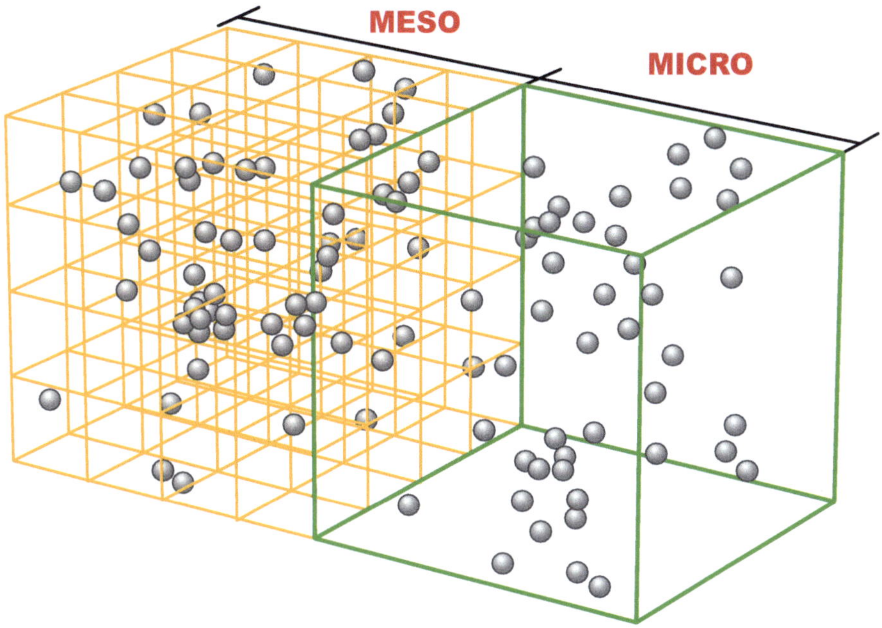

Fig. 2.78 Simulation snapshot shows a hybrid environment with initially homogeneous distribution of diffusing molecules (The left section is mesoscopic, with molecules quantified in one of 64 cubes (subvolumes), whereas the right section is microscopic, where molecules are monitored individually. One hundred diffusing molecules are exhibited)

- The online documentation comprises installation and usage instructions for the latest version, detailed descriptions of all configuration options, and numerous sample configuration files. The sample setups illustrate the full capability of AcCoRD [204].
- AcCoRD's interface has been crafted to assist inexperienced users by delivering informative output messages. AcCoRD additionally encompasses post-processing tools created in MATLAB. These tools provide the consolidation of simulation output files, the graphical representation of receiver observations (including time-dependent behavior or empirical distributions at designated intervals), and the visualization of the physical environment (either as static images or compiled into a movie) [204].

Figures 2.78, 2.79, 2.80, and 2.81 depict four sample scenarios that exemplify the principal features of AcCoRD. Figure 2.78 and 2.79 illustrate hybrid settings encompassing both microscopic and mesoscopic regions. Figure 2.80 depicts reversible surface reactions. Figure 2.81 features a unique transmitter and receiver [204].

Aiming to delineate the physical environment, AcCoRD offers considerable freedom for users, it specifies chemical processes and positions sources and molecules' observers. AcCoRD can be characterized as a \sandbox" reaction-diffusion solver from the perspective of MC methods, allowing users to investigate (\play")

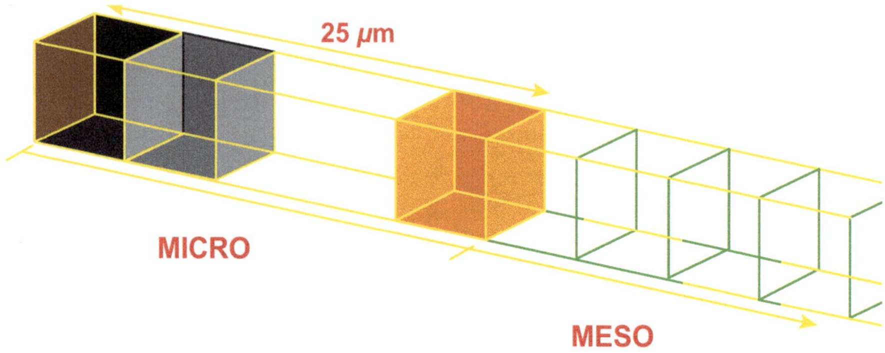

Fig. 2.79 A segment of a hybrid environment including a reactive surface (The environment consists of a rod of 0.5 um in length. One end, depicted in brown, features an absorbent surface. Three viewers examine the molecules located within 5 μm segments of the rod, depicted in brown, black, gray, and orange. The depicted version features a hybrid interface between microscopic and mesoscopic areas, located 25 μm from the absorbing end, where the cubes with blue outlines represent mesoscopic subvolumes)

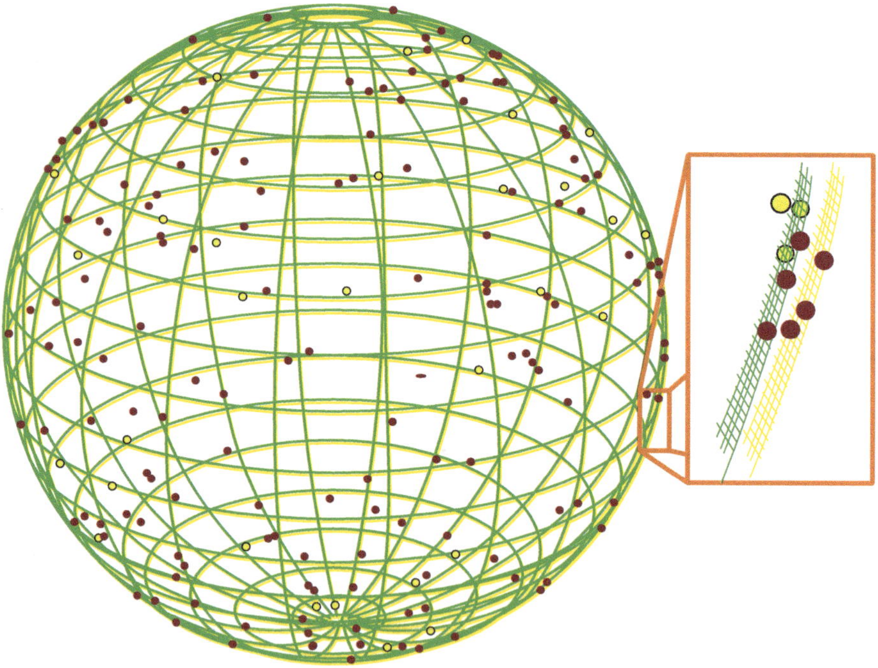

Fig. 2.80 Simulation snapshot shows an environment featuring two enormous concentric spheres with radii of 120 μm and 122 μm (Molecules are positioned in the space between the two spheres. This environment serves as a spherical counterpart to the one-dimensional system. The inner and outer spherical surfaces are delineated with green and yellow edges, respectively. In the depicted version, molecules can probabilistically permeate the inner surface, which functions as a membrane. The molecules shift from red to yellow throughout their transition within the inner sphere)

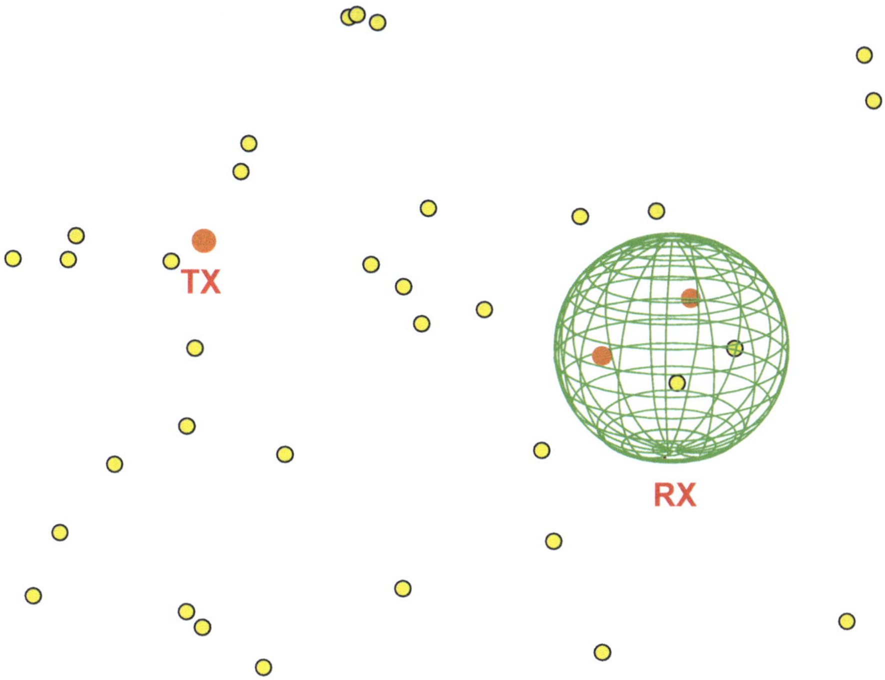

Fig. 2.81 Snapshot of a basic communication simulation environment (Molecules are emitted by a point source and detected at a spherical receptor (radius 1 μm) positioned 5 μm from the source. In the depicted version, the released molecules (yellow) can irreversibly adhere to the receiver's absorbing surface, resulting in a red coloration upon binding. One molecule that seems to be within the receiver is, in fact, located behind it)

and construct their systems. This communication analysis program may simulate data modulation at transmitters and produce channel statistics at observers by executing multiple runs of simulations. The primary contribution concerning the referenced simulator is [204]:

- Encourage the use of simulations: So far, the majority of authors in this subject have used Monte Carlo simulations or numerical assessments of their findings to validate their work. In this case, it is assumed that the channel behavior's time-dependent probability distribution is known. Ignoring simulations might lead to unproven hypotheses about the model's accuracy over a range of parameter values [204].
- Improve the understanding and representation of known reaction-diffusion configurations and related channel reactions. This is especially important for making this multidisciplinary field more accessible and encouraging new researchers to get involved without having to develop their simulation tools [204].
- Offer a framework for assessing transceiver designs and verifying novel analyses [204].

The following paragraphs discuss briefly AcCoRD applications.

Diffusive MC is a biomimetic communication paradigm that utilizes the Brownian motion of specialized information-carrying molecules for signal transmission. The attributes of the communication channel in a diffusive MC system are significantly influenced by the transmitter and receiver mechanisms, the propagation medium and its limits, and the interactions with the molecules. The study in [206] examines the diffusive MC channel within a spherical shell, characterized by inner and outer surfaces that may exhibit partial or complete absorption or reflection, with any boundary surface potentially serving as the diffusive MC receiver. Geometry and boundary conditions can be employed to describe a range of practically significant scenarios for MC systems. The general parametric solution of the partial differential equation for these instances is obtained in [206], and the authors in [206] provide formulas for the resultant channel response. The outcomes of [206] are delineated by numerical analysis using particle-based AcCoRD simulations.

Diffusion-based MC is a bio-inspired system that uses random walk diffusive molecules as information carriers between the transmitter and the receiver. A primary problem of that system is ISI, induced by channel memory and characterized by a pronounced tail in the impulse response. Most studies in this domain have suggested employing enzymes to catalyze the breakdown of residual molecules, thereby alleviating ISI and enhancing the data rate; however, the enzymes will also diminish signal strength by degrading the molecules that convey the information. Reference [207] proposes employing non-enzymatic processes to selectively degrade incoming molecules, thereby amplifying the received signal and simultaneously reducing ISI, which enhances signal strength and achievable throughput. The referenced study examined photolysis reactions that utilize light to rapidly decompose compounds; it investigated the best timing of light emission by 3D stochastic simulations with the AcCoRD simulator. Simulation results indicate an enhancement of the received signal when employing non-enzymatic reactions in contrast to enzymatic systems. The efficacy of the suggested technique was assessed by analyzing the interference-to-total received molecules [207].

Diffusive MC depends on the Brownian motion of certain molecules for information transmission. The signal propagation characteristics in a diffusive MC channel are dictated by the geometries of the environment, receiver, and transmitter, along with their physical and chemical properties, which establish the boundary and initial conditions for the partial differential equation governing the diffusion process [208]. In [208], the directional signal propagation properties of a point source over, or on, a spherical surface—whether reflecting, totally absorptive, or partially absorptive—are examined. The general solution of the partial differential equation for this instance is obtained in [208] and validated by particle-based simulations. Their findings can be utilized to ascertain the signal propagation characteristics throughout a diverse array of practically significant scenarios, wherein the influence of system parameters on signal propagation directivity is examined. AcCoRD is employed for this analysis [208].

MC necessitates low-complexity symbol recognition techniques to address the numerous sources of uncertainty intrinsic to these channels. Therefore, reference

[209] proposes two variants of a high-performance asynchronous peak detection technique for a receiver that conducts separate observations. The initial variant exhibits little complexity and quantifies the maximum observation within a sampling interval. The second variant incorporates decision feedback to alleviate ISI. While the algorithm does not necessitate synchronization between the transmitter and receiver, results indicate that the bit error performance of symbol-by-symbol detection utilizing the first variation surpasses that of employing a single sample with a predetermined sampling period. The second variation demonstrates performance equivalent to that of an energy detector. Both implementations of the method exhibit superior resilience to temporal offsets compared to existing detectors, with the simulator for the aforementioned scenarios being AcCoRD [209].

The AcCoRD simulator is utilized in [210] to analyze their proposed sequential average probability ratio test-based strategy, assessing the efficacy of the entire recurrent neural networks approach using the sole realistic data generated from particle-based simulations in practical application scenarios. Consequently, in [210], a simulated animal cell environment and a simplified blood vessel environment are presented in AcCoRD. Chemical signals are employed in nature for intercellular and intracellular communication at the micro- and nanoscale levels. Investigating communication processes within the intricate cellular environment is essential, given that cells are the primary structural and functional units of life. Abnormality detection with diffusive MC is examined in a fundamental blood artery context, while Reference [210] presents an MC system aimed at exploring this detection within a cellular setting. Figure 2.82 and 2.83 illustrate a basic depiction of the cellular environment and the AcCoRD configuration, respectively [210].

Therefore, the MCs system is represented for simulation in [210], choosing a typical animal cell has a diameter of 5–20 μm. The receiver is chosen as the endoplasmic reticulum (ER) in the cell with the size of 2–3 μm. ER is considered as receiver because the ER is the cell's biggest organelle and is involved in protein synthesis and transport, protein folding, lipid and steroid production, glucose metabolism, and calcium storage. The obstacles represent other agents in the cell environment such as small organelles like lysosomes with the size of 0.5 μm, A is the maximum number of molecules for each pulse, is chosen according to the number of small messenger molecules, such as adenosine monophosphate (AMP), and diffuses to act on signaling process. These molecules are called second messengers, such as molecules cyclic AMP, cyclic guanosine monophosphate (GMP), inositol triphosphate, diacylglycerol, and calcium. The diffusion coefficient of the second messengers in the cytosol is in the range of 100–500 μ^2/s. In the signal generation using particle-based simulator AcCoRD, the diffusion time step is set to 5 μm, the symbol length values are established to be significantly longer than the period at which the proportion of absorbed molecules attains its maximum, hence minimizing the presence of residual molecules that contribute to ISI [210].

Simulations utilizing AcCoRD are shown in [211], which elucidate that information transmission via chemical molecules is a fundamental aspect of biology over many distance scales and has garnered recent attention in bioengineering and communication. The aggregate signal strength at the receiver, determined by the

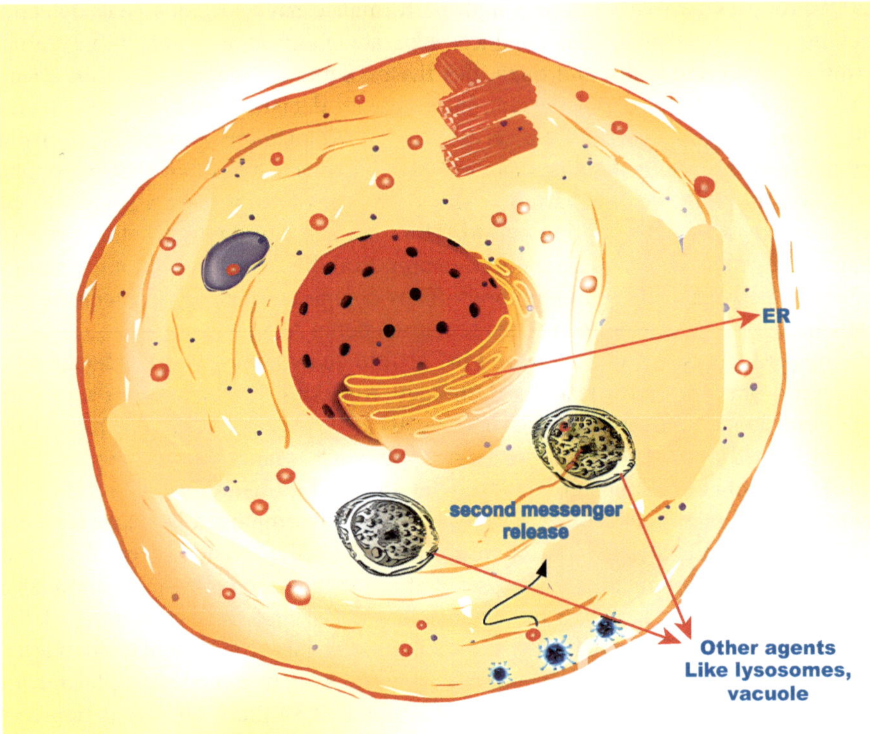

Fig. 2.82 Simulation setup mimicking a basic animal cell environment

anticipated quantity of detected molecules within the receiver, influenced by several transmitters at varying distances (e.g., due to mobility), can significantly affect the reliability and effectiveness of the MC system. Modeling the aggregate signal from numerous diffusion sources presents significant computational and analytical difficulties. Consequently, Reference [211] introduces a manageable analytical model for the aggregate signal strength resulting from randomly positioned transmitters, whose locations are represented as a homogeneous Poisson point process in three-dimensional space. Utilizing stochastic geometry, the authors in Ref. [211] derive analytical formulas for the anticipated quantity of observed molecules at both a fully absorbing receiver and a passive receiver. Their findings indicate that the aggregate signal intensity at both receiver types escalates correspondingly with the augmentation of transmitter density. The suggested paradigm significantly streamlines the investigation of large-scale molecular systems in communication and biological applications [211].

In [212], AcCoRD is used in simulations. This article examines the topic of communication through molecular diffusion, a widely used tactic that occurs naturally and travels quickly over distances of less than a millimeter. The transmitter is typically assumed to be a point source in existing closed-form analyses of the diffusion

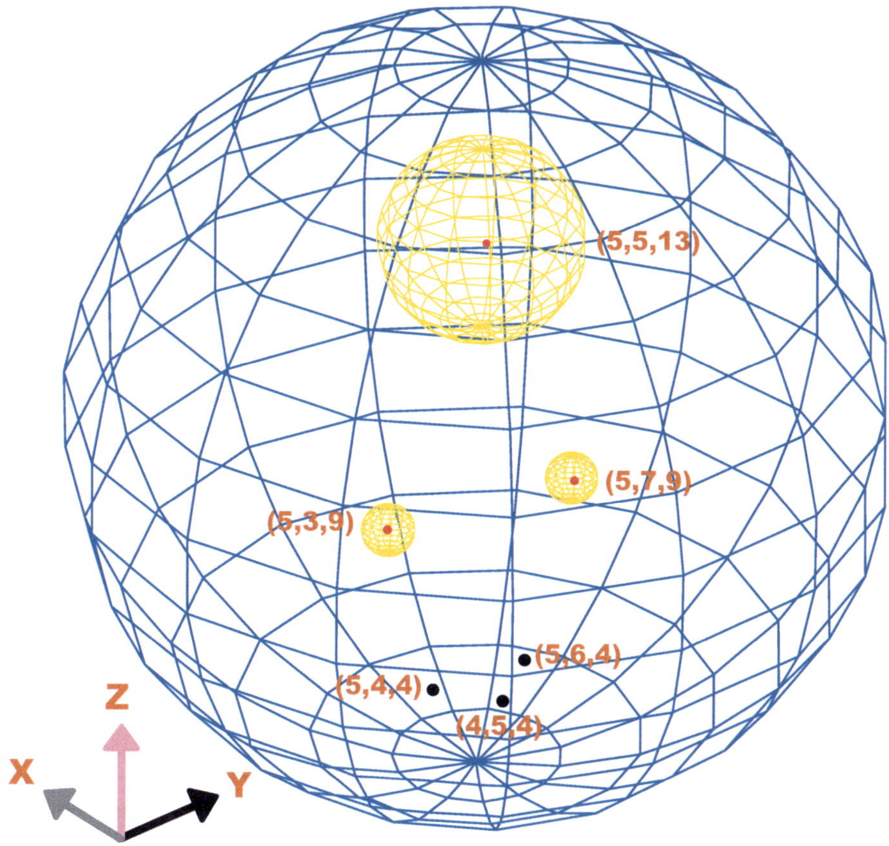

Fig. 2.83 Basic animal cell representation in AcCoRD

channel impulse response. Channel impulse responses for spherical transmitters with either an absorbing or passive receiver are obtained in [212]. For a one-dimensional environment, the resultant channel impulse responses are in closed-form; for a three-dimensional environment, they can be determined using numerical integration. In order to quantify the correctness of the point transmitter assumption, it is explicitly defined and compared to the resulting spherical transmitter impulse responses. When the distance between a transmitter and its receiver is tiny in relation to the emitter's size, the spherical transmitter model performs significantly better than the point transmitter assumption. Using the MC simulation tool AcCoRD, tiny particle-based simulations are used to validate the results derived. Through modeling, a spherical transmitter variation is also taken into consideration, in which molecules are released from the surface of a solid sphere [212].

Reference [213] also makes use of AcCoRD. It mentions that the primary obstacle facing bio-inspired diffusion-based MCs is ISI. Biological systems reduce this ISI by the breakdown of the leftover molecules from a prior transmission. The majority of articles recommend using enzymes to catalyze the breakdown of molecules. Enzymes also break down the molecules that contain information, significantly reducing the signal intensity. The utilization of photolysis processes, which rapidly change the released molecules so they can no longer be identified after being detected, is suggested in [213]. The best timing of the light's emission will enable the receiver to pick up as many molecules as possible, boosting the signal's intensity and mitigating ISI. The expectation of the quantity of molecules detected at the receiver is expressed with a lower bound. A bit error probability expression is also developed, and simulation results verify both expressions, demonstrating a discernible improvement with photolysis reactions. The resulting bit error probability and the interference-to-total-received-molecules metric are used to assess the effectiveness of the suggested approach [213].

An analytical comparison of active and passive receiver models in diffusive MC is conducted in [214]. Molecules are absorbed upon colliding with the receiver's surface in the active model. In the passive paradigm, the receiver functions as a virtual border that does not influence molecular behavior. Two methodologies are delineated in [214] to determine transformations between the receiver signals. Two models for an unbounded diffusion-only MC system with a spherical receiver are consolidated. As time progresses in the three-dimensional system, the transformation functions maintain constant scaling factors, rendering the receiver models practically comparable. Methods are introduced to facilitate the transformation of stochastic simulations via AcCoRD, which serve to validate the transformations and illustrate that the simulation of a passive receiver may be executed with more efficiency and accuracy compared to the direct simulation of an absorbing receiver [214].

An approach is introduced in [205] to compute the transition rates between neighboring mesoscopic subvolumes under conditions of flow and diffusion. These rates can be incorporated into stochastic simulations of reaction-diffusion systems that adopt a mesoscopic method, which partitions the environment into homogeneous subvolumes and employs the spatial stochastic simulation algorithm. The rates are obtained by integrating Fick's second law over a single subvolume in one dimension and are demonstrated to be applicable in three dimensions. The suggested algorithm adjusts the calculated rates to guarantee their physical significance and is executed within the AcCoRD simulator. Simulations employing the suggested methodology are juxtaposed with a simplistic mesoscopic approach, microscopic simulations that monitor each molecule, and analytical outcomes that are exact in one dimension and approximate in three dimensions. By selecting subvolumes that are adequately diminutive, ensuring that the Péclet number related to a subvolume is significantly below two, the precision of the proposed method aligns with that of the microscopic method, thereby facilitating the simulation of advection–reaction–diffusion systems using the spatial stochastic simulation algorithm [205].

In [215], the AcCoRD simulator is employed to investigate information transmission via chemical molecules, as it is a fundamental aspect of biology over many distance scales and has garnered current attention in bioengineering and communication theory. Possible applications encompass cooperative networks comprising numerous uncomplicated devices that may be randomly situated (e.g., as a result of mobility). [215] introduces the inaugural manageable analytical model for the aggregate signal intensity resulting from randomly positioned transmitters in a three-dimensional large-scale MC system, regardless of the presence of deterioration in the propagation environment. Transmitter positions in an infinite and uniform fluid are represented as a homogeneous Poisson point process. Utilizing stochastic geometry, mathematical formulas are formulated for the anticipated quantity of molecules absorbed by a fully absorbing receiver or detected by a passive receiver. The bit error probability is calculated based on ON/OFF keying and either a fixed or adaptive judgment threshold. The findings indicate that the aggregate signal strength escalates in direct correlation with transmitter density, and the minimal bit error probability can be enhanced with the incorporation of molecular degradation. Moreover, the system analysis can be extrapolated to other receiver designs and many performance metrics in extensive MCs systems [215].

Path loss constitutes a primary challenge in mobile communications. The transmission of information by molecules by natural diffusion results in the quantity of molecules received being inversely proportional to the square of the distance between the transmitter and receiver, significantly affecting the strength of the received signal. The implementation of an MIMO approach can enhance the performance of MCs by augmenting the data rate [216]. The receiver utilized in molecular MIMO communications is examined in [216]. Research in [216] concentrates on three critical characteristics for receiver design: channel distance, inter-detector spacing, and detector diameter. To enhance the design of a 3×3 MIMO receiver, the authors in [216] employed the AcCoRD simulator to generate 3D stochastic simulations for each scenario. The simulation results were assessed by analyzing the error probability and the quantity of molecules indicative of signal strength. In [216], two optimization issues are suggested to optimize the selection of receiver parameters, along with two techniques to address these problems. The research indicates that a prudent selection of the three parameter combinations might enhance MIMO receiver design, hence reducing error probability and augmenting MC performance [216].

A unique nanoscale architecture for executing matrix multiplications is suggested in [217]. Matrix multiplications in [217] constitute fundamental operations in machine learning algorithms, hence facilitating their implementation at the nanoscale, such as within the human body in the IoBNT. The process relies on molecular exchange across interconnected compartments and the incorporation of chemical reactions in certain compartments. The matrix entries are exclusively determined by the volumes of the compartment. A comprehensive mathematical characterization of the system's stochastic and dynamic behavior is proposed in [217]. In [217], the authors establish principles for the proposed architecture and evaluate the approach using particle-based simulations with AcCoRD.

In [218], the detection of anomalies in a fluid medium is examined using MC-based network of nanoscale sensors. This task involves the perception, identification, and communication of anomalous alterations in the environment that may indicate a disorder or an unusual occurrence. Distributed detection challenges are very important, particularly in applications including health monitoring, disease diagnosis, targeted medicine delivery, environmental sensing and monitoring, contamination detection and removal, and environmental remediation. In [218], a sequential probability ratio test-based method for decision fusion in diffusive MC distributed detection is proposed. The suggested method yields substantial reductions in the average sample size needed for decision-making relative to fixed-sample size alternatives, hence significantly enhancing the average decision delay. In the examined dispersed detection scenarios, the authors in [218] report a reduction of up to 50% in the number of samples necessary for decision fusion, utilizing the digital simulation tool AcCoRD [218].

The development of appropriate analytical models is a crucial phase in the design and study of MC systems. Nonetheless, numerous current models exhibit restricted applicability in real-world situations owing to different simplifications (e.g., the presumption of an unbounded environment) [219]. A realistic model for particle diffusion within a limited sphere and particle transport across a semi-permeable border is established in [219]. This model is applicable for diverse purposes, including the modeling of inter- and intracellular communication or the release mechanisms of drug carriers. The suggested analytical model utilizes a transfer function technique, facilitating rapid numerical assessment and offering insights into the influence of pertinent MC system elements. The suggested resolution of the constrained spherical diffusion issue in the formulation in [219] employs a state-space description, incorporating the semi-permeable border through a feedback loop. Particle-based simulations with AcCoRD validate the suggested modeling methodology.

In [220], AcCoRD is employed for the modeling of biological systems. Precise analytical models of biological environments, such as cells or blood arteries, provide numerous benefits in the study and design of microchannel systems. Functional transformations represent a promising modeling approach for the diffusion of particles within cells or their transit across a cellular chain. [220] examines the modeling of particle diffusion within a sphere under impedance boundary conditions, which serves as a basic representation of diffusion in cells enclosed by a semi-permeable membrane [220].

2.3.10 Nervous NaNoNetwork Simulator (N^4Sim)

Nervous NaNoNetwork Simulator[45] (N^4Sim) reduces the issues with modern neural network simulators on the one hand and helps with the ICT analysis of the nervous system on the other hand. Reference [221] describes a practical way to deal with this problem and improve the scalability of the N4Sim simulator, as synaptic connection is one of the most complex processes in simulator design.

[45] GitHub - nafiturgut/N4Sim

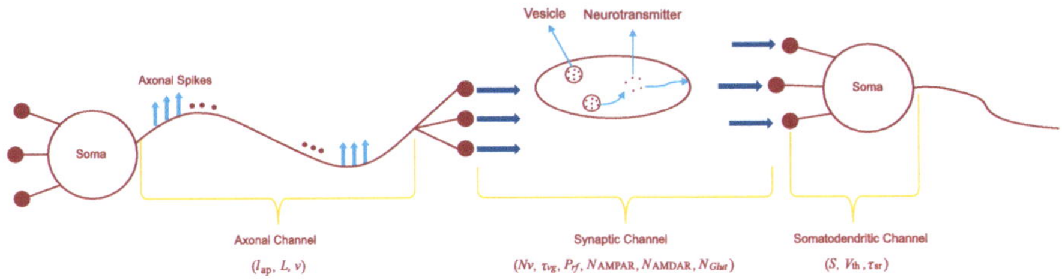

Fig. 2.84 Neural network model (Respective channel parameters are explained in Table 2.2) of N⁴Sim

Axons, which are thin, branching extensions from the soma, carry electrical signals from the presynaptic neuron's soma, or cell body, to the synapse, where they are transformed into chemical signals by releasing neurotransmitters into the synaptic cleft. This is how biological neural networks (BNNs) are made up of neurons connected at synapses. Receptors on the postsynaptic membrane at the synaptic cleft then pick up the chemical signals, allowing ions to enter the postsynaptic neuron and completing the chemical-to-electrical conversion process. The dendrites, which are tubular projections from the soma, and the soma itself at the postsynaptic end are where synapses are mostly found. The axonal, synaptic, and somatodendritic channels are the main communication routes that make up a neuro-spike communication network. The neural networks to be investigated with N^4Sim in connection to these channels are examined in the ensuing paragraphs; Figure 2.84 presents the network's graphical form, whereas Table 2.2 [221] gives a description of the channel parameter simulations and their nominal values.

- Axonal channel: N^4Sim axons are primitively modeled by two parameters, their length and conduction velocity, and the speed at which an action potential (AP) propagates through the axon, following the groundbreaking basic cable model of axons. The former is determined by consideration of distance between the corresponding neurons and the average curvature along the axon, which is proportional to the increase of axon length compared to interneuron distance, and is chosen according to experimental estimates in literature. The latter is also experimentally measured for a large variety of neurons, and in N^4Sim, unless otherwise specified, the conduction velocity of an axon is randomly assigned according to measured values. In its initial phase, N^4Sim is also designed very simplistically in terms of geometrical considerations of axons. In particular, between two given neurons there is at most one axon per direction, there is no branching, and each axon terminates with a single synapse [221].

  N^4Sim also models transmission failure in an axon, where the probability of a transmission failure is characterized by the axonal propagation length constant l_{ap} in units of length. The probability that the axon will fail to successfully propagate a signal is assumed to be exponentially distributed, that is, for a given axon of length l, the probability of axonal propagation failure (APF), $P_{APF}(l)$ is given by Eq. (2.32) [221].

Table 2.2 N[4]Sim channel parameters and the notional values that are used in descriptions and simulations. A uniform random selection process is used to select parameters with a range of values [221]

Channel	Description	Parameter	Nominal value
Axonal channel	Axonal propagation length constant	l_{ap}	0.5 mm
	Length of axon	L	8–200 mm
	Action potential propagation speed in axon	v	0.79 m/s
Synaptic channel	Number of available vesicles	N_v	0–25
	Vesicle generation time constant	τ_{vg}	4 ms
	Release failure probability of single vesicle	P_{rf}	0.1–0.5
	Number of α-amino-3-hydroxy-5--methyl-4-isoxazolepropionic acid (AMPA) receptor	N_{AMPAR}	100–160
	Number of N-methyl-D-aspartate (NMDA) receptor	N_{NMDAR}	20–60
	Number of glutamate neurotransmitters per vesicle	N_{Glut}	4800–5200
Somatodendritic channel at the axonal hillock	Charge injection elastance at the synapse on the voltage	S	10^{-11} F^{-1}
	Threshold membrane voltage for instigating an action potential	V_{th}	15 mV
	Relaxation time constant for the membrane voltage	τ_{sr}	0.265–2.65 ms

$$P_{APF} = \frac{1}{l_{ap}} \int_0^l e^{-\frac{x}{l_{ap}}} dx = 1 - e^{-\frac{x}{l_{ap}}} \tag{2.32}$$

Most often, temporal memoryless events are modeled using the exponential distribution. Assuming that the failure of propagation within a certain length interval is independent of propagation failures in the rest of the axon, it may be shown to be relevant in the context of axonal propagation. Equation (2.32) makes it easy to determine that when the axon's length trends toward infinity, the APF gets closer to certainty [221].

- Synaptic channel: Upon the arrival of an AP at the axonal synapse, it may induce the release of neurotransmitters into the synaptic cleft, subsequently engaging with the membrane receptors of the postsynaptic neuron. Neurotransmitters are contained in vesicles within the synapse, and their release entails the fusion of the vesicular membrane with the cell membrane, leading to the expulsion of their contents into the synaptic cleft. Synapses generally house several vesicles, which are perpetually regenerated since each vesicle release depletes a vesicle along with its contents. Following the release of neurotransmitters into the synaptic cleft, a diffusion-reaction process occurs, wherein neurotransmitters diffuse across the synaptic cleft and interact with the membrane receptors of the postsynaptic neuron to enable ion influx. Consequently, simulating the synaptic channel necessitates consideration of both vesicle replenishment and release mechanisms, in addition to the diffusion-reaction process within the synaptic cleft [221].

In N[4]Sim, each synapse has a number of available vesicles at a given time t, $N_V(t)$. The following rules are used to update this value: The synapse generates vesicles with an exponential distribution of time constant τ_{vg}; it can be derived again assuming that vesicle generation is a sequential memoryless process. If t_0 is defined as the time of last vesicle generation; then, the time of next vesicle generation, $t_1 = t_0 + \Delta t$, is found by sampling Δt from the exponential distribution in Eq. (2.32) and the number of vesicles is updated following Eq. (2.34) [221].

$$f_{vg}(t) = \frac{1}{\tau_{vg}} e^{-\frac{t}{\tau_{vg}}},\ t \in [0,\infty) \tag{2.33}$$

$$N_V(t_1) = N_V(t_0) + 1 \tag{2.34}$$

Depending on whether the synapse is able to release a vesicle (or vesicles) in response to an incoming AP, this number may drop. Every available vehicle is considered to be released according to the same independent Bernoulli distribution in N[4]Sim, with a release failure probability of P_{rf}. This suggests that the binomial distribution, shown in Eq. (2.35) [221], is followed by the total number of released vesicles, N_{rv}.

$$\Pr(N_{rv} = k) = \binom{N_V}{k} (1 - P_{rf})^k P_{rf}^{N_V - k}, 0 \le k \le N_V \tag{2.35}$$

where N_V represents the number of available vesicles in the synapse at the time of AP arrival. As Reference [221] demonstrates, the assumptions of memoryless vesicle production and independent vesicle release are common in the literature. When a vesicle is successfully released, the synaptic channel's final phase—more precisely, the diffusion-reaction process inside the synaptic cleft—begins. This approach involves the slow and resource-intensive accurate simulation of MC methods. Rapid algorithms that mimic the expected behavior of MC dynamics at a synapse have been developed, but they are still ineffective for incorporation into a large-scale network simulator.

- Because of this, it seems hard to incorporate MC dynamics at synapses into large-scale neural network models. In order to solve this problem, N[4]Sim makes use of Look-Up Tables (LUTs), which serve as databases that hold the input/output connections of the diffusion-reaction mechanisms connected to various synapse types, including glutamatergic and GABAergic. A thorough comparison of the novel Sim with earlier neural network simulators is provided in [221], which also explains how MC is included at synapses using the LUT approach.

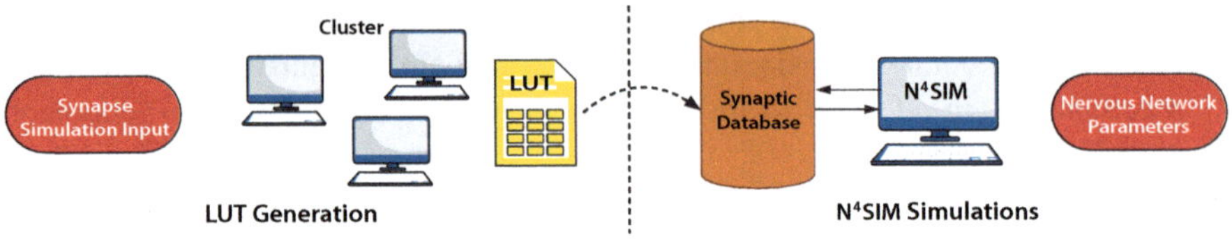

Fig. 2.85 N⁴Sim uses one-time precalculated LUTs for fast simulations involving MC at synapses

– LUTs for MC in N⁴Sim at synapses: It is not feasible to directly integrate simulations that faithfully depict MC at synapses into a large-scale neural network simulator because of the significant time and computer resources required. In order to overcome this difficulty, N⁴Sim employs LUTs, which, for a given synapse type, contain the charge injected into the postsynaptic neuron according to specified synaptic parameters, such as synaptic geometry, diffusion coefficient in the cleft, number of neurotransmitters released, and number of receptors at the PSD, covering the permitted physiological ranges [221].

  The size of the LUT for the designated synapse type is determined by discretizing the input parameter space into a finite number of points. Using information from the LUT, linear interpolation or extrapolation is used to compute the synaptic output at any point in parameter space. Because of its implementation cost and the adequate accuracy and speed of the used linear inter/extrapolation, which requires few arithmetic operations, [221] does not use a machine learning technique to generalize the LUT data throughout the full parameter space.

  Our method's main benefit is that it eliminates the need to repeat the laborious diffusion-reaction mechanism at synapses for every simulation. The creation of the LUT is a single process that does not have to be repeated during the N⁴Sim runtime, as schematically shown in Fig. 2.85. Apart from its main benefit of speed, this approach also adds flexibility to N4Sim, enabling temporary users to submit LUTs for other kinds of synapses (or extended synaptic parameters) they develop. These LUTs may subsequently be included into the synaptic database [221].

  Therefore, the range of synapses included in the N⁴Sim synaptic database may also increase as the number of devoted N⁴Sim users grows. Accordingly, [221] establishes the LUT for glutamatergic synapses. A deterministic probabilistic emulation methodology is used in [221], with accuracy on par with traditional Monte Carlo simulation techniques. It uses a minimal amount of processing resources to do this: For each synaptic configuration, or per LUT entry, the used emulation algorithm and a single Monte Carlo experiment roughly take the same amount of computation time (approximately 1000 seconds for 10 *ms* of real time) on the same workstation (21 GHz); however, many Monte Carlo experiments are required to capture the expected dynamics found by the emulation algorithm [221].

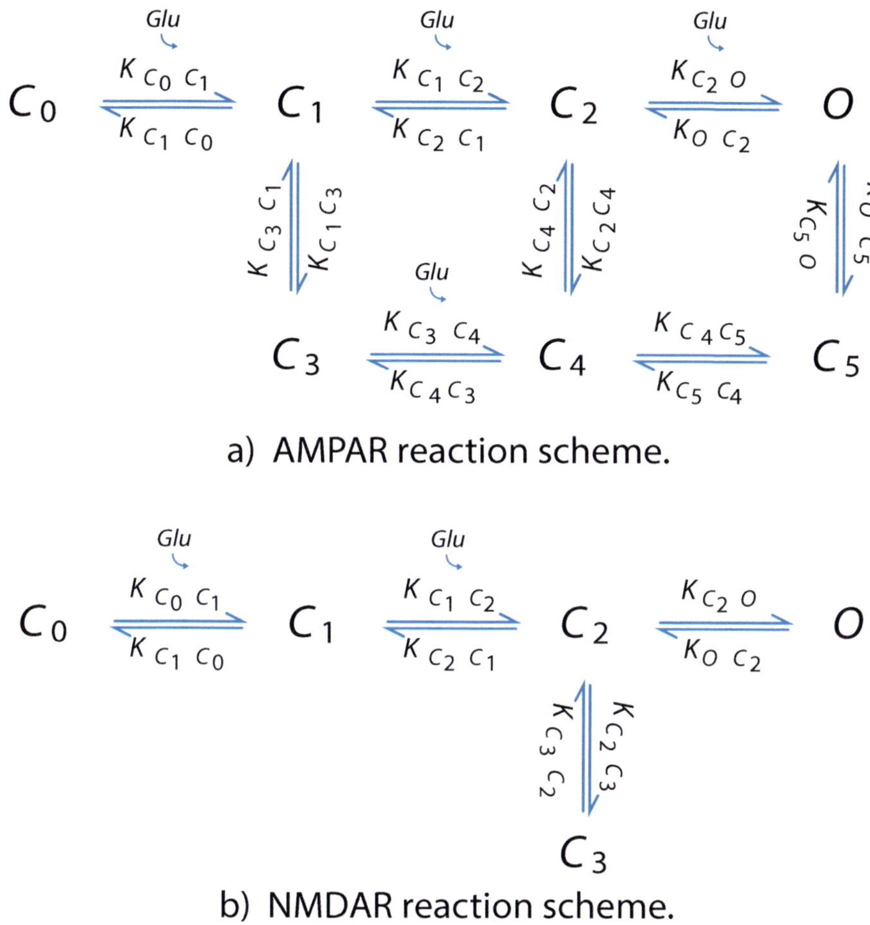

Fig. 2.86 Reaction schemes for the kinetic models of (**a**) AMPARs and (**b**) NMDARs

Complex neurotransmitter-receptor response schemes, like the one for AMPA and NMDA receptors in Fig. 2.86, may be correctly represented by the emulation technique in terms of predicted MC dynamics in a synapse [221].

– Advantages of MC in synapses: There are several benefits of including MC at synapses in N⁴Sim. The use of MC at synapses not only improves N⁴Sim's accuracy but also makes it easier to study the molecular processes behind synaptic plasticity, a critical feature of BNNs that determines learning principles. Understanding the molecular underpinnings of the mechanisms behind synaptic plasticity may help create novel treatments for many neurological conditions that cause learning disabilities. It may also provide important information about how data is processed in BNNs, which may help create a common framework for neuromorphic computing [221].

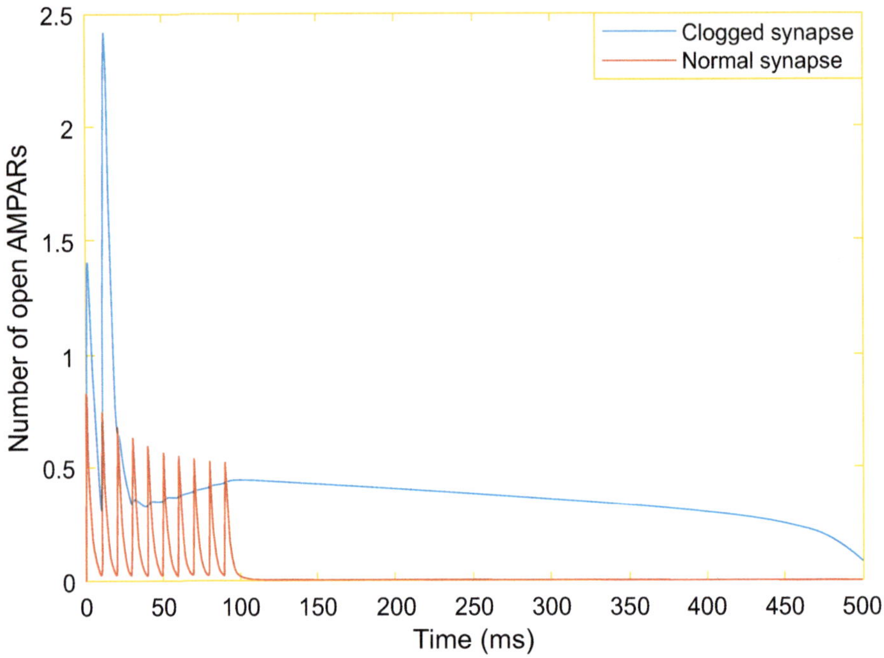

Fig. 2.87 Emulated responses of normal and clogged glutamatergic synapses to 100 Hz impulse train stimulus of 10 pulses length

Furthermore, by using appropriate boundary conditions for the synaptic cleft's horizontal boundary, the model described in [222] can simulate a number of situations, such as a congested synapse, in which the buildup of big molecules at the horizontal cleft boundary prevents glutamate molecules from diffusing out of the cleft. The contrasting reactions of blocked and normal glutamatergic synapses to a 100 Hz stimulus train with 10 pulses are shown in Fig. 2.87. Obstacle-induced glutamate molecule buildup in the synaptic cleft severely limits the amount of impulses that may reach the postsynaptic neuron, which may result in excitotoxicity-induced neuronal death. Because these scenarios are applicable, N^4Sim may be able to predict diseases of the nervous system, such as Alzheimer's disease, in which the buildup of beta-amyloid plaques is associated with synaptic blockage [221].

- Somatodendritic channel: It is composed of the dendrites and the soma of the neuron, where Na^+ and Ca^{++} ions that enter this channel due to the ionic influx mediated by the synapses disperse within the channel, effectively changing local membrane potentials throughout. For the generation of AP, the membrane voltage at the axonal hillock, where the soma opens to the axon, is determinant, which, if exceeds a certain threshold voltage V_{th}, triggers the AP generation mechanism [221].

Fast and accurate mathematical models, desired for scalability to large network calculations, that capture dynamics of ion concentrations inside the somatodendritic channel and corresponding fluctuations in local membrane voltages have been proposed; however, they all specialize to specific types of neuron morphologies, and more importantly, they are not nearly fast enough. Thus, [221] has employed a much simpler model, in which it is assumed the effect of charge injection from the synapse to the membrane voltage at the axonal hillock is instantaneous, and characterized by a linear relationship, where [221] defines the effective elastance (inverse of capacitance) of the synapse as the constant of proportionality, which corresponds to the voltage change at the hillock per unit charge injected at the synapse. The effect of dissipation of charge in soma, which is mediated by the ion channels on the cell membrane that drive the neuron back to its resting state, is modeled by an exponential relaxation (with time constant τ_{SR}) of the membrane potential at the axonal hillock to its resting potential [221].

- N^4Sim the contribution metric: The impact of other neurons' activity on the membrane potential at the neuron's axonal hillock often starts a spike in that neuron. The ratio of the associated voltage contribution at the time of spiking to the total voltage contribution needed to elicit a spike from the neuron's resting state, specifically the difference between the resting and threshold voltages of that neuron, can be used to represent $c(s, r)$, the contribution of incoming spikes, s, to the resultant spike, r. The term "at the spiking" instant describes the computation of incoming spike contributions, which also takes into consideration the decay that occurs prior to the contributing neuron spiking [221]. [221] only takes into account the partial contribution up to the threshold voltage when calculating the last spike's contribution, considering the rest to be wasted. Therefore, one can identify the contributions of the spikes that cause each spike to occur, except the user-defined input spikes that initiate the network [221].

Considering excitatory synapses, which generate excitatory postsynaptic potentials (EPSPs), i.e., positive voltage differences at the membrane of the postsynaptic neuron, then, $c(s, r) \in [0, 1]$ and $\sum_{s \in G_r} c(s,r) = 1$, where G_r is the set of generator or parent spikes of the spike r [221].

The proper mathematical structures for analyzing the relationships between individual spikes and groups of spikes, or brain activity, are directed acyclic graphs (DAGs). Finite sets of nodes connected by directed edges that lack cycles are known as DAGs; starting from any node and moving along the directed edges prevents one from returning to the starting node. In our case, the nodes of the directed acyclic graph of spikes (DAGoS) are spikes, and the directed edges between nodes are defined by the contribution connection previously stated, where each edge is linked to the contribution's numerical value as shown in [221]. Since the directed edges in this graph are arranged according to a causal process, which implies a temporal sequence among the relevant nodes, and because the edges are directed in an ascending temporal order, it is clear that this directed graph is acyclic. In a DAG, an ordered sequence of directed edges defines a path π of length $k \geq 1$, i.e., $\pi = \{e_1, \ldots, e_k\}$, where any two consecutive edges e_i and e_{i+1} intersect at a node s being e_i directed toward and e_{i+1} directed away from s; this can be denoted as $\vec{e}_i = \overleftarrow{e}_{i+1} = s$ [221].

For convenience, $\pi = \{s_1, \ldots, s_{k+1}\}$ is an equivalent alternative notation for a path of length $k \geq 1$ [221] where the edges e_i of the previous notation connect s_i to s_{i+1}, $1 \leq i \leq k$. Reference [221] states that a node r is reachable by the node s, or s reaches r, if there is a path from s to r. The given definition of contribution between two spikes s and r in DAGoS, $c(s, r)$, applies only when s and r are directly connected by a single edge directed from s to r. In [221] now is chosen to extend the definition to apply between any pair of spikes in DAGoS. Given a simulation of duration T, let $\mathcal{S}$ and $\mathcal{U}$ denote the set of all spikes and user-defined spikes that occur during the simulation, respectively, $\mathscr{P}$ denote the set of all paths in the associated DAGoS determined by contribution relations revealed by the simulation. For any $s \in \mathcal{S}$, let DAG of contributors (DAGoC) of s be the sub-DAG of DAGoS composed of all spikes that can reach to s and the directed edges along which they reach to s [221].

To start with, [221] sets $c(s, s) = 0$ for all $s \in \mathcal{S}$ implying spikes have no self-contribution, and [221] sets $c(s, r) = 0$, if s does not belong to the DAGoC of r, signifying the non-existence of contribution paths. Since user-defined spikes are not contributed by any other spike, for any $r \in \mathcal{U}$ DAGoC of r is only composed of r itself, and consequently $c(s, r) = 0$ for all $s \in \mathcal{S}$. Finally, given any $r \in \mathcal{S} \setminus \mathcal{U}$, if s belongs to DAGoC of r, then the contribution of s to r is found as $c(s,r) := \sum_{\pi \epsilon \mathcal{P}_{s,r}} c_\pi$, where $\mathcal{P}_{s,r} \subset \mathcal{P}$ is the set of all paths from s to r, and c_π is the contribution along the path π. For a path $\pi = \{e_1, \ldots, e_k\}$in $\mathscr{P}$ of length $k \geq 1$, c_π is defined as $c_\pi := \prod_{i=1}^{k} c_{e_i}$, where $c_{e_i} := c(\overleftarrow{e}_i, \vec{e}_i)$ is the contribution over the edge e_i of its prior node to its posterior. This completes the bidirectional definition of contribution between any pair of spikes in DAGoS [221], as contributions over edges are previously established.

The idea of contribution to have a meaning between two sets of spikes, or brain activity, is then expanded in [221]. Given two neural activities S, $R \subset \mathcal{S}$ let us denote by $\mathcal{P}_{S,R}$ the set of all paths emanating from S to R, that is all paths $\pi \in \mathcal{P}$ with the property that (using alternative notation) $\pi(1) \in S$, $\pi(k+1) \in R$ and $\pi(i) \in \mathcal{S} \setminus S$ for all $i \in \{2, \ldots, k\}$, where $\pi(j)$ stands for the jth entry of $(k+2)$-tuple path π, $j \in \{0, \ldots, k+1\}$. Then, the contribution of spikes from S to R is defined as $c(S,R) := \sum_{\pi \epsilon \mathcal{P}_{s,r}} c_\pi$, where c_π, the contribution along π, is defined as similar to before. To put it another way, the set of pathways that radiate from S to R are those that start in S, never return, and end in R, where many trips to R are permitted [221].

The justification for distinguishing the contribution of two neural activities along particular paths is the knowledge that the contribution along a path that passes through several elements of S is already captured by the contribution along a path that is obtained by cutting the original path off at its beginning, so that the new path starts from S and does not return to S later. The reason for the overlap is that both routes end at the same node in $\boldsymbol{R}$, one of which involves several trips to S and the other of which starts in S and doesn't return. In contrast, numerous visits and their shortened paths do not cross in their contributions for the destination set $\boldsymbol{R}$ since they invariably end at different nodes in $\boldsymbol{R}$ [221].

Both the neural activity R is independent of the neural activity S if $c(S, R) = 0$, and they are mutually independent if the converse also holds, as claimed in [221]. The concept of contribution between neuronal activities and/or groups of neurons may be defined after the contribution between any two neural activities has been established. Let $\mathcal{N}$ denote the set of all neurons in our neural network, and for each $n \in \mathcal{N}$, let $\mathcal{S}_n \subset \mathcal{S}$ denote the set of spikes that belong to n. Then, for any two groups of neurons $M, N \subset \mathcal{N}$ in [221] is defined the contribution of the activity of neurons in M to the activity of those in N as $c(M,N) := c\left(\bigcup_{n \in M} S_n, \bigcup_{n \in N} S_n\right)$ [221].

Self-contribution of a neural activity S is defined as $c(S) := c(S, S)$, which captures the amount of crosstalk inside the neural activity S. In BNNs, dispersed neural clusters of high activity are known to operate as separate functional blocks that process neural signals. In practice, these clusters are distinguished from the rest of the network by maximizing the density of crosstalk. Thus, in a similar spirit, for a given group of neurons N, in [221] is defined self-contribution density $c_d(N)$ as $c_d(N) := \frac{c(N)}{|N|} = \frac{c(N,N)}{|N|}$, where $|N|$ is the number of neurons in N. Groups of neurons that locally optimize self-contribution density are called functional blocks of a particular network by [221].

Calculating this measure adds little burden to simulation time because spike-to-spike contributions are continuously computed during an N^4Sim simulation. Similar to the synaptic weights and thresholds used in ANNs, the contribution metric in a way represents the strength of synaptic connections while accounting for resting and threshold potentials. The contribution metric is also sensitive to changes in spike timing because the somatodendritic channel of N^4Sim accounts for the time decay of voltage contributions from previous spikes. This adds interest to the metric and saves neural network characterization from the static image of synaptic weights and ANNs threshold values. Therefore, even if two neurons or spikes have the same relationship to the network in terms of connection and strength, a spike or neuron that is "temporally well-coupled" with others would contribute more to network activity than a "temporally ill-coupled" one [221].

- N^4Sim delay: The sum of the delays for many distinct processes, including axonal transmission, neurotransmitter release, synaptic diffusion and binding processes, and somatodendritic ion propagation, determines the total delay in the transmission of an AP across a neuron. The majority of the previously mentioned individual delays can be disregarded in favor of axonal delays as it makes up the majority of the delay in neurons with axons that are many orders of magnitude bigger than the synapse, as is typically the case. Hence, for a neuron with an axonal length of L and conduction velocity v, the transmission delay is taken as $r = \frac{L}{v}$ [221].

The total delay of a branch made up of several neurons linked in series one after the other is equal to the sum of the delays of the individual neurons that make up the branch. For a branch with j neurons, its delay is given as $R = \sum_{i=1}^{j} r_i$, where r_1, r_2, …, r_l are the individual delays for the branch neurons. Now, in the case of a network of neurons with multiple paths between input and output terminals, the overall network delay, R_N is a function of the delay for each of the individual paths between the two sides, i.e., $R_N = f(R_1, R_2, \ldots, R_n)$, where $R_1, R_2, \ldots, R_n$ are the delays for n different branches of the network, and *f*() is a function that depends on the application of interest. For instance, in cases where an average network delay is required, *f*() translates as the *mean*() function. Similarly, *min*(), *max*(), and *median*() delays may also be required by some applications [221].

- N^4Sim software design serves as a valuable instrument for researchers in comprehending the fundamentals of neural transmission and synapse activity. To achieve this, N^4Sim is designed to be open source. Because of the simulator's modular design, various parts may function independently, and parallel processing is made easier. According to its creators in [221], N^4Sim is designed to provide multiscale functioning in both time and space, allowing for large-scale simulations of small cerebral areas or the entire brain in addition to simulations at the level of individual synapses [221].

2.3.11 Simulator nanoNS3 (A Network Simulator for Bacterial Nanonetworks)

In [223], nanoNS3 has been designed to tackle the issues associated with constructing a bacterial MC (BMC) network simulator. nanoNS3 is developed and validated through research conducted using genetically modified bacteria in a microfluidic setting. nanoNS3 is constructed upon NS-3, a discrete event-based simulator (NS-3 is examined in this chapter, see Sect. 2.3.5). A discrete-event simulator is optimal for modeling systems characterized by prolonged delays. It monitors the alteration in the condition of events rather than the exact time. Simulating absolute time is costly for extremely slow networks, such as BMC; hence, event-based simulations are more efficient and scalable. BMC networks exhibit exceedingly slow performance characterized by prolonged transmission delays [223].

Consequently, employing NS-3 enhances the temporal efficiency of nanoNS3. Consequently, nanoNS3 emulates bit-level transmission and reception rather than molecular-level transmission and reception, resulting in enhanced computing efficiency. nanoNS3 can replicate the modulation of transmitted bits into pulses at the transmitter and the propagation and attenuation of molecule concentration along a microfluidic channel. The biological reaction of N-acyl homoserine lactone (AHL) diffusion through the biofilm and across the bacterial membrane is reproduced at the receiver end (physical layer). The receiver can ascertain the transmitter's ID based

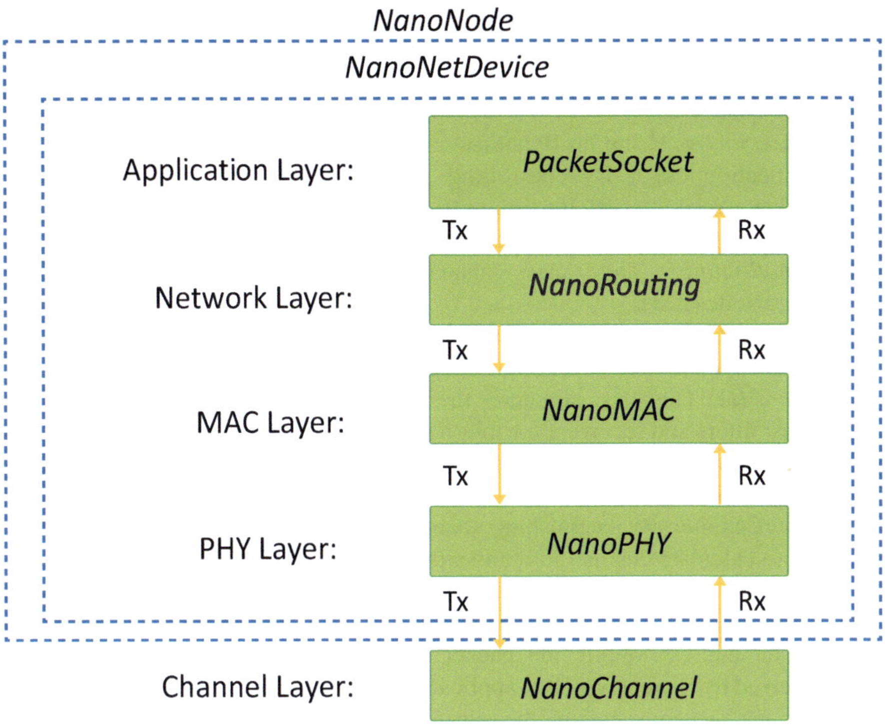

Fig. 2.88 NanoNS3 architecture

on the received concentration at the (MAC) layer. Subsequently, the receiver can demodulate the received signal to retrieve the transmitted information [223].

In order to address the issues of simulation time efficiency and simulating large networks, Reference [223] chooses to develop the nanoNS3 BMC network simulator based on ns-3. There are several noteworthy benefits of using nanoNS3 in ns-3 [223]:

- New algorithms are easily implemented and made available as open-source.
- Excellent computing performance for extensive networks.
- Direct use of NS-3 supporting tools, such as logging and tracing systems, is possible.

Then, subsequently, nanoNS-3 will be described in following paragraphs.

- NanoNS3 network architecture. NS3's overall structure is seen in Fig. 2.88. It lists seven important classes and the structure of the network tiers that correspond to them. Below is a brief discussion of each class's functionality [223].
  - *NanoNetDevice:* It is similar to the network interface card (NIC) and may handle a variety of nanocommunication technologies and related protocols, such as amplitude addressing and diffusive or EM wave-based nanocommunication methods [223].

- *NanoNode*: The physical device can be considered, and various *NanoNetDevices* can be integrated with NanoNode to offer relevant communication technologies and protocols, facilitating intercommunication among NanoNodes [223].
- *PacketSocket*: Using no IP addresses, this class is a simple and creative ns-3 application class. By controlling application-related parameters such as packet arrival interval, maximum transmission packet count, and packet size, it is used to set up user-defined programs for microcommunications [223].
- *NanoRouting:* This class manages message forwarding by each NanoNode [223].
- *NanoMAC*: This class manages channel access of different *NanoNodes*, and it also manages MAC-layer addressing mechanism [223].
- *NanoPHY*: This class simulates the transmission and reception processes of transmitters and receivers for micro signals. This class encompasses modulation, demodulation, error analysis, transfer rate analysis, and receiver response [223].
- *NanoChannel*: By establishing channel conditions, this class makes it possible to calculate channel loss and simulate how transmitted signals propagate and attenuate inside the corresponding microfluidic channel. The packet socket class is related to three categories: packet client, packet server, and packet address. While the packet server class provides how packets are received by the receiver (Rx) application layer, the packet client class describes how packets are sent from the transmitter (Tx) application layer. The setup of addresses for different nodes is specified by the packet address class [223].
- In [223] is integrated the *ns-3 original application class* into nanoNS3, to facilitate the integration of additional original *ns-3 classes* with nanoNS3, such as transport-layer protocols. In NS-3, the transport-layer protocol is integrated with sockets; thus, the authors in [223] incorporate packet sockets into nanoNS3 to facilitate transport-layer extension. Users can adjust the settings for each of the aforementioned classes [223].

• Protocols implemented for nanoNS3: The cited simulator implements some of the basic protocols to simulate bacterial MC networks. The five important models implemented in nanoNS3 are [223]:

  - Receiver response
  - Channel loss
  - On–Off Keying (OOK)
  - Amplitude addressing
  - Transfer rate and error analysis

• Receiver response model: Simulating BMC networks requires precise receiver response modeling. The simulation of BMC networks is the main goal of nanoNS3. This simulator uses a population of genetically modified *E. coli* bacteria as the bacterial receiver, and when they absorb AHL molecules, they produce a GFP. Microfluidic paths link the transceivers, which are housed within a microfluidic device. Green fluorescence is released by the receiver after molecules are transferred from the transmitter via microfluidic channels. The signal received is shown by the receiver bacteria's relative fluorescence. In nanoNS3, the following proposed models are implemented [223]:

$$\frac{dAHL_i}{dt} = k_c\left(AHL_e - AHL_i\right) - k_1 AHL_i^2 LuxR^2 + k_1 C_1 \tag{2.36}$$

$$\frac{dC_1}{dt} = k_1 AHL_i^2 LuxR^2 - k_1 C_1 - k_2 C_2 P_{\text{Lux}} \tag{2.37}$$

$$\frac{dC_2}{dt} = k_2 C_1 P_{\text{Lux}} - k_2 C_2 - k_{tr} C_2 \tag{2.38}$$

$$\frac{dLuxR}{dt} = k_{Lc} - 2C_1 - 2C_2 \tag{2.39}$$

$$\frac{dGFP_i}{dt} = k_{tr} C_2 - k_{Gm} GFP_i - k_{Gd} GFP_i \tag{2.40}$$

$$\frac{dGFP_m}{dt} = k_{Gm} G + i - k_{Gd} GFP_m \tag{2.41}$$

AHL_e and AHL_i are the external and internal concentrations of molecules at the receiver. Lux$R$, C_1, C_2, and GFP_i represent internal parameters of the receiver bacteria. GFP_m represents the concentration of GFP which in turn represents the relative fluorescence at the receiver. Vector $k = [k_C, k_1, k_2, k_{Lc}, k_{tr}, k_{Gd}, k_{Gm}]$ reflects the receiver's process rate constants. A detailed explanation of the equations and their accompanying derivation may be found in [223, 224].

The channel and receiver response in nanoNS3 are modeled using the aforementioned equations. The modulator receives the bits that are generated by the transmitter module. Modulator implemented in nanoNS3 is OOK and hence bits are mapped to rectangular pulses. The rectangular pulses are input to these equations to simulate the GFP response of the receiver. A numerical inverse of these equations is used to sample and quantize the received signal. The output of the sample and quantize modules is then fed to the demodulator to process. This model gives a bit-level response at the receiver [223].

OOK model modulation is the process of varying the properties of a signal to convey the information. Modulation determines the rate of transmission. NS-3 is a packet-level simulator and hence allows users to change modulation by changing the transmission rate. In nanoNS3 is implemented bit-level simulation and hence implement a module for modulation that maps each input bit to a signal to be transmitted. OOK is one of the simplest modulation techniques and majority of works on BMC assume OOK as the modulation technique. OOK transmits a rectangular pulse of amplitude/concentration A for a duration of T_1 time units to send bit 1 and no signal for T_2 time units to send bit 0. OOK module in nanoNS3 generates a rectangular pulse of a given amplitude and a given duration. Then, the generated rectangular pulse is fed to the channel model [223].

- Channel loss model: A crucial part of BMC simulators is the microfluidic channel loss model, which sheds light on how signal concentration is reduced during microfluidic channel propagation. The fluid is moving via a microfluidic channel in diffusion-based flow-induced MC (FMC), which connects areas of molecular transmitter and receiver, such as bacterial habitat, by acting as a communication channel. In accordance with [224], an analytical investigation of the molecules' propagation in the form of an impulse response is carried out in [225] while taking the physical system factors into account.

  The goal of the propagation loss model is to determine the channel loss effects caused on the molecular signal with respect to the distance, fluid flow parameters (pressure drop, flow velocity, microfluidic channel geometry, and fluid type), and type of the molecule (diffusion constant). According to [223] in [225], channel loss models for the basic microfluidic channel shapes (straight and turning) and cross-sections (rectangular, square, elliptical, and circular) are developed incorporating the characteristics of the fluid flow and mass transport in the microfluidic channels; nanoNS3 implements the models presented in [225]. In [223] is illustrated rectangular cross-section microfluidic channel loss model. The governing set of equations for channel loss in a rectangular microfluidic channel is given as [223, 225]:

$$G_{\mathrm{rect}} = \frac{h^3 w}{12\mu l} * \left(1 - 0.63\frac{h}{w}\right) \tag{2.42}$$

$$u_{\mathrm{rect}} = G_{\mathrm{rect}}{}^{*} \Delta p \tag{2.43}$$

$$\tau_{\mathrm{rect}} = \frac{l}{u_{\mathrm{rect}}} \tag{2.44}$$

$$TF_{\mathrm{rect}} = e^{-\left(k^2 D + jku_{\mathrm{rect}}\right) * \tau_{\mathrm{rect}}} \tag{2.45}$$

where the hydraulic conductance of the microfluidic channel and area-averaged flow rate are denoted by G_{rect}, u_{rect}, respectively. G_{rect} is a function of channel cross-section shape, dimensions, and viscosity of the fluid (μ). u_{rect} is a function of pressure drop (Δp) and G_{rect}; τ_{rect} and TF_{rect} are the delay and attenuation of channel, respectively. h and w represent the channel height and width; l and D are the lengths of the straight channel and Taylor dispersion adjusted diffusion constant. All microfluidic channel loss models are explained in depth in [225].

- Transfer rate and error analysis model: The theoretical limits of the information transfer rate and associated error probabilities of bacterial MC are estimated using a theoretical analysis model implemented in nanoNS3 [223, 226] based on the framework of nanoNS3 and based on the uncertainty of communications, i.e., transmission, propagation, and reception.

  According [223, 226] considers an MC setting in which the diffusion channel inputs and outputs are OOK with the binary set $\{0, 1\}$. The effects of uncertainty in the production of molecules, channel parameters, and reception process on the overall noise of the communication are considered. This model can be used to study the theoretical limits of the information transfer rate in terms of the number of bacteria per node, noise level, and maximum molecule production levels. Also, it can be utilized to analyze the achievable rates and the error probabilities of *M-ary* schemes [223]. In nanoNS3, the models presented in [226] are implemented.
- Amplitude addressing model: The channel and physical layer of BMC networks are modeled using the aforementioned models (receiver response, channel loss, transfer rate, error analysis, and OOK). For a network with various sources to be fair and minimize collisions, an MAC protocol is necessary. MAC protocols used in wired or wireless networks as implemented in NS-3 increase the delay and decrease the throughput in a super-slow network like the BMC network. [227] considers a multiple sources and single receiver topology and proposes an MAC protocol to improve the BMC network throughput. Multiple sources transmitting to a single receiver is a typical sensing network scenario.

  [227] proposes a local addressing mechanism which implicitly performs MAC, and it is implemented in nanoNS3. [227] also analyzes the advantages and disadvantages of various addressing mechanisms and proposes Amplitude Addressing for BMC networks. Amplitude Addressing assigns distinct amplitudes to each user in a BMC network and each user uses the assigned amplitude with OOK to transmit information. The receiver receives the sum of the transmitted amplitudes which are then resolved to identify the individual amplitude, thus solving addressing and MAC in the local network. [223] further implements two different receivers, viz., load-aware and load-unaware. Load-aware receiver has an estimate of input load of the transmitter. Therefore, the decoding table (decoding table is used to decode the transmitter ID based on the received signal) is a function of the input load of the transmitter. On the other hand, load-unaware receiver assumes that the input is uniformly distributed for the transmitter and the decoding table is fixed. A global address is required to map local address to the source. nanoNS3 does not have a global addressing module, but MAC address of *ns-3 nodes* can be used as a global address module in simulations [223].

- The assessment outcomes of nanoNS3 employing both simulation-based and experimental-based validation for the five models previously stated are shown in [223]. An approach for validating nanoNS3 using the previously specified methods is also provided in [223]. Findings from experiments and the MATLAB analytical model used in [224] are utilized to validate the simulation findings of nanoNS3 for the receiver response model in [223]. NanoNS3 is verified for the amplitude addressing model in [223] using the Python simulator utilized in [227]. NanoNS3 is verified for the channel loss model in [223] using the MATLAB analytical model utilized in [225]. NanoNS3 is verified using the MATLAB analytical model used in [227] for the transfer rate and error analysis model. Unless otherwise noted, the carrier signal is a chemical signal, and the transceivers are bacteria. To validate the performance of the aforementioned models in nanoNS3, the following scenarios are used [223]:
  - Scenario with only one Tx and one Rx: One transmitter transmits signals to one recipient in a single link situation. 15 μM is the default transmitted molecular concentration [223].
  - Scenario with many Txs and a single Rx: In this case, a single receiver receives signals from many transmitters. Each transmitter's allotted amplitude is determined by the procedures described in [227]. This scenario is used to validate only the amplitude addressing model [223].
- Receiver response: In [223] is implemented the receiver response model derived in [224]. The set of differential equations presented in [224] defines the *GFP* response of the receiver bacteria to a given concentration of molecules. Then, in [223] is built an Inverse model of the receiver response at the receiver to estimate the molecules received from the receiver *GFP* response. Besides, in [223] is validated nanoNS3 receiver response in two steps:

  First, in [223] is analyzed the numerical inverse response using simulations. A transmitter sends information using OOK, i.e., rectangular pulses of a fixed amplitude and a fixed duration for bit 1 and no signal for bit 0. The rectangular pulses are input to Forward response generating receiver *GFP* response which is then fed to Inverse response derived numerically that estimates the signal transmitted based on the GFP response.
- Second, in [223] is compared the forward receiver response obtained from simulator with the response from experiments. Also, [223] inputs the receiver response from experiments to the Inverse model and compares the estimated signal with the actual transmitted signal.
  - Validation of the receiver response using simulation: The inverse receiver response model is verified using OOK in [223]. The broadcast and received rectangular pulses for the input bits and forward receiver response are shown in Fig. 2.89a, b, respectively. Table 2.3 displays the appropriate simulation parameters. The sent and received rectangular pulses are precisely the same, as can be seen in Fig. 2.89a. This outcome demonstrates that the transmitted

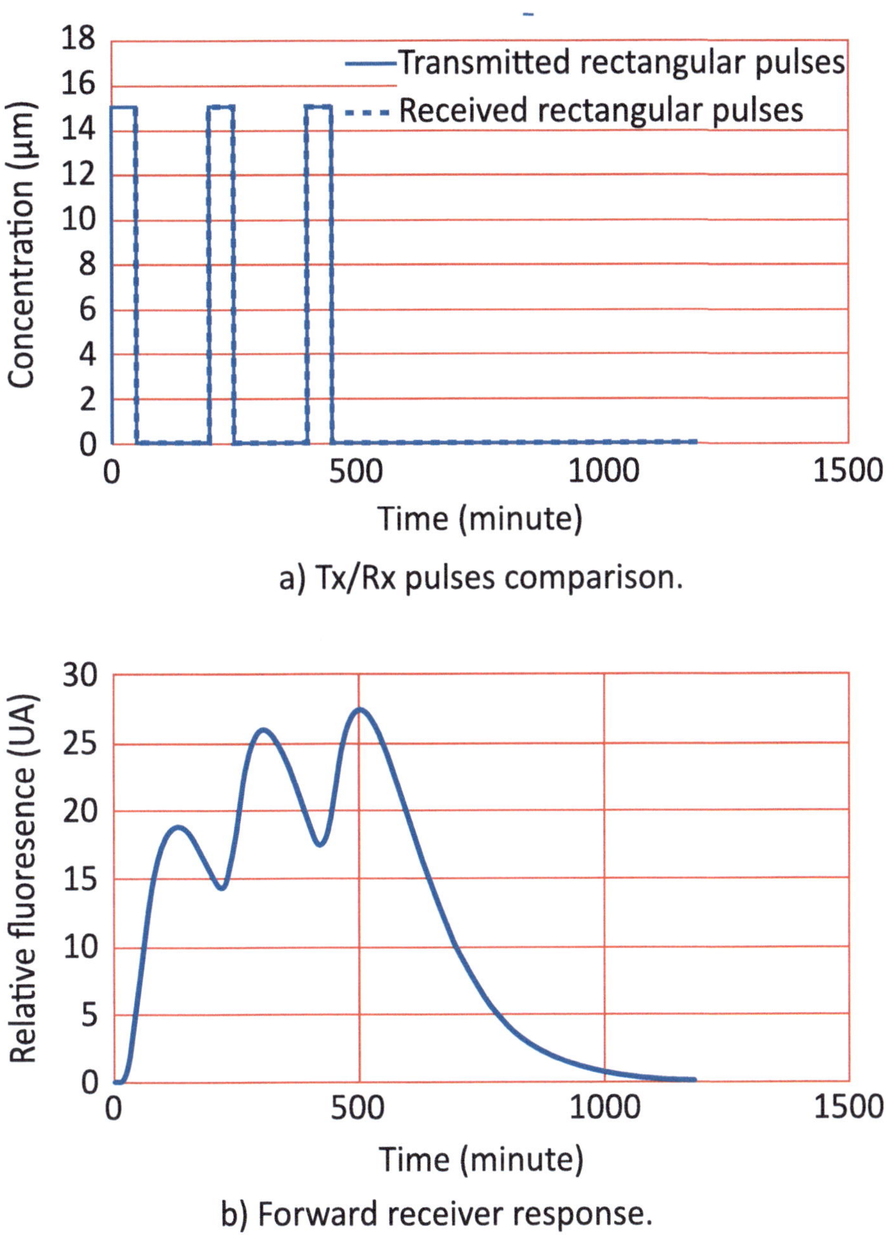

Fig. 2.89 Experimental validation

pulses in nanoNS3 can be recovered by the receiver. Figure 2.89b [223] shows the matching forward receiver response for the transmitted rectangular pulses.

- Validation of receiver response by experimentation: [223] validates the receiver response model based on experimental findings. These results were

Table 2.3 Simulation parameters for receiver response [223]

MC characteristics	Default settings	Parameters for MC	Default settings
Tx sequence	1 000 100 010	k_C	254/60
Modulation	OOK	K_{Gm}	1.8/60
Tx pulses amplitude	15 μM	K_{Gd}	39/60
Tx pulses width	50 min	K_{tr}	1334/60
Threshold	7.5 μM	K_{LC}	1200/60
		K_1	20/60
		K_2	200/60

obtained using the identical experimental setup as described in [224]. As a result, the receiver response from models and tests is compared in [223]. In [223], the receiver response's inverse is also confirmed. In addition to comparing the estimated and real sent signals, the inverse model incorporates the experimental receiver response.

Verification of the reply from the forward receiver: Figure 2.90a shows the rectangular pulses that were broadcast. The forward receiver response modeling and experimental results are shown in Fig. 2.90b. Similar patterns can be seen in the simulation and experimental results, with the simulation faithfully reproducing the peaks of each pulse in the actual data. In the experimental setting, four different Tx sequences, each with ten bits, are set up as Tx sequences. These sequences all show a similar pattern to the data provided [223].

Validation of inverse receiver response: This is confirmed in [223] by the experimental results of the forward receiver response. The real forward receiver response findings are less smooth than the predicted forward receiver response results, as seen in Fig. 2.90b. The experimental data is refined in MATLAB using the Loess smoothing algorithm (with a 10% span) to remove noise and system defects. Figure 2.91a shows the pertinent experimental data. The transmitted bits are then successfully recovered at the receiver end by [223] feeding the smoothed experimental findings into the inverse receiver response model in nanoNS3.

The peaks of the sent and received pulses match, as shown in Fig. 2.91b). [223] calculates the average concentration for the pulse duration of each bit, represented as Ave_i for bit i, in order to demodulate the incoming pulses. The threshold for identifying the bit level is set by the authors in [223] and is half of the transmitted concentration. Equation (2.46) is used to determine the received bit.

$$\text{Received bit} = \begin{cases} 1 & \text{if } Ave_i \geq \text{threshold} \\ 0 & \text{if } Ave_i \leq \text{threshold} \end{cases} \tag{2.46}$$

This method enables [223] to attain a 100% demodulation rate for the demonstrated experimental findings. The mean demodulation rate of four experi-

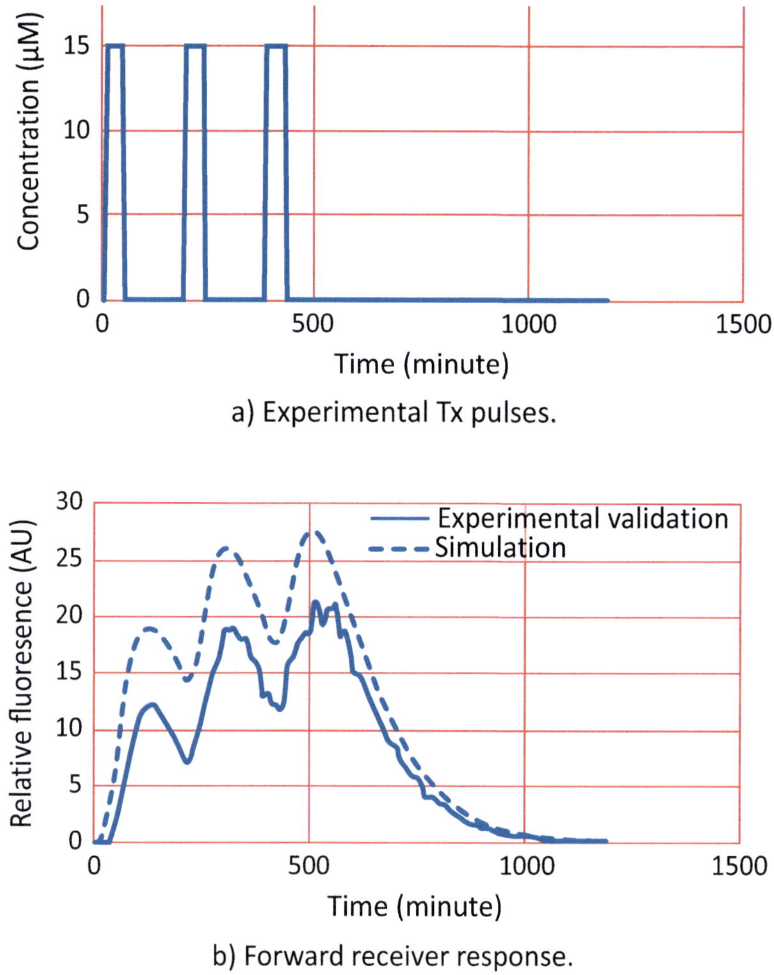

Fig. 2.90 Experimental validation

mental sets attains 92.5%. As a result, nanoNS3 provides an experimental receiver response model, and its simulation results are consistent with the MATLAB analytical model. By improving layer protocols, nanoNS3 can provide enhanced accuracy in simulation performance thanks to the defined model [223].

– Channel loss: A single transmitter and single receiver scenario are used to compare the channel loss model in [223]. Using the analytic channel loss model described in the previous sections, it compares the results using nanoNS3 and MATLAB. The objective is to show that the numerical findings of the analytical channel loss model produced by nanoNS3 match those produced by MATLAB. The details of the default parameter setups are shown in

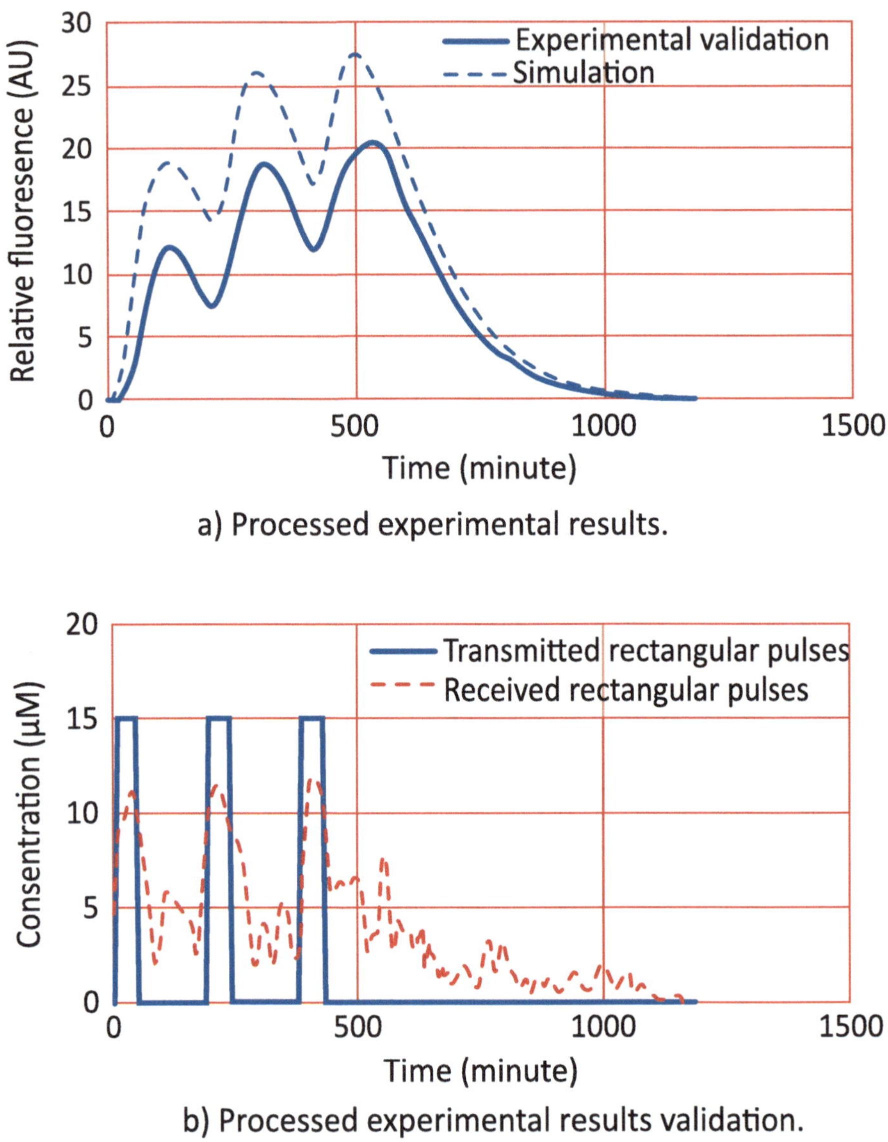

Fig. 2.91 Experimental validation

Table 2.4. The difference between straight and twisting channels' impulse response attenuation as a function of frequency (radians per meter) is shown in Fig. 2.92a, b. For both Fig. 2.92a, b, the analytic model assessment in MATLAB yields the same findings. The microfluidic channel loss model that nanoNS3 provides as a consequence shows that the simulation findings are consistent with the MATLAB analytical model. Using the well-established

Table 2.4. Simulation parameters for channel loss [223]

MC characteristics	Default settings	Parameters for MC	Default settings
Channel shape	Rectangular	D	$10 * 10^{-10}$ m²/s
Turning angle	30°	Δp	500 Pa
Viscosity	10^{-3} Pa s	l	10 mm
		h	6 μm
		w	25 μm

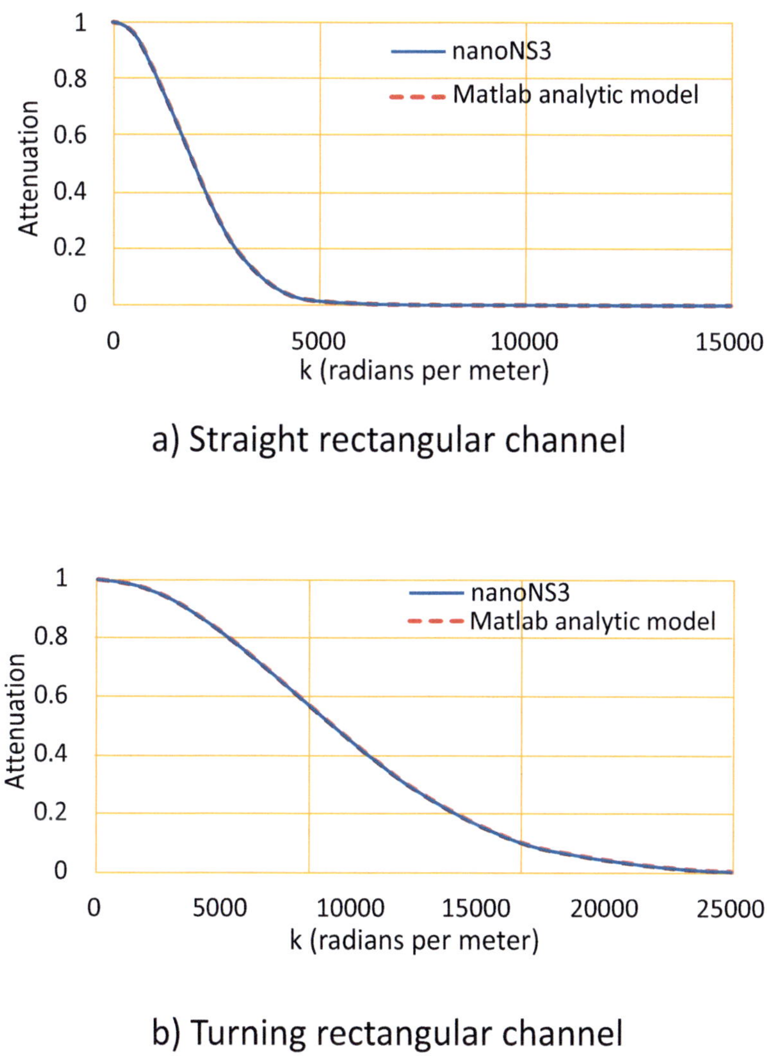

Fig. 2.92 Channel loss validation

Table 2.5 Simulation parameters for capacity and error analysis [223]

MC characteristics	Default settings	Parameters for MC propagation	Default settings
Number of bacterial	100	Square of combined noise variance	0.1
Number of ligand receptors	50	M-ary scheme	16

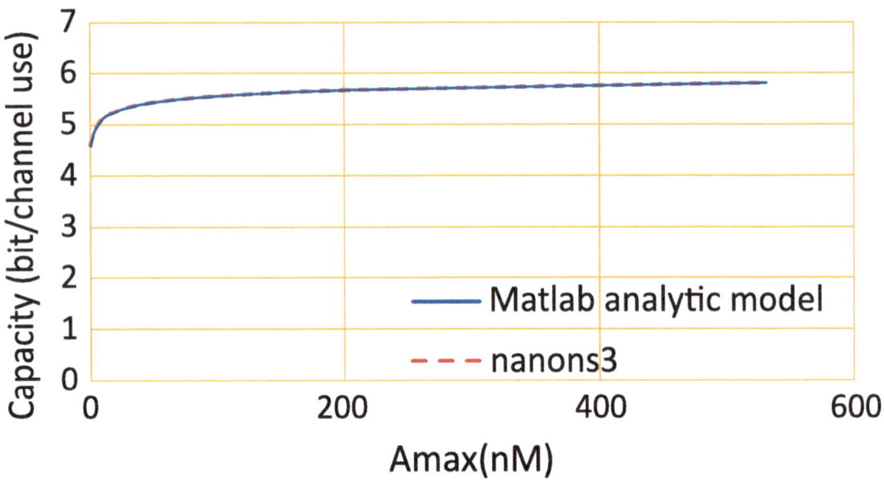

Fig. 2.93 Transfer rate analysis

channel loss model [223], nanoNS3 may easily incorporate additional channel properties and other channel loss models.

– Transfer rate and error analysis: A single transmitter and single receiver scenario is used in [223] to assess the transmission rate and error analysis model. As such, it compares the results obtained by nanoNS3 to the MATLAB simulation mentioned in [226]. The objective is to show that the MATLAB analytical model's numerical results and nanoNS3's numerical results match. The details of the default parameter setups are shown in Table 2.5. The connection between channel capacity and the highest molecular concentration is shown in Fig. 2.93 [223].

In the context of an M-ary system, Fig. 2.94a, b shows how the maximum concentration of molecules varies with the channel capacity and associated error rate. As shown in Figs. 2.93 and 2.94a, b, the theoretical limitations for both capacity and error rate generated by nanoNS3 match the findings from the MATLAB analytic model. In this way, nanoNS3 provides a theoretical analytical model, and its simulation results are consistent with the MATLAB analytical model. The transfer rate and error probability associated with certain modulations may be estimated using the theoretical analytical model [223].

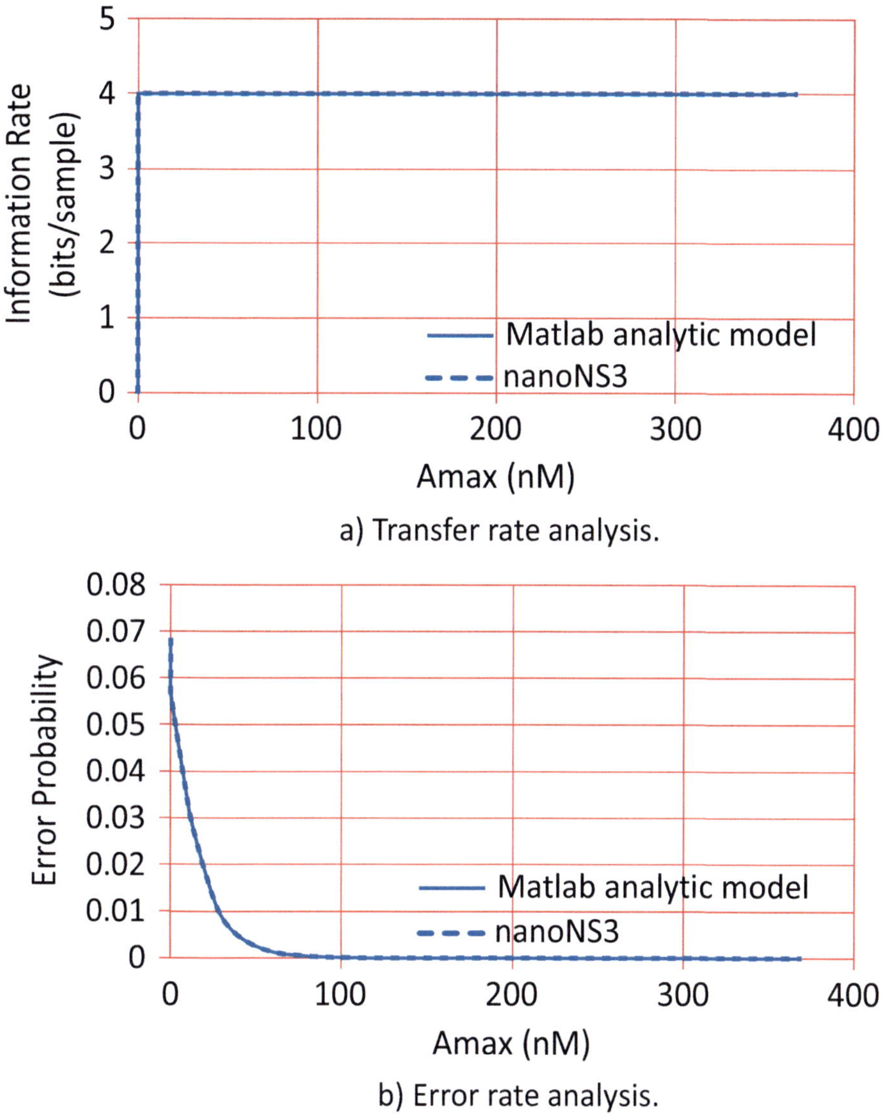

Fig. 2.94 Transfer rate and error rate analysis for corresponding modulation

Amplitude Addressing: The amplitude addressing approach is validated in a scenario with several transmitters and a single receiver in [223]. Subsequently, [223] evaluates the performance of nanoNS3 in comparison to the custom-built Python simulator utilized in [227]. The aim is to demonstrate that the simulation outcomes of nanoNS3 correspond with those of the previously mentioned Python simulator. Table 2.6 [223] presents the specifics of the default parameter settings.

Table 2.6 Simulation parameters for amplitude addressing [223]

MC characteristics at transmitter end	Default settings (value or type)	Default settings (type or unit)
Number of Tx bits	100	-
Amplitude assignment mechanism	Integer	Binary
Tx pulses	50	min
Tx pulses interval	20	min
p_t	0.5	–
Max amplitude	15	μM
Number of Tx users	5	–

Two instances of the amplitude assignment process are briefly described in [223]. In the integer amplitude assignment technique, each user is allocated a distinct amplitude corresponding to the node ID. For instance, amplitudes {1, 2, 3} are allocated to transmitters with IDs {1, 2, 3} in a one-to-one correlation. In the binary amplitude assignment process, amplitudes {1, 2, 4, 8} are allocated to transmitters with IDs {1, 2, 3, 4} in a one-to-one correlation [223]. The demodulation accuracy for load-unaware addressing at the receiver is shown in Fig. 2.95a, which corresponds to various input loads for integer amplitude assignment. The performance of nanoNS3 substantially resembles that of the specially designed Python simulator mentioned in [116], as seen in Fig. 2.95a. Using binary amplitude assignment, Fig. 2.95b shows the decoding accuracy at the receiver for various input loads. The Python simulator's and nanoNS3's simulation outcomes are very comparable. In [223], the effectiveness of load-aware and load-unaware receivers is also contrasted.

Figure 2.96 illustrates the accurate demodulation rate employing both load-aware and load-unaware receivers with binary amplitude assignment. It is evident that when the receiver is cognizant of the transmitter's load, the accuracy of the demodulation rate improves in comparison to a receiver that is uninformed of the load. The amplitude sequences employed for assessment are elaborated upon in [116]. Consequently, nanoNS3 offers an amplitude addressing technique, indicating that its simulation results align with those of the Python simulator. The performance of upper layer protocols, such as routing algorithms, can be examined based on the established addressing model [223].

According to [228], an efficient addressing and multiple-access control method is necessary to improve the system's throughput performance in a communication network with notable processing delays. [228] focuses on source addressing inside bacterial communication networks that include numerous sources and a single receiver. They suggest a method called amplitude-division multiple access, which uses the transmitted signal's amplitude as the source's identification. [228] uses genetically modified *Escherichia coli* bacteria in a microfluidic device to demonstrate the viability of using amplitude for addressing. Using a variety of addressing

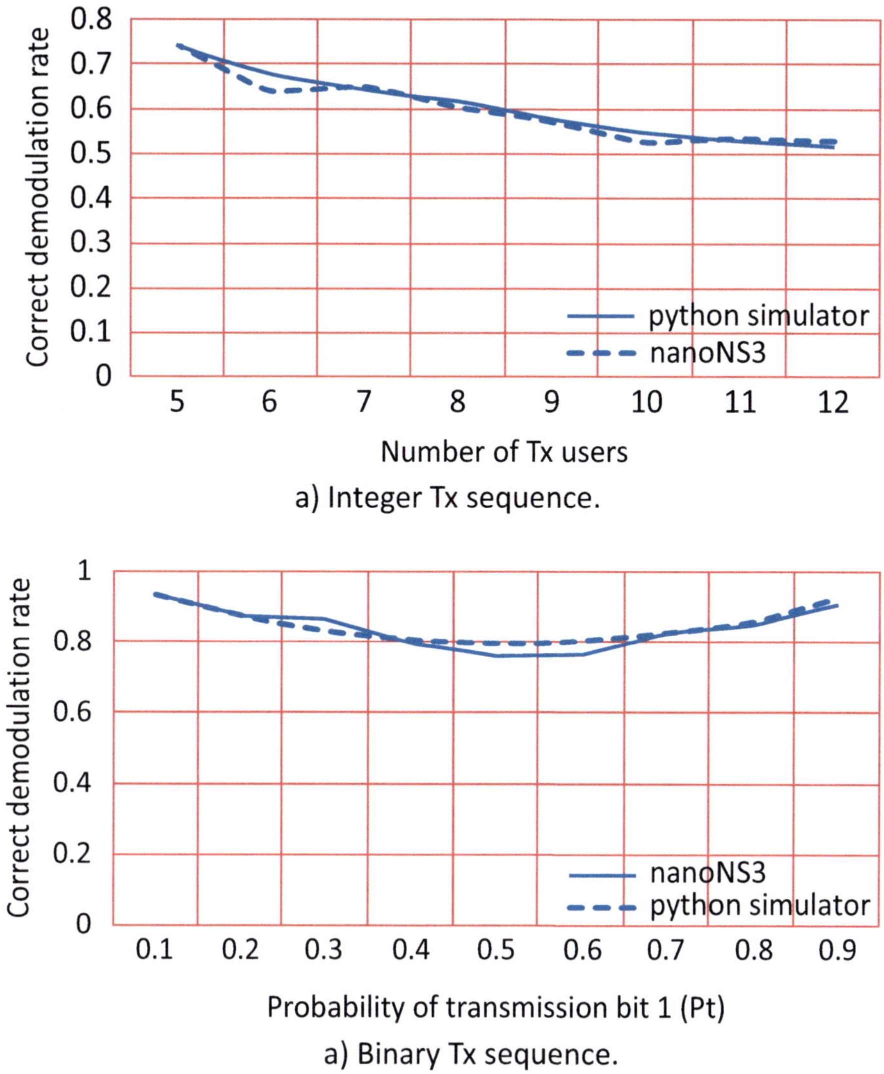

Fig. 2.95 Amplitude addressing validation

algorithms, the authors in [228] assess the network's performance and suggest an amplitude sequence and a low-complexity receiver design that lowers mistakes in source address identification during collisions. The problem of multiple-access control is naturally solved by amplitude-division multiple access, as demonstrated by [228]. NanoNS3 is the simulator in this case.

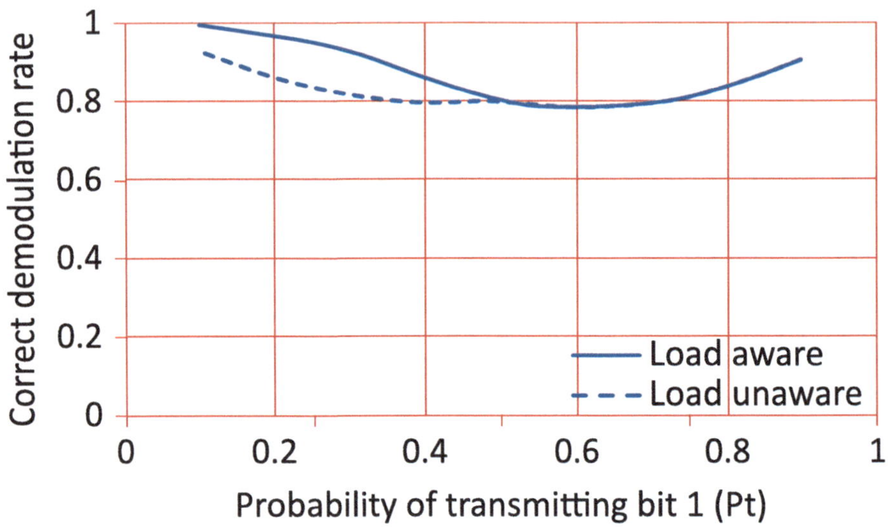

Fig. 2.96 Load-unaware vs. load-aware receiver

2.4 Other Simulators for Analyzing Molecular Communications Systems

In this section, we briefly describe some cases of the use of MCs simulators, and thus, the following investigations represent just a few examples of generic simulators, given that in literature every case of analysis of MCs systems is done through communicational models and digital tools, i.e., with the use of computational programs.

- BloodVoyagerS (BVS), a framework for modeling medical nanonetworks that includes a body simulator with a nanonetwork simulator, is provided in [229]. The requirements for a body model that supports our prototype implementation of the body simulator BloodVoyagerS within the network simulator NS-3 are given by the authors in [229]. The BloodVoyagerS study shows that the framework successfully navigates nanobots throughout the cardiovascular system simulation. After around seven minutes, the nanobot dispersion reaches a dynamic equilibrium. In order to investigate the movement and communication of nanobots in medical applications, the prototype shows promise in providing a more realistic full-body simulation. In order to simulate the flow and dispersion of nanodevices throughout the Human Circulatory System (HCS), BloodVoyagerS is utilized [229].

  Consequently, the referenced simulator presents a prospective application in the deployment of nanosensors within the HCS to identify biomarkers generated by anomalies, especially malignant tissues. These nanosensors are designed to detect and react to certain biomarkers associated with malignant tissue irregu-

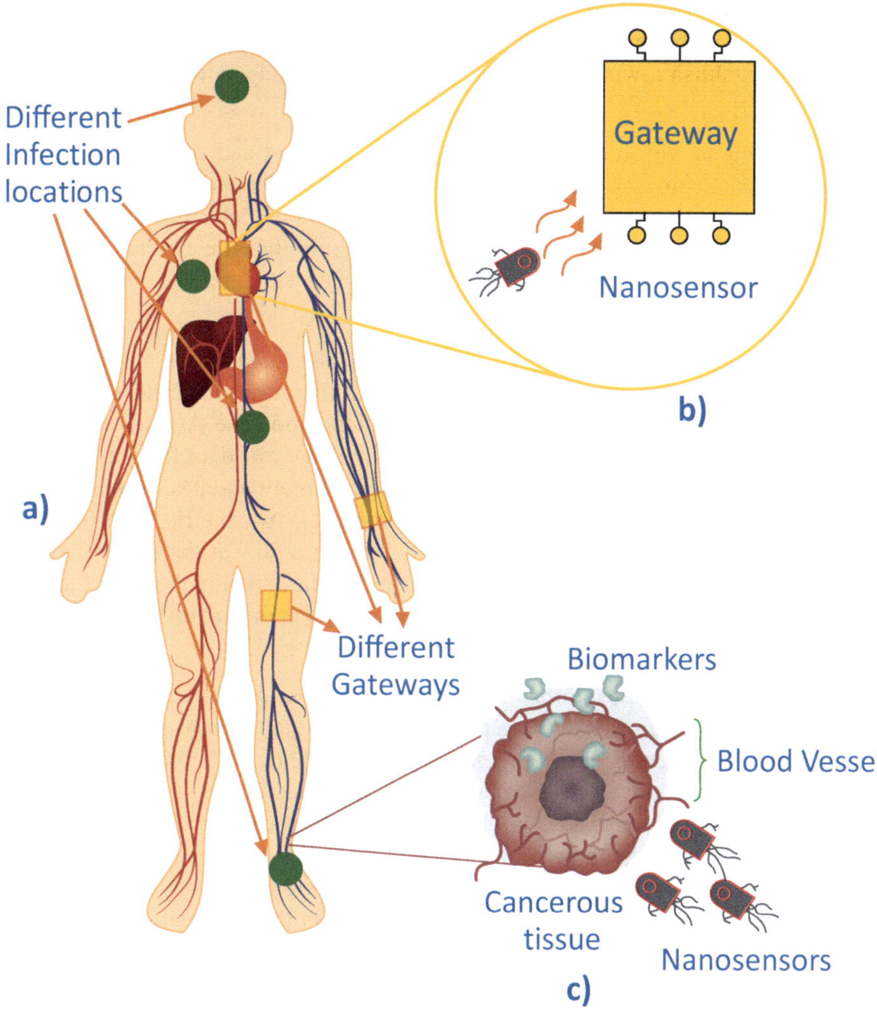

Fig. 2.97 Nanocommunication-based abnormality detection system. (**a**) HCS. (**b**) Communication between nanosensor and the gateway. (**c**) Nanosensors detecting biomarkers at the target location

larities. The injected nanosensors observe the bloodstream, gather data from nearby bodily areas, and transmit their findings to an external monitoring device via a gateway, as seen in Fig. 2.97 [229].

The deployment of IoBNT exemplifies the utilization of a gateway to transmit reported data to healthcare practitioners. A biohybrid implant consists of 3D engineered skeletal muscle that can perform molecular-level sensing and establish a wireless connection according to [229]. The gadget employs a synthesis of 3D-printed scaffolding, an implant antenna, and bioengineered skeletal muscle tissue to facilitate an innovative method of bio-sensing. A target molecule activates the bio-nanosensor, resulting in the reconfiguration of the antenna. The design and performance evaluation of a wireless implant antenna for in-body sensors establish a wireless connection with a wearable antenna (external gateway) [229].

First, it aligns with real-world constraints where cost and resource limitations might make deploying multiple gateways infeasible. Second, it allows us to precisely analyze the impact of a gateway's placement on detection time and ratio, offering insights into optimization. Finally, it simplifies the evaluation process, enabling us to focus on this crucial design aspect without additional complexities [229].

In this context, [229] examines the strategic positioning of the gateway across many sites, with the objective of ensuring that this positioning does not undermine either the detection time or the detection ratio. The Age of Information (AoI) measure is employed to evaluate the freshness of data from a source (in this case, the infection location) and to analyze timely updates, considering the limitations of the MC channel. Utilizing the peak AoI metric, [229] acquires insights into the temporal dynamics of information flow at the receiver (in this case, the gateway). In healthcare contexts, particularly within the dynamic realm of the circulatory system, possessing the most current information is essential for timely reactions to emergent infections or anomalies [229].

Consequently, [229] examines the temporal aspects and detection ratios of the anomalies. The nanosensors are prone to exponential decline over time due to degradation processes in the human body. Thus, the degradation of nanosensors throughout their movement creates ambiguities in identifying anomalies, posing a barrier to the early diagnosis of illnesses, such as cancer. Consequently, it is essential to comprehend the impact of nanosensor degradation on the detection ratio. This understanding is crucial for acknowledging the importance of integrating the lifespan and stability of nanosensors in the formulation of strategies for anomaly monitoring and diagnostics. Therefore, to examine the prompt detection of anomalies, [229] explores the dynamics of AoI in relation to the degradation of nanosensors.

The AoI-based abnormality detection system model essentially includes the following components [229]: (1) Nanosensors: Nanosensors continuously monitor the bloodstream and are engineered to identify specific biomarkers emitted by the site of infection. Upon encountering biomarkers, the nanosensor activates and prepares to submit the finding to the gateway. The detection is recorded at the gateway, and the nanosensor is reset to obtain updated information regarding the infection site [229]. (2) Gateway: A gateway is an on-body computing device placed at different locations on the human body, such as a wearable antenna. It collects data from nanosensors and analyzes this data, alerting external monitoring devices about potential health risks [229]. (3) Communication: The nanosensors transmit data to the gateway within the human body via established intrabody communication links, such as terahertz or ultrasonic channels.

Reference [229] posits that the communication channel is devoid of errors, and they regard the nanosensors as moving passively within the HCS. The nanosensors circulating inside the arteries activate upon identifying the biomarkers emitted by the site of infection. These mobile nanosensors indicate the presence of an infection upon encountering the designated gateway along their trajectory. The authors in [229] employ data produced by the BVS architecture, which models the movement of nanosensors inside the bloodstream throughout the principal vessels of the HCS. Every vessel and organ within the simulator is allocated a unique identifier. The unprocessed data from BVS indicates the global location of nanosensors intermittently traversing vasculature within the HCS [229].

- The NS-3 module BloodVoyagerS (BVS) simulates the global movement of nanobots inside the human body. For BVS result files, there is a cross-platform visualization tool called Blood-VoyagerS Visualizer (BVS-Vis). The simulated nanobots' temporal dispersion inside a three-dimensional cardiovascular model is depicted using BVS-Vis. Rotating, translating, zooming, and navigating through different temporal phases are all possible with this animation. For a quick inspection of simulation data, post-processing can create a heat map showing the concentration of nanobots. BVS-Vis may be accessed as a whole Docker-compose setup or as a webpage. It allows for quick analysis of simulation data and significantly improves workflow with BVS [230].
- Conversely, [231] introduces NanoMedView, a mobile health (mHealth) application. NanoMedView offers and administers a platform for health data derived from an in-body nanonetwork. NanoMedView advances the application of nanotechnology in medicine, providing developers with a platform for in-body communication data transmission and facilitating interactions between patients and physicians. The initial phase situates the NanoMedView concept within the technical framework of the IoNT, delineates the anticipated functionalities, and introduces the fundamental network architecture for message transmission as outlined in [231]. The subsequent description outlines the distinct processes in the user centered design (UCD)-based development process to elucidate the resultant application [231].

The notion of nanodevices and networks is not universally applicable. Nanodevices can be classified as biological, artificial, or hybrid structures. For medical applications, however, their size should not surpass 5 μm; otherwise, they will be unable to traverse capillaries, the tiniest blood vessels in the human body. In numerous scenarios, it is expected that highly resource-constrained nanodevices will navigate the human bloodstream or embed themselves within it to function as sensors that measure and report values, act as actuators that release active substances, or, for instance, aggregate around malignant cells, such as tumors, to isolate and eliminate them. Nevertheless, actuation decisions are typically not straightforward for a single nanodevice, as most medical diagnostic procedures necessitate quantitative measurements rather than qualitative assessments based on a solitary particle. The concentration of the substance is critical, requiring multiple independent measurements from various nanosensors. The

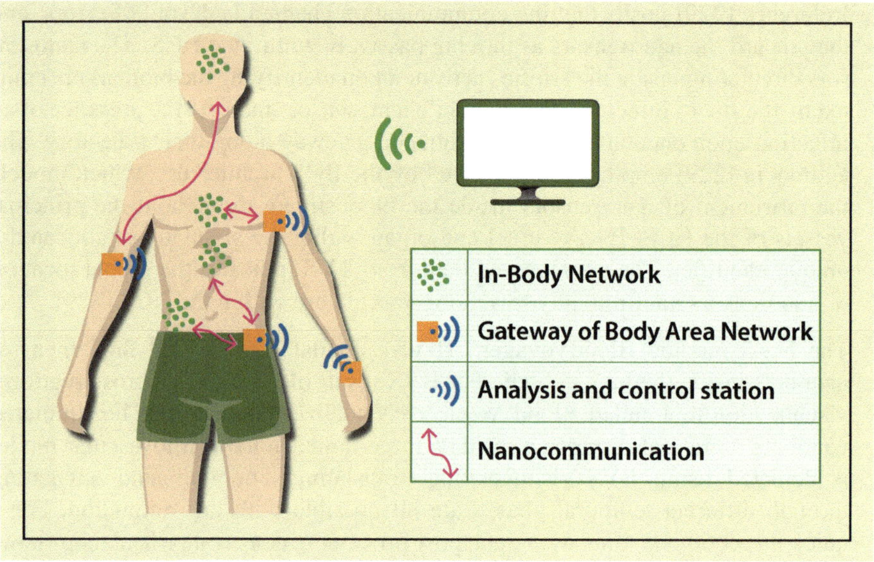

Fig. 2.98 An IoNT network architecture connects nanodevices with gateways to access an external control station

decision to act may also rely on various chemical concentrations, additional patient data, or may be overly complex for a singular, resource-constrained nanodevice [231].

Therefore, in order to aggregate the data collected in the body and process it on-site in a majority consensus decision or forward it for further analysis to more complex devices, databases, and finally present it to human users or medical experts, communication processes are required for the processing of diagnostic data and the implementation of any instructions for action. In this manner, data from nanodevices is transmitted to analysis and control stations like smartphones from an in-body network through gateways, which are more potent stationary microelectronics in or on the body (Fig. 2.98) [231].

This analysis and control station's software instance is called NanoMedView (Fig. 2.99) [231]. It gathers up-to-date measured data from the in-body nanonetwork and provides it to users through a GUI, therefore informing them of acute harmful conditions or their current health status. Pure blood values may be included in this data, but it may also be enhanced with useful metadata that is provided by the application or the data itself. For instance, blood readings can be given together with useful action recommendations in the event of overstock or undersupply, along with limit information for both [231].

- Conversely, in certain circumstances, wireless communication by EM radiation is impractical or too hazardous. MCs operate by the transmitter emitting particles, therefore presenting a possible alternative. A carrier media, such as gas or liquid, transports particles through space. Particles are detected by receivers,

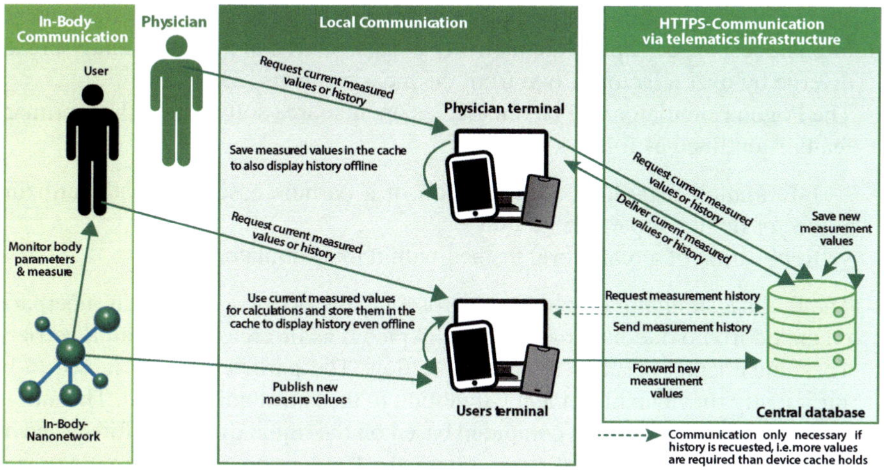

Fig. 2.99 System architecture of the distributed system in which NanoMedView acts as a distribution-transparent component and monitoring system on any mobile end device

which retrieve the sent data. In a large-scale industrial MC system, pre-existing pipe systems may be utilized for the injection of magnetic particles. Existing simulations for this either fail to account for the intricate flow within the pipe system or exhibit suboptimal performance [232].

The "Pogona" simulator, published in [232], CFD to model liquid flow within the piping system. The flow is input into the simulator to forecast particle movement. Furthermore, models for particle injection and sensing are proposed. A scene approach is presented to enhance performance and flexibility by enabling the design of scenarios comprising separate elements. The prototype of Pogona is examined using basic scenarios, which are then compared to analytical outcomes and established effective implementations. Simulation scenarios that correspond to an established testbed were developed, and their results were compared to empirical measurements. A minimal level of consensus was attained, indicating potential for enhancement. My investigation indicates that the interpolation algorithm utilized for the CFD flow output is significant. Nonetheless, identifying appropriate interpolation algorithms continues to be an unresolved research challenge [232].

Pogona employs transmitter and receiver models, accommodates changeable scenes, and utilizes computationally efficient data structures. The accuracy of the simulator was effectively verified for simplified situations with established outcomes. The fluid simulation result was also validated. Furthermore, the integration algorithm was evaluated against a validated implementation. Measurement data for an experimental testbed in Erlangen was acquired for validation reasons. The testbed employs magnetic particles introduced into a water tube, which are detected by a susceptometer at the receiver. The testbed was simulated, and its output was compared to the susceptometer data. A minimal concordance was attained between simulator output and empirical findings. While the channel impulse response shape is qualitatively consistent, the anticipated delay spreads diverge by over a factor of two from the measurements [232].

- The Pogona simulator will be released as open-source software, with its primary features outlined as follows [233]:

  - Information regarding our execution of a particle-based simulator utilizing vector fields for particle motion
  - Replication of a real-world testbed within this simulator

  The Pogona simulator employs a microscopic methodology in which superparamagnetic iron oxide nanoparticles are depicted as discrete information carriers. The fundamental premise is ideal advection: The particles are presumed to be sufficiently tiny to exhibit identical motion to the surrounding liquid. The mobility of particles can then be computed based on the liquid dynamics forecasted by fluid simulation. The fluid flow is precalculated using a CFD simulator and exported as a vector field. At each simulation time step, the flow at the particle's present location is assessed and utilized to direct the particle's movement to its new place. Information is transmitted across a network of tubes by generating particles at the transmitter and detecting their existence at the receiver [233].
- Moreover, it is essential to note that contemporary population genomic datasets possess significant potential for elucidating the evolutionary processes occurring in natural populations; however, a vital prerequisite for achieving this objective is the capacity to model realistic evolutionary scenarios and forecast their anticipated patterns in genomic data. SLiM 5, as reported in [234], is an evolutionary simulation framework that integrates a robust and rapid engine for forward population genetic simulations with the capacity to model many complicated evolutionary scenarios. SLiM attains this versatility via scriptability, enabling control over numerous facets of the simulated evolutionary scenarios through a straightforward R-like scripting language known as Eidos. SLiM features a GUI for constructing simulations, interactive runtime management, and dynamic presentation of simulation results, hence enabling efficient model building through rapid prototyping and visual debugging [234].
- Additionally, concerning software development for multicellular systems, Reference [235] presents a modeling and simulation framework designed for extensive parallel simulation of multicellular systems, incorporating spatially resolved stochastic kinetics at the individual cell level. Through the application of operator-splitting the authors in [235] separate the simulation of reaction-diffusion kinetics within cells from the simulation of molecular interactions between cells at their edges. This decoupling utilizes the intrinsic scale separation in the foundational model to facilitate extensively horizontally scalable parallel simulation, appropriate for execution on heterogeneous, distributed computing infrastructures, including public and private clouds. This framework's modular structure enables the integration of any existing single-cell simulation software with any cell-signaling simulator [235].

- The research in [236] is crucial for understanding this illness through MCs, especially in light of the tremendous progress mankind has made throughout the epidemic in recent years. Airborne pathogen transmission mechanisms play a critical role in the spread of infectious diseases, such as COVID-19. In order to model and statistically investigate airborne pathogen transmission through pathogen-laden particles in turbulent channels from an MC viewpoint, Reference [236] suggests a CFD technique.

  The Reynolds-averaged Navier-Stokes (RANS) equations are used in combination with the realizable $k{-}\varepsilon$ model and the discrete random walk model to represent the turbulent dispersion of droplets and aerosols and the turbulent flows produced by coughing, respectively. A CFD simulator runs simulations to obtain statistical information about the number of particles received. To ascertain the likelihood of infection and to statistically characterize the turbulence effect on reception, the data undergo post-processing. As demonstrated by a multimodal distribution that is a weighted sum of the normal and Weibull distributions, the results show that turbulence has an uneven impact on the risk of infection. Furthermore, depending on the air velocity, the turbulent MCs channel is defined by a multimodal distribution, namely a sum of weighted normal distributions or stable distributions [236].

  One important way that pathogens (such as bacteria and viruses) spread infectious illnesses like COVID-19 and influenza is by airborne transmission. Infectious illnesses spread by respiratory activities such as speaking, coughing, sneezing, and breathing that release pathogen-laden particles (droplets and aerosols). Because of the increased initial emission velocity of particles from the mouth, coughing and sneezing produce turbulent flows. Although there are analytical approaches in the literature on fluid dynamics, these models frequently oversimplify the effects of turbulence. Although they come with a higher computational cost, CFD models are often used to simulate airborne transmission through droplets, improving the realism of turbulent flow and airflow-particle interactions [236].

  Additionally, Reference [236] presents a CFD methodology to simulate the airborne transmission of cough droplets and aerosols, alongside turbulent fluxes between an infectious individual (transmitter, TX) and a vulnerable individual (receiver, RX). This method models the channel as a turbulent two-phase flow medium, consisting of turbulent airflows (continuous phase) and particle movements (discrete phase) that interact with these airflows and RX. The turbulence is represented using the RANS equations in conjunction with the realizable $k{-}\varepsilon$ model, while the turbulent dispersion of particles is monitored in the CFD simulator through the discrete random walk model.

  Extensive CFD simulations are conducted to observe and characterize the impact of turbulence on airborne transmission. Consequently, statistical data about the acquired particles are gathered and utilized to calculate the risk of infection. Statistical research in [236] indicates that estimating the impact of turbulence on infections is complex, as the probability density function (pdf) of received particles is represented by multimodal distributions, namely a weighted combination

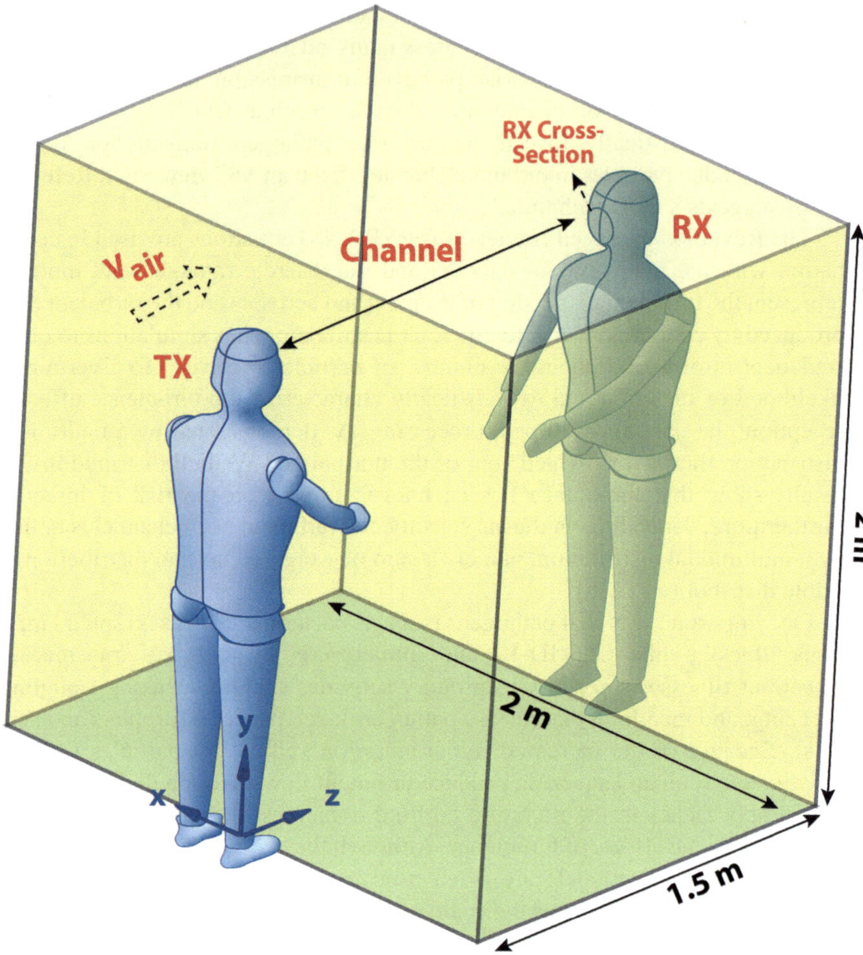

Fig. 2.100 Airborne pathogen transmission scenario between two humans

of normal and Weibull distributions. Furthermore, it is demonstrated that aerosols are more efficacious than larger droplets at a distance of 1.5 m, and an increase in ambient airflow elevates the likelihood of infection. Furthermore, the end-to-end impulse response of the turbulent air-based MCs channel is delineated. While for lower air velocities ($v_{air} = 0.1\frac{m}{s}$ and $v_{air} = 0.3\frac{m}{s}$), its pdf is multimodal, i.e., weighted sum of normal distributions, it has a uni-modal stable distribution for $v_{air} = 0.5\frac{m}{s}$ [236].

The aforementioned research, namely in [236], elaborates on a 3D system model depicting a scenario in which the transmitter emits pathogen-laden spherical particles through coughing toward the receiver in a room with airflow, as represented in Fig. 2.100. As seen in Fig. 2.101, particles emitted from the trans-

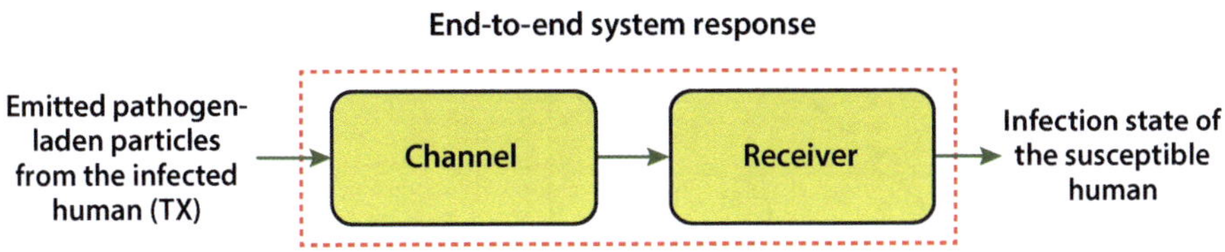

Fig. 2.101 End-to-end system model

mitter (TX) are regarded as an impulsive input signal that propagates through the channel and is detected by the receiver (RX). Therefore, the system's end-to-end response is represented as the output, specifically the infection status of the susceptible individual. The comprehensive system model encompasses the receiver model, the turbulent two-phase channel model, and the characterization of emitted particles, as detailed in [236].

Finally, in [236], a CFD method that incorporates turbulent airflows and the turbulent dispersion of droplets and aerosols is used to mimic the airborne transmission of viruses generated by coughing between two people. Turbulence produces multimodal distributions for the probability density function of particles acquired by susceptible individuals, according to statistical data derived from CFD simulations. Furthermore, numerical results in [236] show that higher air velocity increases the risk of infection. Epidemiological models may effectively incorporate indoor airborne transmission by using expressions of derived likelihood of infection. The air-based turbulence MCs system's overall system response is described. With the exception of greater ambient air velocities, the system response displays multimodal distributions. The realistic design of MC systems with turbulent flows may be achieved by the characterization of turbulent MCs channels. Future research will broaden this analysis to include respiratory and mask settings [236].

- Continuing with another scenario, [237] examines the issue of receiver modeling in MCs systems. Thus, Reference [237] examines the diffusive MC channel between a transmitting nanomachine and a receiving nanomachine within a fluid medium. The information molecules emitted by the transmitter nanomachine into the environment may undergo deterioration in the channel through a first-order degradation reaction, while those that arrive at the receiver nanomachine can engage in a reversible bimolecular reaction with receptor proteins of the receiver.

Consequently, [237] differentiates between two possibilities. In the initial scenario, it is presumed that the complete surface of the receiver is adorned with receptor molecules. A closed-form analytical expression for the predicted received signal at the receiver has been derived, specifically the anticipated number of active receptors on the receiver's surface. In the second scenario, the situation is examined when the quantity of receptor molecules is finite, and the evenly distributed receptor molecules only partially cover the receiving surface.

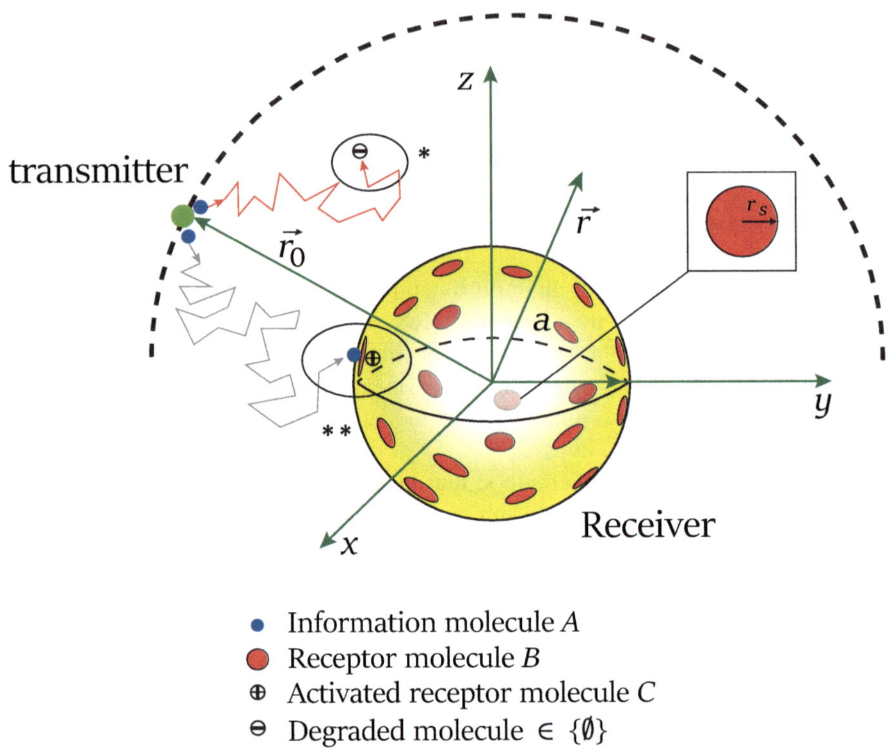

Fig. 2.102 Schematic diagram of the considered system model, where receiver and transmitter are shown as a yellow sphere in the center of the coordinate system and a green dot, respectively [The surface of the receiver is partially covered with type B receptor molecules, shown in blue color, where each receptor is modeled as a circular patch with radius r_s. Two sample trajectories of an A information molecule (denoted by a blue dot) that result in a degraded molecule (circle with a "−" inside) and an activated receptor molecule C (circle with a "+" inside) are shown as red and grey arrows, respectively, where * y ** obeys Eqs. (2.47) and (2.48).]

[237] indicates that the anticipated received signal for this situation can be precisely approximated by the expected received signal of the first scenario, with suitable adjustments to the forward reaction rate constant. The precision of the obtained analytical results is validated using Brownian motion particle-based simulations of the examined environment, demonstrating the influence of receptor occupancy on the generated analytical outcomes [237].

Thus, in [237] is introduced the system model considered with an unbounded three-dimensional fluid environment with constant temperature and viscosity. It is assumed that the transmitter is a point source placed at position $\vec{r_0} = (x_0, y_0, z_0)$ and the receiver is spherical in shape with a fixed radius a and placed at the center of the coordinate system, as it is indicated in Fig. 2.102. Besides, it is also assumed that the transmitter uses one specific type of molecule, denoted by A, for sending information to the receiver. Hence, information molecules are referred as A molecules. Furthermore, another assumption is that the transmitter instantaneously releases a fixed number of A molecules, N_A, at time $t_0 = 0$ into the environment. After their release, the information molecules diffuse in the environment into all directions with constant diffusion coefficient D_A. Also, other assumptions consist of the diffusion processes of different information molecules that are independent of each other, and the A molecules can be degraded throughout the environment via a reaction mechanism of the form [237]:

$$A \xrightarrow{k_d} \varnothing \tag{2.47}$$

where k_d is the degradation reaction constant in s^{-1}, and $\varnothing$ is a species of molecules which is not recognized by the receiver. Equation (2.46) models a first-order reaction but can also be used to approximate higher-order reactions [237].

Some of the A molecules released by the transmitter may reach the receiver surface and *reversibly* react with a B molecule to form an activated receptor (also referred to as ligand receptor complex molecule), denoted by C, via a reaction mechanism of the form [237]:

$$A + B \underset{k_b}{\overset{k_f}{\rightleftarrows}} C \tag{2.48}$$

where k_f and k_b are the microscopic forward reaction constant in $molecule^{-1}m^3s^{-1}$ and the microscopic backward reaction constant in s^{-1}, respectively [237].

In nature, different cell-surface receptors produce different responses inside the cell via different signaling pathways once they are activated. Then in [237] is considered the formation of the C ligand-receptor complex molecules as the *received signal* at the receiver, which could be used for detection of information sent by the transmitter. Two cases regarding the coverage of the receiver surface by receptor B molecules are analyzed. In the first case, this is done in order to make the analysis of the expected received signal at the receiver analytically tractable. In the second case, it is relaxed assumption, mentioned in the following paragraphs [237].

A models each receptor individually. In particular, it is assumed that the receiver surface is *partially* covered with M uniformly distributed receptor B molecules, where it is modeled each receptor B molecule as a circular patch with radius r_s. However, in order to keep the analysis mathematically tractable [237].

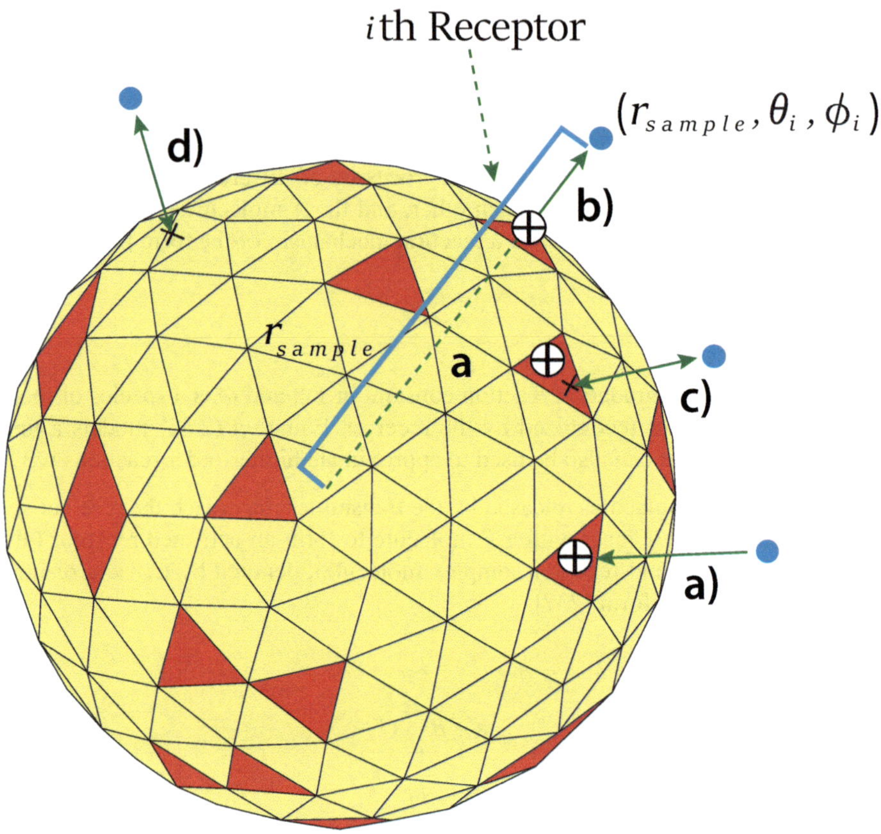

Fig. 2.103 Schematic diagram of the considered triangle mesh grid for constructing the receiver, where the receptor and non-receptor parts of the receiver surface are shown with red and yellow triangles, respectively. The A information molecules and activated receptor C molecules are depicted as red circle and white circle with a "+" inside, respectively. One sample displacement of an A molecule that leads to the production of a C molecule is shown as an arrow marked with (**a**). The production of an *A* molecule after occurrence of a backward reaction is shown as an arrow marked with (**b**). The reflections of an *A* molecule upon contact with an occupied receptor and the part of the receiver that is not covered by receptors is shown with double arrows marked with (**c**) and (**d**), respectively

To continue examining the MCs systems presented in [237], a Brownian motion particle-based simulation is conducted, wherein the exact positions of all individual molecules are monitored throughout the simulation environment. The implemented simulation technique progresses in discrete increments of Δt seconds. To collaboratively replicate the reactions, specifically Eqs. (2.47) and (2.48), together with the diffusion of molecules, the simulator referenced in [237] employs a variety of methods to address first-order and second-order reversible reactions. Additionally, a triangle mesh grid is utilized for the construction of the receiver's surface in [237] (Fig. 2.103). Before the simulation, M

triangles, seen in red in Fig. 2.103, are randomly selected from all triangles forming the surface of the receiver, representing the set of M receptors.

- The integration of microfluidics and MCs has attracted considerable interest within the framework of the growing IoBNT paradigm. Microfluidic systems provide distinct functionalities for the manipulation of fluids at the nanoscale, facilitating exact regulation of molecular interactions. Potential applications of microfluidic MCs in the IoBNT encompass healthcare, environmental monitoring, and biotechnology [198]. The MPPIC solver in OpenFOAM, a prevalent open-source tool for CFD simulations, was first designed for applications in sedimentation, separation, and granular flows. Conversely, the viability of employing the MPPIC solver in OpenFOAM for simulating microfluidic MCs within the framework of the IoBNT is examined in [198].

  Reference [198] presents a simulation framework for investigating the complex interactions between microfluidics and MCs; the analyzed OpenFOAM simulation methodology may serve as a basis for future research in microfluidic MCs simulation. Specifically, [198] contrasts its OpenFOAM simulations with an analytical model and a particle-based simulation methodology. In a uniform particle release on the transmitter side and a flow-dominated environment, simulation results in [198] indicate that the CIR from both the analytical and particle-based approaches closely resembles the results from OpenFOAM simulations. As the shape extends in the direction of fluid flow, the OpenFOAM simulations align with the analytical model, particularly near the tail of the CIR. Furthermore, [198] juxtaposes the OpenFOAM simulation outcomes with empirical measurements obtained from a fluidic MCs testbed at the millimeter scale. Researchers in [198] ascertain that the OpenFOAM simulation outcomes closely align with the experimental observations for the flow-dominated domain, exhibiting a root-mean-square error (RMSE) below 10% [198].

  Hence, [198] investigates the use of OpenFOAM for the simulation of microfluidic MCs. More specifically, [198] investigates the MPPIC solver, which is well-known in the OpenFOAM community as a two-phase solver. The main contribution in [198] is to investigate OpenFOAM with the MPPIC solver as a new tool for the simulation of microfluidic MCs. Thus, the simulation results with OpenFOAM uses the two-phase MPPIC solver for a flow-dominated regime with an analytical channel model and with Particle-Based Simulations (PBSs). Specifically, the error between the simulation results with OpenFOAM (using MPPIC) and the analytical model as well as the PBS are evaluated. In addition, [198] utilizes an experimental testbed setup for conducting experimental measurements, which is compared to the OpenFOAM simulation results that are obtained using the MPPIC solver in the millimeter scale. To facilitate the use of OpenFOAM with the MPPIC two-phase solver for microfluidic MCs simulations, [198] provides a publicly accessible repository. This repository offers a template (folder structure of scripts) for conducting microfluidic MCs simulations using the Open-FOAM MPPIC. In addition, the repository provides Python code for the post-processing of OpenFOAM simulation results [198].

As mentioned previously, the use of MPPIC in [198] represents a *two-phase* solver for microfluidic MCs simulation. A *one-phase* solver, such as icoFoam, is designed to handle the dynamics of a single phase, e.g., a fluid phase, focusing on solving the incompressible Navier-Stokes equations. *One-phase* solvers are suitable for fluid flow simulations where interactions with other phases are absent. Conversely, a *two-phase* solver, such as MPPICFoam, can simulate the interactions between two distinct phases. A two-phase solver can accurately account for interaction phenomena, e.g., fluid-solid interactions, making it a powerful tool in CFD simulations [198].

Therefore, a *two-phase* solver is more appropriate for microfluidic MCs simulations (than a one-phase solver), especially when the interactions between the fluid and solid phases become relevant. As groundwork toward enabling simulation of microfluidic MCs simulation with phase interactions, [198] introduces and evaluates OpenFOAM with the two-phase MPPIC solver for microfluidic MCs simulation in this methods article. The authors in [198] do not examine specific phase interactions; rather, they enable the simulation evaluation of microfluidic MCs with such phase interactions [198].

To verify simulation results, i.e., to compare the OpenFOAM simulation results to experimental testbed measurements, Fig. 2.104 compares the experimentally obtained and simulated data using OpenFOAM in [198]. They consider the flow-dominated regime. Clearly, the two CIRs are close to each other, and the peaks match well. There are slight differences in the course of the tail. The experimentally obtained data exhibit a higher number of received particles, especially in the tail's run-out region, compared to the simulation results. The difference can be quantified with an RMSE of 9.89%, which signifies that the average difference between the experimentally obtained data and the simulation data is below 10% of the data range [198].

- The concept of a complex framework, as stated in [238], is primarily intended to represent and mimic the structures and processes of the human body, as well as the interactions of nanobots with it. Medical nanobots are anticipated to operate within the body, such as identifying and eliminating cancer cells. They are typically assigned a significant role in a future health system centered on precision medicine. It is imperative to simulate the application of nanobots within the human body prior to their real deployment. Nonetheless, only with a full-body simulation architecture, such as the Medical Holistic Simulation Architecture (MEHLISSA), can significant results be attained. Since [238] accurately simulates the human body, it enables dependable assessments about the efficacy and efficiency of nanobot utilization in vivo.
- The design, implementation, and evaluation of a general-purpose simulation platform for multicellular MCs systems are reported in [239], referencing [2, 240]. The platform, constructed on an agent-based model, provides the capabil-

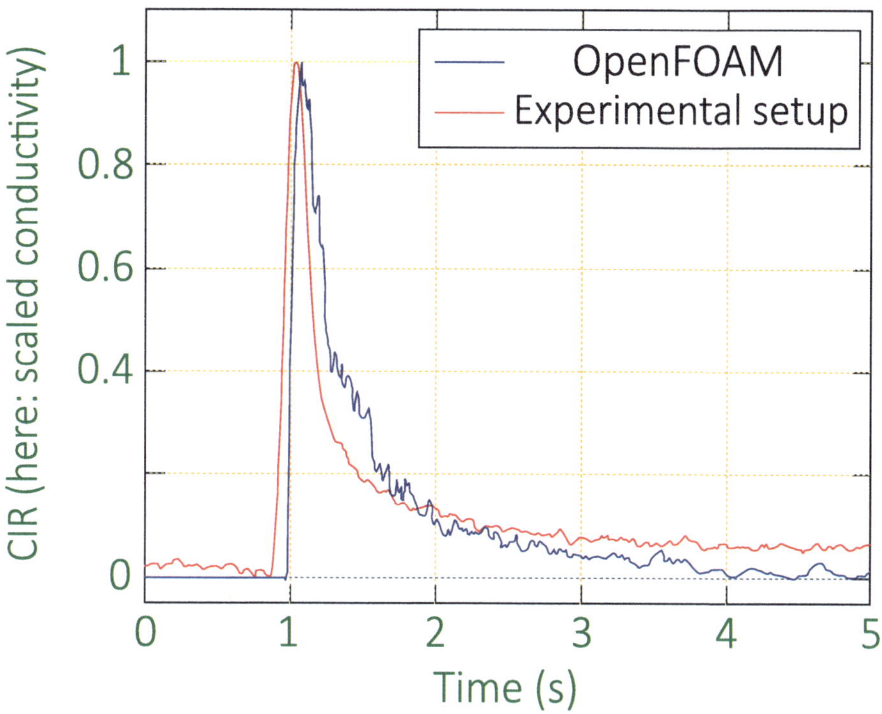

Fig. 2.104 Comparison of the OpenFOAM simulation results versus the measurement data obtained from the experimental testbed setup for a background flow rate of 3.1 mL/min, i.e., flow-dominated regime

ity to simulate various multicellular systems, including cancer spheroids and vascular-like networks. It integrates efficient algorithms, including the Cell-List and Barnes-Hut methods, for computing cell-cell interaction forces and accommodates dynamic processes, such as cell division, growth, and apoptosis. The simulation platform is an essential instrument for enhancing study in multicellular systems and comprehending their collective behavior. This simulator is developed in C++, selected for its superior processing speed. Large-scale simulations entail a significant time expenditure and necessitate great processing speed. Moreover, the aforementioned simulator employed Python as the visualizer due to its prevalence as a programming language and the availability of various open-source modules, including those for data visualization.

The simulator in [239] employs an agent-based approach, where cells represented as discrete entities or agents interact with one another and exhibit collective behavior. A key feature of this simulator is the modeling of cell movement driven by cell-cell interaction forces, making the computation of these forces a critical aspect. To address this, it provides multiple algorithms for calculating these forces. The simulator is designed with flexibility in mind, enabling the simulation of a wide variety of multicellular MCs systems.

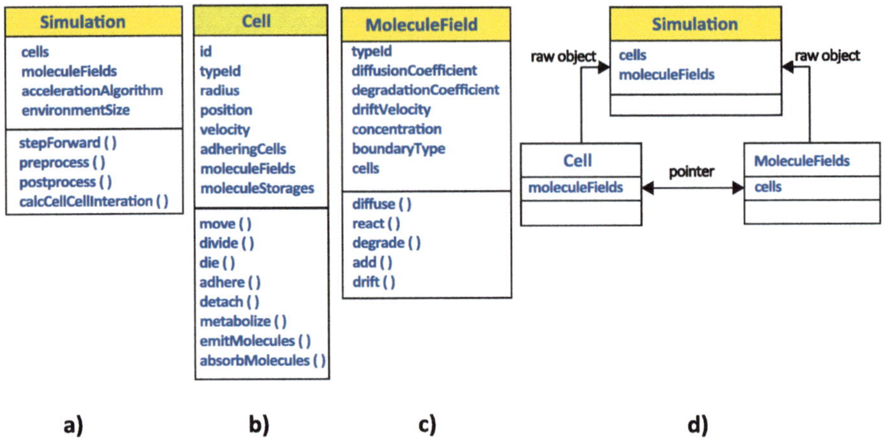

Fig. 2.105 (**a**) The *Simulation* class. (**b**) The *Cell* class. (**c**) The *MoleculeField* class. (**d**) Relationships between the cited classes

- Hence, in [239], a multicellular MC system is defined as a group (or groups) of biological or synthetic cells that interact through MC. These systems can be naturally occurring in biological organisms or artificially engineered for specific applications. Examples of multicellular MCs systems include the following:
  - Calcium signaling systems: Glial cells propagate calcium signals between each other via cell-cell junctions.
  - Vascular networks: Endothelial cells diffuse vascular endothelial growth factor (VEGF) to attract one another, leading to the formation of a vascular network.
  - *Dictyostelium* slugs: *Dictyostelium* cells adhere to each other and use cyclic adenosine monophosphate (cAMP) signaling to form a slug that moves through its environment.
  - Tumor spheroids: Cancer cells grow and form a three-dimensional spherical structure through biochemical and biophysical interactions with each other and their microenvironment.
  - Pattern formation systems: Bacterial cells communicate by diffusing acyl-homoserine lactone (AHL) molecules, creating spatial patterns of gene expression.
  - Cell-based drug delivery systems: Drug-carrying cells communicate to locate tumors and deliver therapeutic molecules to target locations.
  - Cell-based computational systems: Engineered cells use diffusive molecules to perform Boolean logic computations.

  In this context, to develop a general-purpose simulation platform, in [239], an object-oriented approach is employed with designs for simulating multicellular MC systems using three main classes: *Simulation*, *Cell*, and *MoleculeField* classes (Fig. 2.105c) with their relationships illustrated in the indicated figures.

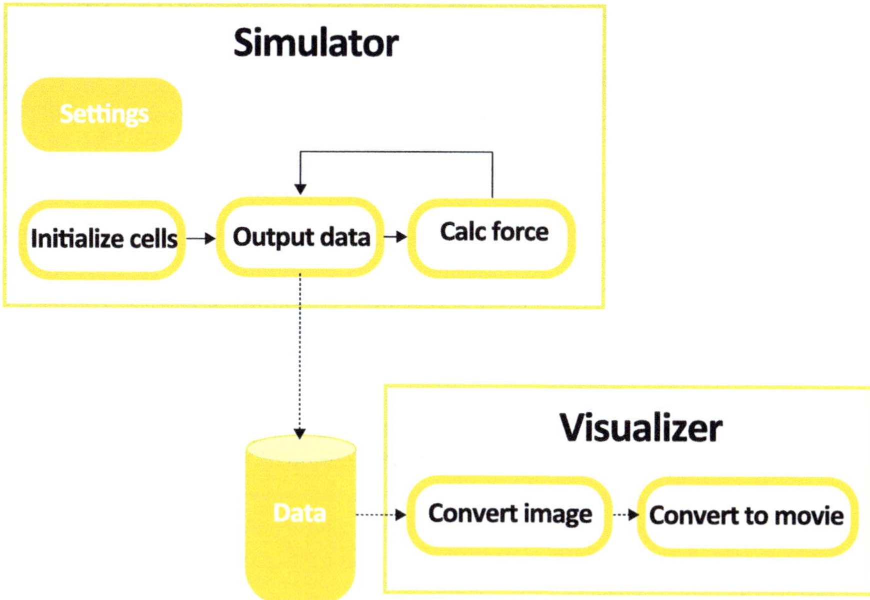

Fig. 2.106 Development of the simulation platform (The blue-framed rectangles denote the primary procedures within the simulator or visualizer. The blue-filled rectangle denotes the configuration area for parameter values, whereas the cylinder symbolizes data. The solid arrows denote processing processes, whereas the dashed arrows indicate data flows. The user-editable processes are populated with dots. Users may modify the settings as well.)

As mentioned above, the architecture of the simulator in [239] is based on the object-oriented design, where cells are represented as agents that interact with one another and their environment to exhibit collective behavior. Such multiagent cell simulators, as demonstrated in the field of biology, are highly flexible and extensible. They allow users to model various interactions, including cell-cell and cell-environment interactions, as well as dynamic cell behaviors and environmental changes.

As mentioned previously, a key feature of simulator in [239] is its ability to model and compute cell movement based on simple cell-cell interaction forces. This simple force-based model reflects a stepwise approach to understanding complex systems. Starting with a simple model allows us to focus on uncovering the fundamental rules governing system behavior [239].

Such force-based models have been shown to successfully replicate key aspects of biological systems, as demonstrated in studies of *Caenorhabditis elegans* embryogenesis, *Dictyostelium discoideum* cell sorting, cancer cell network formation, cancer cell spheroid growth, and angiogenesis. Understanding the behavior of a simple model provides a strong foundation for progressively introducing additional complexity. Figure 2.106 shows the design of the simulation platform, which consists of three main modules: a simulator, a visualizer, and data storage. The simulator is capable of simulating multicellular MC systems employing significant algorithms. The data storage retains simulation results in text format. The visualizer imports simulation results from the data repository and produces movie files that depict cellular movement. The platform design features a decoupled simulator and visualizer, allowing users to visualize simulation findings with their own visualizer or import them into other systems for additional analysis [239].

The simulator comprises three primary procedures: cell initialization, data output, and cell mechanics computation. The cell initialization process imports parameter values from the configuration file to execute simulations. The parameters encompass the dimensions of the simulation field, the duration of the time step, and the beginning cell count. The procedure establishes the initial cell placements based on these parameters. It additionally initializes the cells' state variables to the default values provided in the settings file. The data output procedure produces an output file that contains simulation results, including the positioning information of cells, in text format, and saves the file in data storage. The cell mechanics computation algorithm determines the forces exerted on each cell from cell-cell interactions, modifies the cell's velocity, and updates cell locations [239].

The visualizer functions as an effective instrument that aids users in comprehending simulation outcomes. The visualizer comprises two primary functions: one for transforming text data from storage into a sequence of images (PNG files) and the other for converting a sequence of images into a video file (MP4 format). The visualizer is independent of the simulator, allowing users to create and utilize custom visualizers tailored to their research needs. The data repository archives simulation outcomes as text files. A text file encompasses data regarding the cells (e.g., state variables of the Cell class in Fig. 2.105) at a particular moment in time [239].

A simulation is conducted as illustrated in Fig. 2.106. The simulator initially imports parameter values from the Settings file. The simulator subsequently invokes the Initialize cells method. Throughout the simulation, the simulator invokes the Output Data process depicted in Fig. 2.106 at a specified output interval to produce data files and store them in the data repository. The simulator invokes the *Calc force* algorithm at each simulation step. Upon conclusion of the simulation, the visualizer invokes the turn to image method to transform text files into image files (PNG format), and the Convert to movie process to turn image files into a movie file [239].

- In [241] is presented BSim 2.0, a radically new version of Bacteria Simulator (BSim), a computational Agent-based models (ABM)[46] framework for modeling bacterial dynamics in standard experimental settings, including microfluidic che-

[46] ABMs provide a number of advantages relative to traditional continuum modeling approaches, permitting incorporation of great detail and realism into simulations, allowing in silico tracking of single-cell behaviors and correlation of these with emergent effects at the macroscopic level [237].

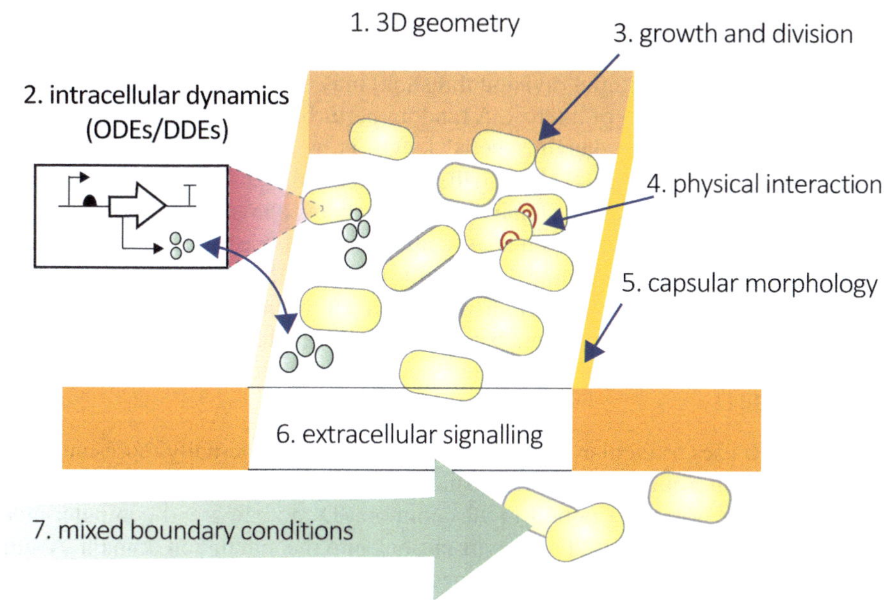

Fig. 2.107 New features of BSim 2.0

mostats. This is achieved by the use of novel methodologies, including cells with capsular geometry that may physically and chemically interact, a realistic model of cellular proliferation, a delay differential equation solver, and authentic environmental geometries [241].

- BSim 2.0 is an innovative agent-based modeling tool for bacterial populations that incorporates all functionalities of BSim4 while substantially enhancing its capabilities by including the following features [241]:
  - Intracellular dynamics characterized by delayed differential equations (DDEs) and ODEs.
  - Realistic 3D cell shape executed by growth, division, and intercellular contact.
  - Authentic physical properties of the three-dimensional simulation environment, including microfluidic chemostat attributes if necessary (Fig. 2.107).

The open-source BSim 2.0 framework's architecture is developed in Java and is intended to be modular, providing an extensible base of common components for simulating common bacterial behaviors associated with cellular development. Cells are created as separate Java objects with per-cell methods that allow properties to be changed based on physical interactions and internal gene-regulatory network dynamics to be integrated. According to the literature, a per-cell ordinary differential equation model of rod elongation with time may be used to explain capsular cell proliferation. Division takes place once the progenitor cell has risen over a fixed threshold length. In accordance with the global simulation time step, the cell length is calculated using an Euler technique [241].

In addition to the cell radius, minimum and maximum length, and elongation rate being chosen according to the traits of the species being imitated, the beginning cell length and maximum division threshold may be precisely adjusted to match the biological system of interest. A random perturbation with an amplitude ranging from ±5% to the daughter cells' length is used to disturb axial symmetry during division by slightly altering the daughter cells' positions and lengths. The mechanics of E. coli cells, which are commonly used in microfluidic studies, are represented as capsular volumes. According to available research, cell-cell contact forces are computed using a semirigid-body technique. An overlap-dependent volume exclusion force is computed for each pair of intersecting cells to account for any nonphysical crossovers between cells. To reduce overlaps at every junction, all cell locations are simultaneously modified using the computed forces [241].

BSim 2.0 uses triangular-element meshes or boxes to internally represent geometry, such in a microfluidic chemostat. The Wavefront object format, which is available as an export option in all common 3D modeling and computer-aided design tools, may be used to import meshes into the simulation domain. A similar rigid-body approach to cell-cell mechanics is used to implement cell-geometry interactions using methods developed in literature. Possible crossings between geometry and cells are assessed at each time step. When overlap occurs, the position constraint relaxation phase outlined in cell-cell mechanics [241] applies an exclusion force to the cells that is proportionate to the interpenetration distance.

The definition of true population-level variation in system properties and the interplay between intracellular and extracellular mechanics is made possible by BSim 2.0. In contrast to existing ABM for cell populations, BSim 2.0 allows for the representation of gene-regulatory networks using delayed differential equations [241].

Besides, in [242] is designed and engineer multicellular consortia with spatial segregation to realize an 8-CSK transceiver—covering both modulation and demodulation—in a fully biological framework using simple and reusable single input single-output cells. The mentioned proposal design demonstrates the scalability of this framework while maintaining the advantages of distributed computation, minimal genetic manipulation, and signal orthogonality. Additionally, in [242] is derived the mathematical framework of 8-CSK based on modular building blocks, enabling an accurate theoretical characterization of the system. Simulation results from the agent-based simulator—BSim—validate the feasibility of their extended design and demonstrate strong agreement with the theoretical analysis, highlighting the robustness and applicability of their CSK framework to higher-order modulation schemes.

- In [242] is proposed and analyzed an octuple Concentration Shift Keying (8-CSK) system implemented using spatially segregated multicellular consortia. The proposed framework leverages reusable engineered cells that emulate fundamental logic functions, including ID gate, NOT gate, and Thresholding unit and cascaded them, together with a propagation block, to form modular transmitter and receiver compartments. To systematically analyze the system, in [242] an analytical framework is developed that decomposes it into four interconnected building blocks, enabling a tractable analysis of molecular signal propagation and detection. Furthermore, an agent-based simulation, coupled with a particle-based approach, is performed to assess the accuracy of the developed analytical expressions. The simulation results obtained from BSim demonstrated that the system exhibited the expected dynamic response to input signals, aligning closely with the theoretical analyses [242].

- Gro is a cellular programming language developed in Klavins Lab for simulating intercellular communication and colony expansion [243]. This program simulates microbial consortia and multicellular biocircuits as a synthetic biology prototyping tool. As a result, reference [243] present several improvements to increase the simulator's effectiveness, make it easier to use, and provide further features. Compared to the original version, the modified gro is between one and two orders of magnitude faster. In around ten minutes, it may grow microbial colonies with concentrations as high as 10^5. Using a new pushing algorithm, a new library called *CellEngine* improves the precision of spatial physical interactions between proliferating and dividing cells. CellPro is a genetic library based on Probabilistic Timed Automata that uses computationally efficient and simplified digital proteins to describe the kinetics of gene expression.

  Furthermore, based on the idea that proteins affect biological activity, reference [243] suggests a more effective language defining layer, *ProSpec*. Another library, *CellNutrient*, integrates nutrient absorption and growth activities based on Monod. A collection called *CellSignals* was created to improve and increase the regulation of intercellular signaling. Another local cell-cell communication mechanism that was included in the simulator was bacterial conjugation. These enhancements show the versatility and potential outreach of this version of Gro [243].

- [244] introduces CardioTwin-XAI, a novel Artificial Intelligence-driven Consumer-Centric Digital Twins framework designed for early detection, risk prediction, and personalized management of coronary artery disease (CAD) and related conditions, such as atherosclerosis, myocardial infarction, and heart failure. The proposed system integrates IoT-enabled wearable sensors, continuous hemodynamic profiling, and federated multimodal learning to build a dynamic, patient-specific digital replica. The key innovation of CardioTwin-XAI is its unified integration of explainable AI (XAI) with privacy-preserving federated learning, wherein interpretable attribution mechanisms are embedded within distributed model updates, enabling transparent predictions without compromising data confidentiality across healthcare institutions.

  According to [244], a hybrid transformer-Graph neural networks (GNN) architecture captures complex spatiotemporal cardiovascular dynamics, while differential privacy and blockchain-based data sharing ensure regulatory compliance. Evaluation on multi-institutional datasets demonstrates 92.6% accuracy in CAD prediction with strong privacy guarantees. The framework enables population-scale deployment from edge devices to cloud infrastructure while empowering patients through multilevel explainable AI that bridges clinical insights with consumer understanding.

- Smoldyn is a spatial and stochastic biochemical simulator. It treats each molecule of interest as an individual particle in continuous space, simulating molecular diffusion, molecule-membrane interactions, and chemical reactions, all with good accuracy. Smoldyn now supports two types of rule-based modeling. These are a wildcard method, which is very convenient, and the BioNetGen package with extensions for spatial simulation, which is better for complicated models. Smoldyn also includes new algorithms for simulating the diffusion of surface-bound molecules and molecules with excluded volume. Both are exact in the limit of short time steps and reasonably good with longer steps. In addition, Smoldyn supports single-molecule tracking simulations. Finally, the Smoldyn source code can be accessed through a C/C ++ language library interface [245]. Smoldyn is a computer program for cell-scale biochemical simulations. It simulates each molecule of interest individually to capture natural stochasticity and to yield nanometer-scale spatial resolution. It treats other molecules implicitly, enabling it to simulate hundreds of thousands of molecules over several minutes of real time. Simulated molecules diffuse, react, are confined by surfaces, and bind to membranes much as they would in a real biological system. Smoldyn is easy to use and easy to install. It is more accurate and faster than other particle-based simulators. Smoldyn's features include simulations in 1, 2, or 3 dimensions, a "virtual experimenter" who can manipulate or measure the simulated system, molecules with excluded volume, rule-based modeling, and Python bindings. The latest version of Smoldyn is 2.75 released July 30, 2025 [245].

References

1. L. Jiao et al., Nature-inspired intelligent computing: a comprehensive survey. Research **7**, 0442 (2024)
2. S. Imanaka, T. Saiki, T. Nakano, Accelerating multicellular molecular communication simulations using Barnes-Hut Algorithm, in *2024 IEEE 4th International Conference on Electronic Communications, Internet of Things and Big Data (ICEIB)*, (2024), pp. 233–235
3. S. Dokos, *Modelling Organs, Tissues, Cells and Devices: Using Matlab and Comsol Multiphysics* (Springer, 2017)
4. Y. Ni et al., Pre-training with fractional denoising to enhance molecular property prediction (2024), Preprint at https://arxiv.org/abs/2407.11086
5. C. Yang et al., The role of a molecular informatics platform to support next generation risk assessment. Comput. Toxicol. **26**, 100272 (2023)
6. M. Zoofaghari et al., In silico model for tumor diagnosis based on bloodstream penetrating extracellular vesicles, in *Proceedings of the 10th ACM International Conference on Nanoscale Computing and Communication*, (2023), pp. 129–135
7. E. De Maria, *Systems Biology Modelling and Analysis: Formal Bioinformatics Methods and Tools* (John Wiley & Sons, 2022)
8. L. Giannantoni, R. Bardini, A. Savino, S. Di Carlo, Biology System Description Language (BiSDL): a modeling language for the design of multicellular synthetic biological systems. BMC Bioinform. **25**(1), 166 (2024)

9. M.G. Durmaz et al., Engineering genetic circuits in receiver cells for diffusion-based molecular data communications. bioRxiv (2023). https://doi.org/10.1101/2023.04.18.536609
10. J. Shin, V. Porubsky, J. Carothers, H.M. Sauro, Standards, dissemination, and best practices in systems biology. Curr. Opin. Biotechnol. **81**, 102922 (2023)
11. K. Tiwari et al., Reproducibility in systems biology modelling. Mol. Syst. Biol. **17**(2), e9982 (2021)
12. C.R.C. Rêgo et al., SimStack: an intuitive workflow framework. Front. Mater. **9**, 877597 (2022)
13. D. Waltemath, Management of simulation studies in computational biology, in *Invited Presentations, Junior Research Groups and Research Highlights at GCB 2015*, (2015)
14. J.K. Medley, A.P. Goldberg, J.R. Karr, Guidelines for reproducibly building and simulating systems biology models. IEEE Trans. Biomed. Eng. **63**(10), 2015–2020 (2016)
15. L.T. Tatka, L.P. Smith, J.L. Hellerstein, H.M. Sauro, Adapting modeling and simulation credibility standards to computational systems biology. J. Transl. Med. **21**(1), 501 (2023)
16. A.M. Uhrmacher et al., Context, composition, automation, and communication-the C2AC roadmap for modeling and simulation. ACM Trans. Model. Comput. Simul. (2023)
17. T. Warnke, T. Helms, A.M. Uhrmacher, Reproducible and flexible simulation experiments with ML-Rules and SESSL. Bioinformatics **34**(8), 1424–1427 (2018)
18. S. Arsene et al., In silico clinical trials: is it possible? in *High Performance Computing for Drug Discovery and Biomedicine*, (Springer, 2023), pp. 51–99
19. A. Galal, X. Hesselbach, Nano-networks communication architecture: modeling and functions. Nano Commun. Netw. **17**, 45–62 (2018)
20. J. Long, L. Zhang, M. Wen, J. Li, K. Wang, Computation and communication cooperation for molecular microvascular networks. IEEE Commun. Lett. (2024)
21. K. Chowdhury, IEEE TCSIM Newsletter. Available at https://tc.computer.org/tcsim-demo/wp-content/uploads/sites/20/2019/07/tcsim-issue-22-july-2016.pdf
22. H. Luo, Y. Huang, B. Xiong, H. Jiang, L. Lin, Neural communication based on the oscillatory characteristics of membrane potential, in *ICC 2024-IEEE International Conference on Communications*, (2024), pp. 4832–4837
23. E. Dedu, M. Asghari, Ray tracing routing using packet reception timing in dense nanonetworks. Comput. Netw. **254**, 110753 (2024)
24. Y. Cevallos et al., *Molecular Communications: An Analysis from Networking Theories Perspective* (Springer Nature, 2023)
25. L. Felicetti, M. Femminella, G. Reali, P. Liò, Applications of molecular communications to medicine: a survey. Nano Commun. Netw. **7**, 27–45 (2016)
26. S. Sicari, A. Rizzardi, G. Piro, A. Coen-Porisini, L.A. Grieco, Beyond the smart things: towards the definition and the performance assessment of a secure architecture for the Internet of Nano-Things. Comput. Netw. **162**, 106856 (2019)
27. Y. Zhou, Y. Chen, Latticed channel model of touchable communication over capillary microcirculation network. IEEE Trans. Nanobiosci. **18**(4), 669–678 (2019)
28. L. Felicetti, M. Femminella, G. Reali, A simulation tool for nanoscale biological networks. Nano Commun. Netw. **3**(1), 2–18 (2012)
29. N. Farsad, W. Guo, A.W. Eckford, Tabletop molecular communication: Text messages through chemical signals. PLoS One **8**(12), e82935 (2013)
30. F. Rossi, S. Ristori, A. Abou-Hassan, Multiscale approach for tuning communication among chemical oscillators confined in biomimetic microcompartments. Acc. Chem. Res. **57**(18), 1854–1859 (2024)
31. J. Suzuki, H. Budiman, T.A. Carr, J.H. DeBlois, A simulation framework for neuron-based molecular communication. Procedia Comput. Sci. **24**, 103–113 (2013)
32. M. Hamidović, S. Angerbauer, D. Bi, Y. Deng, T. Tugcu, W. Haselmayr, Microfluidic systems for molecular communications: a review from theory to practice. IEEE Trans. Mol. Biol. Multiscale Commun. (2024)
33. B.A. Shams, *A Simulation Framework for Performance Analysis of Molecular Nano Communication Networks* (Eastern Mediterranean University (EMU)-Doğu Akdeniz Üniversitesi (DAÜ), 2014)

34. P. Stroobant et al., Fast simulation of interacting carriers in nanosimulators, in *Proceedings of the 5th ACM International Conference on Nanoscale Computing and Communication*, (2018), pp. 1–6
35. D. Martins, *Computational Synthetic Biology for Molecular Communications* (Waterford Institute of Technology, 2019)
36. J. Jung et al., GENESIS: a hybrid-parallel and multi-scale molecular dynamics simulator with enhanced sampling algorithms for biomolecular and cellular simulations. Wiley Interdiscip. Rev. Comput. Mol. Sci. **5**(4), 310–323 (2015)
37. S. Bankler, "Modeling and Simulation of Realistic Molecular Communication Systems/submitted by Stefan Bankler," 2022.
38. Y. Wang, A. Noel, N. Yang, A Novel $ A~ Priori $ Simulation Algorithm for absorbing receivers in diffusion-based molecular communication systems. IEEE Trans. Nanobiosci. **18**(3), 437–447 (2019)
39. F. Bronner, F. Dressler, Towards mastering complex particle movement and tracking in molecular communication simulation, in *Proceedings of the Sixth Annual ACM International Conference on Nanoscale Computing and Communication*, (2019), pp. 1–2
40. B.A. Bilgin, O.B. Akan, A fast algorithm for analysis of molecular communication in artificial synapse. IEEE Trans. Nanobiosci. **16**(6), 408–417 (2017)
41. K. Ioannis, C. Aristotelis, M. Ilias, K. Fragiskos, *SBML-Bridge: An XML Pathway Converting Tool*
42. S. McKeever, D. Johnson, The role of markup for enabling interoperability in health informatics. Front. Physiol. **6**, 152 (2015)
43. D. Köhn, N. Le Novere, SED-ML--an XML format for the implementation of the MIASE guidelines, in *International Conference on Computational Methods in Systems Biology*, (2008), pp. 176–190
44. L.P. Smith et al., The simulation experiment description markup language (SED-ML): language specification for level 1 version 5. J. Integr. Bioinform. **21**(1), 20240008 (2024)
45. M. Scharm, O. Wolkenhauer, D. Waltemath, An algorithm to detect and communicate the differences in computational models describing biological systems. Bioinformatics **32**(4), 563–570 (2016)
46. T. Yu, C. Bradley, O. Sinnen, Hardware acceleration of biomedical models with OpenCMISS and CellML, in *2013 International Conference on Field-Programmable Technology (FPT)*, (2013), pp. 370–373
47. J. Sweet-Jones, M. Ahmad, A.C.R. Martin, Antibody markup language (AbML)—a notation language for antibody-based drug formats and software for creating and rendering AbML (abYdraw). MAbs **14**(1), 2101183 (2022)
48. M. König et al., Specifications of standards in systems and synthetic biology: status and developments in 2022 and the COMBINE meeting 2022. J. Integr. Bioinform. **20**(1), 20230004 (2023)
49. M. Golebiewski et al., *Specifications of Standards in Systems and Synthetic Biology: Status, Developments, and Tools in 2024* (De Gruyter, 2024)
50. M. Hucka et al., The systems biology markup language (SBML): language specification for level 3 version 2 core. J. Integr. Bioinform. **15**(1), 20170081 (2018)
51. K.S. Elnaggar et al., A guide in synthetic biology: designing genetic circuits and their applications in stem cells. SynBio **3**(3), 11 (2025)
52. L. Buecherl et al., A decade of SBOL Visual: growing adoption of a diagram standard for engineering biology. ACS Synth. Biol. **14**(10), 3822–3830 (2025)
53. N.J. DiRoberto, Synthetic biology. Geo. L. Tech. Rev. **8**, 287 (2024)
54. H. Li, J. Hu, J. Song, W. Zhou, SynVectorDB: embedding-based retrieval system for synthetic biology parts. Database **2026**, baaf088 (2026)
55. N. Roehner, Z. Zhang, T. Nguyen, C.J. Myers, Generating systems biology markup language models from the synthetic biology open language. ACS Synth. Biol. **4**(8), 873–879 (2015)
56. L. Watanabe, C.J. Myers, Efficient analysis of Systems Biology Markup Language models of cellular populations using arrays. ACS Synth. Biol. **5**(8), 835–841 (2016)

57. L. Smith, H. M. Sauro, An update to the SBML human-readable antimony language (2024), Preprint at https://arxiv.org/abs/2405.15109
58. J. Xu, J. Jiang, H.M. Sauro, SBMLDiagrams: a python package to process and visualize SBML layout and render. Bioinformatics **39**(1), btac730 (2023)
59. B.E. Jardine, L.P. Smith, H.M. Sauro, MakeSBML: a tool for converting between Antimony and SBML. J. Integr. Bioinform. **21**(1), 20240002 (2024)
60. V.L. Porubsky, H.M. Sauro, A practical guide to reproducible modeling for biochemical networks, in *Computational Modeling of Signaling Networks*, (Springer, 2023), pp. 107–138
61. H. Panchiwala, S. Shah, H. Planatscher, M. Zakharchuk, M. König, A. Dräger, The systems biology simulation core library. Bioinformatics **38**(3), 864–865 (2022)
62. J.C. Schaff et al., SBML level 3 package: spatial processes, version 1, release 1. J. Integr. Bioinform. **20**(1), 20220054 (2023)
63. M. Hucka et al., The systems biology markup language (SBML): language specification for level 3 version 2 core release 2. J. Integr. Bioinform. **16**(2) (2019)
64. S.M. Keating et al., SBML Level 3: an extensible format for the exchange and reuse of biological models. Mol. Syst. Biol. **16**(8), e9110 (2020)
65. M. Hucka et al., The systems biology markup language (SBML): a medium for representation and exchange of biochemical network models. Bioinformatics **19**(4), 524–531 (2003)
66. A. Mutsuddy et al., Computational speed-up of large-scale, single-cell model simulations via a fully integrated SBML-based format. Bioinform. Adv. **3**(1), vbad039 (2023)
67. A. Immaneni, “Modelling and Visualizing Selected Molecular Communication Processes in Biological Organisms: A Multi-Layer Perspective,” 2018.
68. K.L. Garner, Principles of synthetic biology. Essays Biochem. **65**(5), 791–811 (2021)
69. H. Baig et al., Synthetic biology open language (SBOL) version 3.0. 0. J. Integr. Bioinform. **17**(2–3), 20200017 (2020)
70. N. Roehner et al., Specifying combinatorial designs with the synthetic biology open language (sbol). ACS Synth. Biol. **8**(7), 1519–1523 (2019)
71. J. Mante et al., Curation principles derived from the analysis of the SBOL iGEM data set. ACS Synth. Biol. **10**(10), 2592–2606 (2021)
72. G. Mısırlı, J. Beal, T.E. Gorochowski, G.-B. Stan, A. Wipat, C.J. Myers, SBOL visual 2 ontology. ACS Synth. Biol. **9**(4), 972–977 (2020)
73. U. Urquiza-García, T. Zieliński, A.J. Millar, Better research by efficient sharing: evaluation of free management platforms for synthetic biology designs. Synth. Biol. **4**(1), ysz016 (2019)
74. J. Mante et al., Synthetic biology knowledge system. ACS Synth. Biol. **10**(9), 2276–2285 (2021)
75. J.V. Mante, *Promotion of Data Reuse in Synthetic Biology* (University of Colorado at Boulder, 2022)
76. L. Buecherl, C.J. Myers, Engineering genetic circuits: advancements in genetic design automation tools and standards for synthetic biology. Curr. Opin. Microbiol. **68**, 102155 (2022)
77. Z. Sents, T.E. Stoughton, L. Buecherl, P.J. Thomas, P. Fontanarrosa, C.J. Myers, Synbiosuite: a tool for improving the workflow for genetic design and modeling. ACS Synth. Biol. **12**(3), 892–897 (2023)
78. L. Clarke, R. Kitney, Developing synthetic biology for industrial biotechnology applications. Biochem. Soc. Trans. **48**(1), 113–122 (2020)
79. G. Mısırlı et al., Constructing synthetic biology workflows in the cloud. Eng. Biol. **1**(1), 61–65 (2017)
80. A. Pandey, *IWBDA 2022: Toward a Modular Future of Synthetic Biology Driven by Bio-Design Automation* (ACS Publications, 2024)
81. J.W. Yeoh et al., *SynBiopython: an open-source software library for synthetic biology* (Oxford University Press, 2021)
82. B. Calles, Á. Goñi-Moreno, V. de Lorenzo, Digitalizing heterologous gene expression in Gram-negative bacteria with a portable ON/OFF module. Mol. Syst. Biol. **15**(12), e8777 (2019)
83. C. Wagner, U. Urquiza-Garcia, M.D. Zurbriggen, H.M. Beyer, GMOCU: digital documentation, management, and biological risk assessment of genetic parts. Adv. Biol. **8**(4), 2300529 (2024)

84. I.S. de Murieta, M. Bultelle, R.I. Kitney, Data model for biopart datasheets. Eng. Biol. **2**(1), 7–18 (2018)
85. G. Vidal, C. Vitalis, T. Matúte, I. Núñez, F. Federici, T.J. Rudge, Genetic network design automation with LOICA, in *Synthetic Biology: Methods and Protocols*, (Springer, 2024), pp. 393–412
86. M. Cashman, J. Firestone, M.B. Cohen, T. Thianniwet, W. Niu, An empirical investigation of organic software product lines. Empir. Softw. Eng. **26**(3), 44 (2021)
87. E. Martínez-García et al., SEVA 4.0: an update of the Standard European Vector Architecture database for advanced analysis and programming of bacterial phenotypes. Nucleic Acids Res. **51**(D1), D1558–D1567 (2023)
88. P. Boeing, M. Leon, D.N. Nesbeth, A. Finkelstein, C.P. Barnes, Towards an aspect-oriented design and modelling framework for synthetic biology. Processes **6**(9), 167 (2018)
89. J. Beal et al., The long journey towards standards for engineering biosystems: are the Molecular Biology and the Biotech communities ready to standardise? EMBO Rep. **21**(5), e50521 (2020)
90. N.R. Zelenka et al., Data hazards in synthetic biology. Synth. Biol. **9**(1), ysae010 (2024)
91. J.A. McLaughlin et al., The synthetic biology open language (SBOL) version 3: simplified data exchange for bioengineering. Front. Bioeng. Biotechnol. **8**, 1009 (2020)
92. L. Buecherl et al., Synthetic biology open language (SBOL) version 3.1. 0. J. Integr. Bioinform. **20**(1), 20220058 (2023)
93. J.Y. Quinn et al., SBOL visual: a graphical language for genetic designs. PLoS Biol. **13**(12), e1002310 (2015)
94. J.A. McLaughlin, M. Pocock, G. Mısırlı, C. Madsen, A. Wipat, VisBOL: web-based tools for synthetic biology design visualization. ACS Synth. Biol. **5**(8), 874–876 (2016)
95. L. Terry, J. Earl, S. Thayer, S. Bridge, C.J. Myers, SBOLCanvas: a visual editor for genetic designs. ACS Synth. Biol. **10**(7), 1792–1796 (2021)
96. H. Baig et al., Synthetic biology open language visual (SBOL visual) version 3.0. J. Integr. Bioinform. **18**(3), 20210013 (2021)
97. C.J. Clark, J. Scott-Brown, T.E. Gorochowski, paraSBOLv: a foundation for standard-compliant genetic design visualization tools. Synth. Biol. **6**(1), ysab022 (2021)
98. T. Zieliński, J. Hay, A. Romanowski, A. Nenninger, A. McCormick, A.J. Millar, SynBio2Easy—a biologist-friendly tool for batch operations on SBOL designs with Excel inputs. Synth. Biol. **7**(1), ysac002 (2022)
99. J. Mante, J. Abam, S.P. Samineni, I.M. Pötzsch, J. Beal, C.J. Myers, Excel--SBOL converter: creating SBOL from excel templates and vice versa. ACS Synth. Biol. **12**(1), 340–346 (2023)
100. T.S.D. Wang, T. Quaiser, E. Shochat, Semantically linking in silico cancer models. Cancer Inform. **13**(s1) (2014)
101. C.M. Lloyd, M.D.B. Halstead, P.F. Nielsen, CellML: its future, present and past. Prog. Biophys. Mol. Biol. **85**(2–3), 433–450 (2004)
102. G. Gricourt, T. Duigou, S. Dérozier, J.-L. Faulon, neo4jsbml: import systems biology markup language data into the graph database Neo4j. PeerJ **12**, e16726 (2024)
103. C.B. Collin et al., Computational models for clinical applications in personalized medicine—guidelines and recommendations for data integration and model validation. J. Pers. Med. **12**(2), 166 (2022)
104. M.A. Walker, R. Madduri, A. Rodriguez, J.L. Greenstein, R.L. Winslow, Models and simulations as a service: exploring the use of galaxy for delivering computational models. Biophys. J. **110**(5), 1038–1043 (2016)
105. M. Clerx et al., CellML 2.0. J. Integr. Bioinform. **17**(2–3), 20200021 (2020)
106. M. Clerx et al., CellML 2.0. 1. J. Integr. Bioinform. **20**(1), 20230003 (2023)
107. B. Maier, D. Göddeke, F. Huber, T. Klotz, O. Röhrle, M. Schulte, OpenDiHu: an efficient and scalable framework for biophysical simulations of the neuromuscular system. J. Comput. Sci. **79**, 102291 (2024)
108. Y. Shi, P. Lawford, D.R. Hose, Construction of lumped-parameter cardiovascular models using the CellML language. J. Med. Eng. Technol. **42**(7), 525–531 (2018)

109. M. Akbarpour Ghazani, M. Pan, K. Tran, A. Rampadarath, D.P. Nickerson, A review of the diverse applications of bond graphs in biology and physiology. Proc. R. Soc. A **480**(2294), 20230807 (2024)
110. M.L. Neal et al., Harmonizing semantic annotations for computational models in biology. Brief. Bioinform. **20**(2), 540–550 (2019)
111. Y. Villota-Narvaez, C. Bleiler, O. Röhrle, Data sharing in modeling and simulation of biomechanical systems in interdisciplinary environments. GAMM-Mitteilungen **47**(2), e202370006 (2024)
112. E. Neufeld, B. Lloyd, B. Schneider, W. Kainz, N. Kuster, Functionalized anatomical models for computational life sciences. Front. Physiol. **9**, 1594 (2018)
113. J. Cooper, M. Scharm, G.R. Mirams, The cardiac electrophysiology web lab. Biophys. J. **110**(2), 292–300 (2016)
114. K.D. Carlson, F. Rothganger, J.B. Aimone, Computational perspectives on adult neurogenesis, in *The Rewiring Brain*, (Academic Press, 2017), pp. 425–441
115. M.M.B.R. Vellasco, Neuronal circuit specification language and tools for modelling the virtual fly brain (Doctoral dissertation, Programa de Pósgraduação em Engenharia Elétrica of the Departamento de Engenharia Elétrica, PUC-Rio) (2015)
116. S. Panagiotou, H. Sidiropoulos, D. Soudris, M. Negrello, C. Strydis, EDEN: a high-performance, general-purpose, NeuroML-based neural simulator. Front. Neuroinform. **16**, 724336 (2022)
117. L. Nayak, A. Dasgupta, R. Das, K. Ghosh, R.K. De, Computational neuroscience and neuroinformatics: Recent progress and resources. J. Biosci. **43**(5), 1037–1054 (2018)
118. F. Schreiber et al., Specifications of standards in systems and synthetic biology: status and developments in 2021. J. Integr. Bioinform. **18**(3), 20210026 (2021)
119. E.E. Tsur, Data models in neuroinformatics, in *Bioinformatics in the Era of Post Genomics and Big Data*, (2018), pp. 146–166
120. I.Y. Abdurakhmonov, *Bioinformatics in the Era of Post Genomics and Big Data* (BoD--Books on Demand, 2018)
121. C.N. Ojiugwo, A.B. Abdallah, C. Thron, Simulation of biological learning with spiking neural networks, in *Implementations and Applications of Machine Learning*, (2020), pp. 207–227
122. O. Awile et al., Modernizing the NEURON simulator for sustainability, portability, and performance. Front. Neuroinform. **16**, 884046 (2022)
123. E. Ezra-Tsur, O. Amsalem, L. Ankri, P. Patil, I. Segev, M. Rivlin-Etzion, Realistic retinal modeling unravels the differential role of excitation and inhibition to starburst amacrine cells in direction selectivity. PLoS Comput. Biol. **17**(12), e1009754 (2021)
124. A. Sinha et al., The NeuroML ecosystem for standardized multi-scale modeling in neuroscience. bioRxiv (2024). https://doi.org/10.1101/2023.12.07.570537
125. R.A. Tikidji-Hamburyan, V. Narayana, Z. Bozkus, T.A. El-Ghazawi, Software for brain network simulations: a comparative study. Front. Neuroinform. **11**, 46 (2017)
126. M. Stimberg, R. Brette, D.F.M. Goodman, Brian 2, an intuitive and efficient neural simulator. eLife **8**, e47314 (2019)
127. K. Dai et al., The SONATA data format for efficient description of large-scale network models. PLoS Comput. Biol. **16**(2), e1007696 (2020)
128. J.E. Pedersen et al., Neuromorphic intermediate representation: a unified instruction set for interoperable brain-inspired computing (2023), Preprint at https://arxiv.org/abs/2311.14641
129. J. Birgiolas, V. Haynes, P. Gleeson, R.C. Gerkin, S.W. Dietrich, S. Crook, NeuroML-DB: Sharing and characterizing data-driven neuroscience models described in NeuroML. PLoS Comput. Biol. **19**(3), e1010941 (2023)
130. L.L. Bologna, A. Tocco, R. Smiriglia, A. Romani, F. Schürmann, M. Migliore, Online interoperable resources for building hippocampal neuron models via the Hippocampus Hub. Front. Neuroinform. **17**, 1271059 (2023)
131. G.P. Sarma et al., OpenWorm: overview and recent advances in integrative biological simulation of Caenorhabditis elegans. Philos. Trans. R. Soc. B **373**(1758), 20170382 (2018)

132. R.C. Gerkin, J. Birgiolas, R.J. Jarvis, C. Omar, S.M. Crook, NeuronUnit: a package for data-driven validation of neuron models using SciUnit. bioRxiv (2019). https://doi.org/10.1101/665331
133. L. Felicetti, M. Femminella, G. Reali, P. Gresele, M. Malvestiti, J.N. Daigle, Modeling CD40-based molecular communications in blood vessels. IEEE Trans. Nanobiosci. **13**(3), 230–243 (2014)
134. L. Felicetti, M. Femminella, G. Reali, Simulation of molecular signaling in blood vessels: software design and application to atherogenesis. Nano Commun. Netw. **4**(3), 98–119 (2013)
135. L. Felicetti, M. Femminella, G. Reali, P. Gresele, M. Malvestiti, Simulating an in vitro experiment on nanoscale communications by using BiNS2. Nano Commun. Netw. **4**(4), 172–180 (2013)
136. L. Felicetti, M. Femminella, G. Reali, T. Nakano, A.V. Vasilakos, TCP-like molecular communications. IEEE J. Sel. Areas Commun. **32**(12), 2354–2367 (2014)
137. L. Felicetti, M. Femminella, G. Reali, A simple and scalable receiver model in molecular communication systems, in *Proceedings of the 3rd ACM International Conference on Nanoscale Computing and Communication*, (2016), pp. 1–2
138. P. Stroobant et al., Parallel algorithms for simulating interacting carriers in nanocommunication. Nano Commun. Netw. **20**, 20–30 (2019)
139. L. Felicetti, M. Femminella, G. Reali, Smart antennas for diffusion-based molecular communications, in *Proceedings of the Second Annual International Conference on Nanoscale Computing and Communication*, (2015), pp. 1–6
140. I. Isik, H.B. Yilmaz, M.E. Tagluk, A preliminary investigation of receiver models in molecular communication via diffusion, in *2017 International Artificial Intelligence and Data Processing Symposium (IDAP)*, (2017), pp. 1–5
141. L. Felicetti, M. Femminella, G. Reali, Congestion control in molecular cyber-physical systems. IEEE Access **5**, 10000–10011 (2017)
142. M. Femminella, G. Reali, A simple queuing model for molecular communications receivers. Sensors **21**(22), 7664 (2021)
143. I. Llatser, D. Demiray, A. Cabellos-Aparicio, D.T. Altilar, E. Alarcón, N3Sim: Simulation framework for diffusion-based molecular communication nanonetworks. Simul. Model. Pract. Theory **42**, 210–222 (2014)
144. I. Pascual Mariñelarena, "N3SIM: Simulator for Diffusion-based Molecular Communication in Nanonetworks," 2011.
145. S. S. Assaf, "Synthetic Biology Guidelines for Diffusion Based Molecular Communication," 2019.
146. S.S. Assaf, S. Salehi, R.G. Cid-Fuentes, J. Solé-Pareta, E. Alarcón, Characterizing the physical influence of neighboring absorbing receivers in molecular communication, in *Proceedings of the 3rd ACM International Conference on Nanoscale Computing and Communication*, (2016), pp. 1–2
147. S.S. Assaf, S. Salehi, R.G. Cid-Fuentes, J. Solé-Pareta, E. Alarcón, Influence of neighboring absorbing receivers upon the inter-symbol interference in a diffusion-based molecular communication system. Nano Commun. Netw. **14**, 40–47 (2017)
148. P. Mašek, J. Hosek, *Body-area Communications: Sci-Fi or Promising Paradigm?*
149. L. Hong, W. Chen, F. Liu, Cooperative molecular communication for nanonetwork, in *2014 Sixth International Conference on Ubiquitous and Future Networks (ICUFN)*, (2014), pp. 369–370
150. S. Salehi, N.S. Moayedian, S.H. Javanmard, E. Alarcón, Lifetime improvement of a multiple transmitter local drug delivery system based on diffusive molecular communication. IEEE Trans. Nanobiosci. **17**(3), 352–360 (2018)
151. S. Salehi, N.S. Moayedian, S.S. Assaf, R.G. Cid-Fuentes, J. Solé-Pareta, E. Alarcón, Releasing rate optimization in a single and multiple transmitter local drug delivery system with limited resources. Nano Commun. Netw. **11**, 114–122 (2017)

152. Q. Zhao, M. Li, J. Luo, Relationship among reaction rate, release rate and efficiency of nanomachine-based targeted drug delivery. Technol. Health Care **25**(6), 1119–1130 (2017)
153. S.S. Assaf, J. Solé-Pareta, E. Alarcón, Characterizing glutamatergic synapse of a diffusion-based molecular communication channel to measure Aβ oriented to Alzheimer's disease, in *Proc. Workshop Mol. Commun.*, (2018), pp. 1–2
154. L. Felicetti, S.S. Assaf, M. Femminella, G. Reali, E. Alarcon, J. Sole-Pareta, The molecular communications markup language (MolComML). Nano Commun. Netw. **16**, 12–25 (2018)
155. L.F. Borges, M.T. Barros, M. Nogueira, Cell signaling error control for reliable molecular communications. Front. Commun. Netw. **5**, 1332379 (2024)
156. N.A. Turgut, B.A. Bilgin, O.B. Akan, N^4 Sim: the first nervous NaNoNetwork simulator with synaptic molecular communications. IEEE Trans. Nanobiosci. **21**(4), 468–481 (2021)
157. M.T. Barros, S. Balasubramaniam, B. Jennings, Comparative end-to-end analysis of ca 2+-signaling-based molecular communication in biological tissues. IEEE Trans. Commun. **63**(12), 5128–5142 (2015)
158. M.T. Barros, S. Balasubramaniam, B. Jennings, Y. Koucheryavy, Transmission protocols for calcium-signaling-based molecular communications in deformable cellular tissue. IEEE Trans. Nanotechnol. **13**(4), 779–788 (2014)
159. E. Akyol, A.B. Ozturk, I. M. Bolhassan, M. Kuscu, Experimental characterization of hydrodynamic gating-based molecular communication transmitter (2024), Preprint at https://arxiv.org/abs/2408.09835
160. M. Kuscu, O.B. Akan, Modeling convection-diffusion-reaction systems for microfluidic molecular communications with surface-based receivers in Internet of Bio-Nano Things. PLoS One **13**(2), e0192202 (2018)
161. D. Bi, Y. Deng, M. Pierobon, A. Nallanathan, Chemical reactions-based microfluidic transmitter and receiver design for molecular communication. IEEE Trans. Commun. **68**(9), 5590–5605 (2020)
162. L. Khaloopour et al., An experimental platform for macro-scale fluidic medium molecular communication. IEEE Trans. Mol. Biol. Multiscale Commun. **5**(3), 163–175 (2019)
163. I.M. Bolhassan, A. Abdali, M. Kuscu, Microfluidic molecular communication transmitter based on hydrodynamic gating. IEEE Trans. Mol. Biol. Multiscale Commun. (2024)
164. M.M. Al-Zubi, A.S. Mohan, P. Plapper, S.H. Ling, Intrabody molecular communication via blood-tissue barrier for Internet of Bio-Nano Things. IEEE Internet Things J. **9**(21), 21802–21810 (2022)
165. W. Guo et al., Molecular physical layer for 6G in wave-denied environments. IEEE Commun. Mag. **59**(5), 33–39 (2021)
166. Y. Shuai, A.G. Maslovskaya, C. Kuttler, Modeling of bacterial communication in the extended range of population dynamics. Математическая биология и биоинформатика **18**(1), 89–104 (2023)
167. S. Pal, N. Islam, S. Misra, VIVID: in vivo end-to-end molecular communication model for COVID-19. IEEE Trans. Mol. Biol. Multiscale Commun. **7**(3), 142–152 (2021)
168. D. Bi, Y. Deng, Microfluidic QCSK transmitter and receiver design for molecular communication. IEEE Trans. Commun. **70**(9), 5837–5852 (2022)
169. S.M. Musa, *Computational finite element methods in nanotechnology* (CRC Press, 2017)
170. T.C. Gasser, *Vascular biomechanics* (Springer, 2021)
171. C. Koca, O. Ergul, M. Civas, and O. B. Akan, "Information theoretic modeling of paranodal regions in myelinated axons," 2021.
172. L. Campanile, M. Gribaudo, M. Iacono, F. Marulli, M. Mastroianni, Computer network simulation with ns-3: a systematic literature review. Electronics (Basel) **9**(2), 272 (2020)
173. M. Viroli, D. Pianini, G. Scaparrotti, P.S. di Laurea, Cross-simulator integration: ns3 as a network simulation back-end for Alchemist
174. A. Mohta, S.I. Ajankar, M.M. Chandane, Network Simulator-3: a review
175. L. Felicetti, M. Femminella, G. Reali, Modeling approaches for simulating molecular communications, in *Encyclopedia of Wireless Networks*, (Springer, 2020), pp. 903–910

176. IEEE recommended practice for nanoscale and molecular communication framework, IEEE Std 1906.1-2015 (2016), pp. 1–64. https://doi.org/10.1109/IEEESTD.2016.7378262
177. IEEE standard data model for nanoscale communication systems. IEEE Std 1906.1.1-2020 (2020), pp. 1–142. https://doi.org/10.1109/IEEESTD.2020.9285373.
178. A. Rana, S. Prajapat, P. Kumar, D. Gautam, C.-M. Chen, Designing a security framework based on hybrid communication in the Internet of Nano Things. IEEE Internet Things J. (2023)
179. R. Fotohi, Securing of Unmanned Aerial Systems (UAS) against security threats using human immune system. Reliab. Eng. Syst. Saf. **193**, 106675 (2020)
180. P. Masek, J. Hosek, D. Kovac, J. Miklica, On simulation techniques for modeling of molecular-based nanodevices' communication in human body environment, in *2015 38th International Conference on Telecommunications and Signal Processing (TSP)*, (2015), pp. 627–631
181. M. Stelzner et al., Precise detection and treatment of human diseases based on nano networking, in *Proceedings of the 11th EAI International Conference on Body Area Networks*, (2016), pp. 58–64
182. D.F. Astudillo Salinas, E.A. Zuñiga Gomez, J.P. Calderon Calderon, L.P. Tello Oquendo, A diffusive propagation model for molecular communications: analysis and implementation in NS-3
183. N. Farsad, H.B. Yilmaz, A. Eckford, C.-B. Chae, W. Guo, A comprehensive survey of recent advancements in molecular communication. IEEE Commun. Surv. Tutor. **18**(3), 1887–1919 (2016)
184. G. Wei, P. Bogdan, R. Marculescu, Efficient modeling and simulation of bacteria-based nanonetworks with BNSim. IEEE J. Sel. Areas Commun. **31**(12), 868–878 (2013)
185. C. Lo, G. Wei, R. Marculescu, Towards autonomous control of molecular communication in populations of bacteria, in *Proceedings of the Second Annual International Conference on Nanoscale Computing and Communication*, (2015), pp. 1–6
186. X. Li et al., A reconfigurable wireless NoC for large scale microbiome community analysis. IEEE Trans. Comput. **66**(10), 1653–1666 (2017)
187. S. Ahmed, J. Hu, S.M.Z.A. Naqvi, Y. Zhang, L. Linze, A.M. Iderawumi, Molecular communication network and its applications in crop sciences. Planta **255**(6), 128 (2022)
188. J. Zhuang et al., Analytical modeling and experimental characterization of chemotaxis in Serratia marcescens. Phys. Rev. E **89**(5), 52704 (2014)
189. H. Koorehdavoudi et al., Multi-fractal characterization of bacterial swimming dynamics: a case study on real and simulated serratia marcescens. Proceedings of the Royal Society A: Mathematical, Physical and Engineering Sciences **473**(2203), 20170154 (2017)
190. F.M. Lovatt, A.R. Hoelzel, Impact on reindeer (Rangifer tarandus) genetic diversity from two parallel population bottlenecks founded from a common source. Evol. Biol. **41**, 240–250 (2014)
191. G. Wei, R. Marculescu, Don't let history repeat itself: optimal multidrug quorum quenching of pathogens network, in *Proceedings of ACM the First Annual International Conference on Nanoscale Computing and Communication*, (2014), pp. 1–9
192. A. Akkaya, G. Genc, T. Tugcu, HLA based architecture for molecular communication simulation. Simul. Model. Pract. Theory **42**, 163–177 (2014)
193. A. Akkaya, T. Tugcu, dMCS: distributed molecular communication simulator, in *Proceedings of the 8th International Conference on Body Area Networks*, (2013), pp. 468–471
194. B. Acar, A. Akkaya, G. Genc, H.B. Yilmaz, M. Şükrü Kuran, T. Tugcu, Understanding communication via diffusion: Simulation design and intricacies, in *Modeling, Methodologies and Tools for Molecular and Nano-scale Communications: Modeling, Methodologies and Tools*, (2017), pp. 139–163
195. H.B. Yilmaz, C.-B. Chae, Simulation study of molecular communication systems with an absorbing receiver: modulation and ISI mitigation techniques. Simul. Model. Pract. Theory **49**, 136–150 (2014)

196. I.F. Akyildiz, J.M. Jornet, M. Pierobon, Nanonetworks: a new frontier in communications. Commun. ACM **54**(11), 84–89 (2011)
197. E. Shitiri, H.B. Yilmaz, H.-S. Cho, A time-slotted molecular communication (TS-MOC): framework and time-slot errors. IEEE Access **7**, 78146–78158 (2019)
198. P. Hofmann, P. Zhou, C. Lee, M. Reisslein, F.H.P. Fitzek, C.-B. Chae, OpenFOAM simulation of microfluidic molecular communications: method and experimental validation. IEEE Access (2024)
199. A.P. Kaur, D.K.K. Randhawa, Comparative analysis of three dimensional channel models in molecular communication via diffusion. Int. J. Adv. Res. Comput. Commun. Eng. **6**, 522–529 (2017)
200. G. Alesiani, M. Magarini, S. Bregni, Confidence interval of peak-based distance estimation in diffusive molecular communication, in *2021 IEEE International Mediterranean Conference on Communications and Networking (MeditCom)*, (2021), pp. 454–459
201. T. Islam, E. Shitiri, H.-S. Cho, In-body sequential multidrug delivery scheme using molecular communication. IEEE Access **10**, 39975–39985 (2022)
202. T. Islam, E. Shitiri, H.-S. Cho, A simultaneous drug release scheme for targeted drug delivery using molecular communications. IEEE Access **8**, 91770–91778 (2020)
203. A.P. Kaur, D.K.K. Randhawa, G.K. Walia, Simulation study of messenger molecule displacement in communication via diffusion. Int. J. Comput. Sci. Eng. **5**(6), 1–6 (2017)
204. A. Noel, K.C. Cheung, R. Schober, D. Makrakis, A. Hafid, Simulating with AcCoRD: actor-based communication via reaction--diffusion. Nano Commun. Netw. **11**, 44–75 (2017)
205. A. Noel, D. Makrakis, Algorithm for mesoscopic advection--diffusion. IEEE Trans. Nanobiosci. **17**(4), 543–554 (2018)
206. M. Öner, E. Sorguven, Channel characteristics for diffusive molecular communications in a spherical shell. IEEE Commun. Lett. **27**(4), 1120–1124 (2022)
207. O.A. Dambri, S. Cherkaoui, Enhancing signal strength and ISI-avoidance of diffusion-based molecular communication, in *2018 14th International Wireless Communications & Mobile Computing Conference (IWCMC)*, (2018), pp. 1–6
208. M. Öner, E. Sorguven, Signal propagation in diffusive molecular communications over a spherical surface. IEEE Commun. Lett. **26**(1), 64–68 (2021)
209. A. Noel, A.W. Eckford, Asynchronous peak detection for demodulation in molecular communication, in *2017 IEEE International Conference on Communications (ICC)*, (2017), pp. 1–6
210. S.N. Solak, *Abnormality Detection Using Molecular Communications Based Nano-scale Sensor Networks* (University of Sussex, 2022)
211. Y. Deng, A. Noel, W. Guo, A. Nallanathan, M. Elkashlan, 3D stochastic geometry model for large-scale molecular communication systems, in *2016 IEEE Global Communications Conference (GLOBECOM)*, (2016), pp. 1–6
212. A. Noel, D. Makrakis, A. Hafid, Channel impulse responses in diffusive molecular communication with spherical transmitters (2016), Preprint at https://arxiv.org/abs/1604.04684
213. O.A. Dambri, S. Cherkaoui, Performance enhancement of diffusion-based molecular communication. IEEE Trans. Nanobiosci. **19**(1), 48–58 (2019)
214. A. Noel, Y. Deng, D. Makrakis, A. Hafid, Active versus passive: receiver model transforms for diffusive molecular communication, in *2016 IEEE Global Communications Conference (GLOBECOM)*, (2016), pp. 1–6
215. Y. Deng, A. Noel, W. Guo, A. Nallanathan, M. Elkashlan, Analyzing large-scale multiuser molecular communication via 3-D stochastic geometry. IEEE Trans. Mol. Biol. Multiscale Commun. **3**(2), 118–133 (2017)
216. O.A. Dambri, A. Abouaomar, S. Cherkaoui, Design optimization of a MIMO receiver for diffusion-based molecular communication, in *2019 IEEE Wireless Communications and Networking Conference (WCNC)*, (2019), pp. 1–6
217. S. Angerbauer, T. Pankratz, F. Enzenhofer, W. Haselmayr, Novel nano-machine architecture for machine learning in the IoBNT, in *GLOBECOM 2023-2023 IEEE Global Communications Conference*, (2023), pp. 1161–1166

218. S. Solak, M. Öner, Sequential decision fusion for abnormality detection via diffusive molecular communications. IEEE Commun. Lett. **25**(3), 825–829 (2020)
219. M. Schäfer, W. Wicke, W. Haselmayr, R. Rabenstein, R. Schober, Spherical diffusion model with semi-permeable boundary: a transfer function approach, in *ICC 2020-2020 IEEE International Conference on Communications (ICC)*, (2020), pp. 1–7
220. M. Schäfer, R. Rabenstein, An analytical model of diffusion in a sphere with semi-permeable boundary, in *Proc. Workshop on Molecular Communications*, (2019), pp. 1–2
221. L.F. Borges, M.T. Barros, M. Nogueira, Toward reliable intra-body molecular communication: an error control perspective. *IEEE Commun. Mag.* **59**(5), 114–120 (2021)
222. T. Khan, B.A. Bilgin, O.B. Akan, Diffusion-based model for synaptic molecular communication channel. *IEEE Trans. Nanobiosci.* **16**(4), 299–308 (2017)
223. Y. Jian et al., nanoNS3: a network simulator for bacterial nanonetworks based on molecular communication. Nano Commun. Netw. **12**, 1–11 (2017)
224. C.M. Austin et al., Modeling and validation of autoinducer-mediated bacterial gene expression in microfluidic environments. Biomicrofluidics **8**(3) (2014)
225. A.O. Bicen, I.F. Akyildiz, System-theoretic analysis and least-squares design of microfluidic channels for flow-induced molecular communication. IEEE Trans. Signal Process. **61**(20), 5000–5013 (2013)
226. A. Einolghozati, M. Sardari, F. Fekri, Design and analysis of wireless communication systems using diffusion-based molecular communication among bacteria. IEEE Trans. Wirel. Commun. **12**(12), 6096–6105 (2013)
227. B. Krishnaswamy, R. Sivakumar, Source addressing and medium access control in bacterial communication networks, in *Proceedings of the Second Annual International Conference on Nanoscale Computing and Communication*, (2015), pp. 1–6
228. B. Krishnaswamy et al., ADMA: amplitude-division multiple access for bacterial communication networks. IEEE Trans. Mol. Biol. Multiscale Commun. **3**(3), 134–149 (2017)
229. R. Geyer, M. Stelzner, F. Büther, S. Ebers, BloodVoyagerS: simulation of the work environment of medical nanobots, in *Proceedings of the 5th ACM International Conference on Nanoscale Computing and Communication*, (2018), pp. 1–6
230. R. Wendt, C. Deter, S. Fischer, BVS-Vis: a web-based visualizer for BloodVoyagerS, in *Proceedings of the 7th ACM International Conference on Nanoscale Computing and Communication*, (2020), pp. 1–2
231. S. Brauer, M. Stelzner, S. Fischer, NanoMedView: a mhealth visualizer app for nano communication networks, in *Proceedings of the 10th ACM International Conference on Nanoscale Computing and Communication*, (2023), pp. 40–45
232. J.P. Drees, *A Computationally Inexpensive and Flexible Simulation Model for Molecular Communication* (Universitätsbibliothek, 2019)
233. J.P. Drees et al., Efficient simulation of macroscopic molecular communication: the Pogona simulator, in *Proceedings of the 7th ACM International Conference on Nanoscale Computing and Communication*, (2020), pp. 1–6
234. B.C. Haller, P.W. Messer, SLiM 2: flexible, interactive forward genetic simulations. Mol. Biol. Evol. **34**(1), 230–240 (2017)
235. A. Coulier, A. Hellander, Orchestral: a lightweight framework for parallel simulations of cell-cell communication, in *2018 IEEE 14th International Conference on e-Science (e-Science)*, (2018), pp. 168–176
236. F. Gulec, F. Dressler, A.W. Eckford, A computational approach for the characterization of airborne pathogen transmission in turbulent molecular communication channels. IEEE Trans. Mol. Biol. Multiscale Commun. **9**(2), 124–134 (2023)
237. A. Ahmadzadeh, H. Arjmandi, A. Burkovski, R. Schober, Comprehensive reactive receiver modeling for diffusive molecular communication systems: reversible binding, molecule degradation, and finite number of receptors. IEEE Trans. Nanobiosci. **15**(7), 713–727 (2016)
238. R. Wendt, S. Fischer, MEHLISSA: a medical holistic simulation architecture for nanonetworks in humans, in *Proceedings of the 7th ACM International Conference on Nanoscale Computing and Communication*, (2020), pp. 1–6

239. T. Saiki, S. Imanaka, S. Kobayashi, T. Nakano, A general-purpose simulation platform for multicellular molecular communication systems. IEEE Trans. Mol. Biol. Multiscale Commun. (2025)
240. T. Saiki, T. Nakano, Design and implementation of a multicellular molecular communication simulator, in *2022 Joint 12th International Conference on Soft Computing and Intelligent Systems and 23rd International Symposium on Advanced Intelligent Systems (SCIS&ISIS)*, (2022), pp. 1–5
241. A. Matyjaszkiewicz et al., BSim 2.0: an advanced agent-based cell simulator. ACS Synth. Biol. **6**(10), 1969–1972 (2017)
242. A. Abdali, R. Jadsadaphongphaibool, D. Bi, M. Kuscu, Y. Deng, High-order CSK realization for MC via spatially distributed multicellular consortia. IEEE Trans. Mol. Biol. Multiscale Commun. (2026)
243. M. Gutiérrez, P. Gregorio-Godoy, G. Perez del Pulgar, L.E. Munoz, S. Sáez, & A. Rodríguez-Patón, A new improved and extended version of the multicell bacterial simulator gro. ACS synthetic biology, **6**(8), 1496–1508 (2017)
244. J. Yang et al., CardioTwin-XAI: a consumer-centric digital twin framework for predictive risk stratification and personalized management of coronary artery disease in healthcare 5.0. IEEE Trans. Consum. Electron. (2026)
245. S.S. Andrews, Smoldyn: particle-based simulation with rule-based modeling, improved molecular interaction and a library interface. Bioinformatics **33**(5), 710–717 (2017)

Chapter 3
Molecular Communications Scenario Analysis Using MolComML and NS-3

3.1 Background Information

MCs create a field of study that integrates the knowledge of two primary disciplines: biology and engineering. This multidisciplinary approach, besides being a useful asset, makes it challenging for researchers to share knowledge. The diversity of biological environments capable of facilitating nanoscale communication has led to various proposals (e.g., neuronal networks, molecular diffusion, and flow-based carrier mobility), which have been examined through distinct research methodologies and instruments (various analytical models, simulators, and laboratory experiments). Consequently, the necessity to combine research activities at the European Union (EU) level has arisen. The CIRCLE project presentation created a beneficial practice in MCs by consolidating many research activities into a unified research framework. The proposal is strategically aligned with EU research objectives, highly interdisciplinary, and open to contributions from various research activities that can help establish a research roadmap for the next years and viable future exploitation plans.

CIRCLE establishes a Europe-wide MCs forum and offers a support infrastructure for the coordination of research throughout Europe. In the intermediate term, it promotes information dissemination through the CIRCLE forum and a specialized web portal.[1] This emphasizes the dissemination of research methodology and simulation code sources. It forms specialized working groups on various research subjects within the MCs sector and formulates strategic roadmaps for academic research

[1] The cited information regarding the CIRCLE project is detailed in the following paragraphs, with specifications taken from the webpage at: CIRCLE: Coordinating European Research on Molecular Communications (Coordinating European Research on Molecular Communications | Ufuk Avrupa; and CIRCLE: Coordinating European Research on Molecular Communications | Internet of Everything (IoE) Group)

Y. Cevallos et al., *Markup Languages and Simulators for Molecular Communications*, Signals and Communication Technology,
https://doi.org/10.1007/978-3-032-33040-6_3

and industrial engagement. CIRCLE advocates for the roadmaps at both the Member State and EU levels to guarantee that MCs' research swiftly aligns with viable market products.

The CIRCLE[2] project seeks to unify disparate research efforts in MCs throughout Europe by establishing a systematic research agenda to expedite convergence and promote pioneering research in the field. A theoretical and experimental study will be conducted on an intrabody sensor network that facilitates the integration of improved tumor treatment methods for diseases, such as cancer. The integrated detection and treatment methodologies encompass circulating tumor cell (CTC) identification in the bloodstream, the injection of genetically engineered bacteria into the tumor mass, and real-time monitoring of tumor progression alongside the bacteria's transcriptome, proteome, and temporal and spatial concentration, thereby facilitating the development of numerous highly disruptive therapeutic techniques. In addition, CIRCLE determines:

- Harmonized research methodologies and techniques within the area of MCs
- Breakthrough advances in MCs research through focused critical mass

Besides, tackling "grand" or societal challenges in CIRCLE Project:

- This project will facilitate scientific advancement in a Key Enabling Technology for future biomedical applications involving biological systems, including humans.
- Industrial Innovation, encompassing advancements in services, products, and processes.
- This project facilitates industry collaboration that influences novel medical solutions.

Objectives in CIRCLE projects can be defined as:

- Objective 1: Unify disparate research domains in MCs throughout Europe by establishing a systematic research agenda via collaborative development and ongoing enhancement of a research roadmap.

  Contemporary research on MCs is diverse and fragmented across various disciplines, with separate research teams following distinct strategic agendas. To enhance European influence in this nascent research domain, a comprehensive strategic research plan must be orchestrated across the Member States and aligned with international best practices. CIRCLE will implement a procedure for the creation and ongoing revision of a Strategic Research Roadmap concerning MCs.
- Objective 2: Facilitate directed learning for novice researchers in the field of MCs by enhancing the efficacy of information acquisition in essential disciplines.

[2] Information related to Coordinating European Research on MCs, which is detailed in the following paragraphs, is taken from the webpage at: Coordinating European Research on Molecular Communications | Ufuk Avrupa

The MCs hypothesis necessitates a distinct array of competencies that span information and communication technologies (ICT), biology, and medicine. This understanding is essential for comprehending the biological processes of communications (BIO), information theory, communication protocol design (ICT), and both experimental and theoretical validation in proposed situations (medicine). Presently, ICT and biology researchers seeking to engage in this field encounter an ineffective procedure for attaining the requisite skill set to proficiently develop information and influence in this domain. CIRCLE will implement a mechanism to build and continuously update a Skills Development Plan for young researchers aiming to enter this highly interdisciplinary and unorthodox research area, thereby significantly enhancing their efficiency in doing so. This will guarantee that Europe expedites scientific progress in the domain of MCs and secures a competitive edge in the future.

- Objective 3: Establish a cohesive network throughout Europe comprising research leaders and collaborators engaged in the field of MCs.
  The arguments here are exactly the same as Objective 1.
- Objective 4: Facilitate the rapid exchange of knowledge and best practices among researchers within Europe and beyond, concentrating on MCs.
  To swiftly achieve critical mass in MCs research, the exchange of knowledge and exemplary experimental procedures and methodologies is vital. CIRCLE establishes the requisite infrastructure and procedures to provide open access to laboratory equipment, the exchange of code repositories and experimental data, prompt publication of results, coordination of laboratory studies, and the sharing of experimental methodology. This purpose seeks to enhance learning among young researchers via collaboration and knowledge exchange.
- Objective 5: Implement a staff exchange initiative across CIRCLE partners, with specific emphasis on young researchers.
  The most effective strategy to enhance the learning of novice researchers in a new field is to collaborate with professionals, acquire best practice procedures and expertise, and gain direct access to laboratory equipment and experimental data. Emerging researchers will have the chance to allocate significant time at an alternative hosting institution, thereby enhancing their knowledge and skills in the domain of MCs.
  Facilitating visits among the various institutes within the consortium would strengthen a systematic collaborative research culture, thereby enhancing the dissemination of knowledge in this burgeoning field of study.
- Objective 6: Reduce the barriers for entry into the area of MCs for high-tech small- and medium-sized Enterprises (SME) via the joint delineation and ongoing enhancement of an industry engagement roadmap.
  As the area of MCs still seems unconventional and high risk, there has yet to be any explicit and strategic interest from high-tech SMEs and entrepreneurs to get involved in this area. There are a number of barriers to entry that must be addressed for this research area to deliver disruptive technological achievements within the next 20 years. These include European policy, legal frameworks, patenting, licensing, and other exploitation mechanisms necessary for industry

engagement. CIRCLE will engage with a wide range of stakeholders including policymakers, medical professionals, legal advisors, and others through the election of targeted expert. Working groups and focused workshops to tackle the major driving factors will ensure that when the research is sufficiently mature, the frameworks are in place for European high-tech SMEs to take advantage of this potentially game-changing technology to develop novel solutions for society.

A new markup language has been established from the aforementioned project, designated as MolComML. It facilitates the integration of interdisciplinary research endeavors. It delineates the fundamental elements of generic MC scenarios, utilizing the inherent flexibility of markup languages, through the integration of numerical analysis and experimental synthesis. MolComML exhibits its mobility and the potential for integrating many platforms to develop intricate simulation tools [1, 2].

On the other side, the IEEE has propelled the 1906.1 and 1906.1.1 standards due to the necessity to have a common framework to simulate and subsequently to use the programming code for other simulations by other researches, to avoid the heterogeneity of noncompatible programming codes and issues with information transmission, propagation, and reception in MCs systems, having as a simulation tool Network Simulator-3 (NS-3, which is studied in Chap. 2, Sect. 2.3.5).

The recommended practice in IEEE 1906.1 standard (the first IEEE nanocommunication standard) provides a definition, common framework, and common parameters for nanoscale communication networks. For justifying the introduction of IEEE 1906.1 standard, IEEE argued that there was not a standard or recommended practice that defines precisely what a nanoscale communication network is, nor how its components should be viewed in a conceptual framework, nor precise definitions of standardized parameters and components for interoperable simulation modules. The lack of standards in these areas has hindered research and development of this technology. The goal of this recommended practice is to provide a general framework that partitions the problem in such a way as to allow research and development to focus on specific, individual components that interoperate with one another. It also allows researchers to focus on specific components while being able to make assumptions about the environment in which those components will reside [3].

According to [3], to date, few scholarly articles purporting to advance nanoscale communication networking endeavor to explicitly define nanoscale communication. The absence of a clearly articulated challenge hinders systematic progress in technology. Consequently, each researcher pursues distinct and frequently disparate objectives, and the absence of a unified framework and definition hinders progress by obstructing the accumulation of prior findings. Establishing a unified concept for nanoscale communication is difficult, mostly due to the necessity for researchers from several disciplines, including information theory, physics, and biology, to attain a mutual comprehension of the subject. To prevent miscommunication of ideas, common terminology and definition are necessary.

Researchers may be distinguished by identical language; for instance, "communication" may signify one concept to a biologist (organisms in direct contact) and a distinct one to an information theorist (transmission of information quantified by information entropy). These varied disciplines frequently possess distinct overarching objectives; for instance, information theorists concentrate on information entropy, whereas biologists emphasize mechanisms, specifically the comprehension and manipulation of systems, including causally interacting phenomena and processes that yield observable outcomes. The IEEE 1906.1 standard delineates a description, a unified structure, and standardized parameters for nanoscale communication networks [3].

As yet, few scholarly articles purporting to advance nanoscale communication networking endeavor to accurately delineate nanoscale communication. The absence of a clearly articulated challenge hinders consistent technological progress. Consequently, each researcher pursues distinct and frequently disparate objectives, and the absence of a unified framework and definition hinders progress by obstructing the accumulation of prior findings [3].

One factor that makes it difficult to establish a universal definition of nanoscale communication is the necessity for experts from various disciplines spanning information theory and physics to biology to arrive at a mutual understanding of the subject; common terminology and definition are required in order to avoid miscommunication of ideas.

IEEE standard 1906.1.1 fortifies and directs the advancement of practical nanoscale communication systems across various media by establishing a unified data model, articulated in Yet Another Next Generation (YANG) language, which adheres to IEEE standard 1906.1–2015 for nanoscale communication systems. The IEEE 1906.1.1 YANG data model establishes a standardized framework for the management and configuration of nanoscale communication systems, aimed at achieving several objectives [4]:

- Helps guide conformance with IEEE Standard 1906.1–2015.
- Use a self-describing data structure for management and control of nanoscale communication systems.
- Embodies essential physics influencing IEEE 1906.1 systems.
- Establishes configuration and management for simulation and analysis.
- Characterizes a self-descriptive data structure utilized in repositories for nanoscale communication experimental data.

A standardized network, along with a management and configuration data model, facilitates the effective comprehension and utilization of nanoscale systems and simulations. The data model addresses the problem caused by the continuing development of many tools for nanoscale communication simulation, emulation, and data analysis that have different interfaces and cannot easily be composed into more useful toolchains [4].

The data model also tackles the issue that nanoscale systems and tools operate under varying assumptions regarding data collection and performance comparison across diverse nanoscale communication channels and technologies. This standardized data model is essential for the precise and equitable comparison of nanoscale communication systems, while also providing documentation that is comprehensible to both humans and machines. This data model tackles the deficiency of YANG data models pertaining to small-scale communication system interactions with nanoscale physics by offering a model for pertinent fundamental physics and the International System of Units [4].

As it seems obvious, the attempts, i.e., CIRCLE project; and IEEE, with MolComML and NS-3, respectively, have propelled initiatives to "standardizes" a universal programming tool to be used commonly by investigation, academia, and industry. This is the main reason to analyze these programs in this chapter.

3.2 Molecular Communication Markup Language (MolComML) Analysis

The tools provided by ICT demonstrate the feasibility for the analysis, design, and subsequent simulations and/or experimentation in the transfer of information in MCs. However, despite the existence of established communication theories from traditional transmission systems, which are applied to biological communication systems, scattered initiatives have emerged globally. These initiatives have primarily focused on developing software tools for programming, simulation, and eventually, the creation of hardware interfaces for experimentation in MCs systems.

Despite the significant scientific contributions these initiatives represent, there has been an incompatibility among these individual efforts, making it very difficult to utilize elements, scenarios, and research findings in the field of MCs interchangeably. Currently, there is a wide array of markup languages and simulators for nanocommunication systems at molecular level (as discussed in Chap. 2), which are generally non-interoperable. Therefore, there is a pressing need for a standard or a general language that enables the universal management of simulation and/or experimentation resources in biological communication systems. Thus, MolComML (according to CIRCLE project) emerges to ensure effectiveness in the reproducibility of results, as it provides compatibility, reproducibility, and information exchange across different platforms, facilitating the comparison and validation of results obtained in biological research laboratories.

The swift advancement of MCs study and the collaborative endeavors of specialists from many fields have prompted the establishment of a unified terminology for articulating and structuring intricate MCs investigations. This necessity has arisen within the context of the CIRCLE project. The European Commission under the Future and Emerging Technology Open program supports this initiative and has coordinated research operations in member countries across the EU, with extra participation from organizations in the USA, Asia, and Australia. This coordination and harmonization effort has facilitated the development of a unified syntax enabling researchers to conduct experiments, gather and disseminate results, and cross-validate findings, regardless of the research instruments employed [1, 2].

In the following, the main objectives of the MolComML format are summarized [5]:

- MolComML delineates the many categories of MCs scenarios across all levels of abstraction.
- MolComML facilitates the utilization of various software tools, and it does not necessitate the modification of models to adhere to disparate file formats.
- MolComML must ensure the persistence of models beyond the duration of the program utilized for their creation.
- MolComML employs a singular language to assess the specified situations using software tools and to create real laboratory investigations. This guarantees replicability and cross-validation of outcomes.
- MolComML facilitates the sharing and publication of models in a format accessible to researchers utilizing various software packages. Ultimately, MolComML should contemplate future development in light of the swift growth of the MCs domain.

The expectation beyond MolComML is to prevent model rewriting and to establish consistent input/output interfaces, hence ensuring experimental reproducibility. For instance, the current MCs simulators, including BiNS2 (examined in Chap. 2, Sect. 2.3.1), N3Sim (a simulation framework for diffusion-based molecular communication nanonetworks, analyzed in Chap. 2, Sect. 2.3.2), and others based on NS-3 (discussed in Chap. 2, Sect. 2.3.5), incorporate functionalities to simulate diffusion-based channels. They can be configured through various interfaces and produce distinct output files, complicating the comparison of simulation results, particularly in intricate scenarios. An effective solution for ensuring result reproducibility, whether derived from numerical analysis or experimental synthesis, is provided by a versatile XML-based markup language known as MolComML [2].

MolComML is intended for setting software tools and is also envisioned as a future addition to organizing wet lab experiments conducted in biological research laboratories, facilitating cross-validation of results and expediting convergence toward the desired outcome. Results from wet lab tests can be utilized to configure both targeted simulations and additional wet experiments, hence ensuring complete compatibility, reproducibility, and interchangeability across several platforms [2].

MolComML has been developed to offer a cohesive framework for characterizing molecular complexes in simulations or experimental laboratory settings. It offers platform-agnostic specification language that may be utilized to specify experiments as well as their output format, encompassing monitored amounts, monitoring frequency, and preservation technique. Subsequently, the experiment designer must choose the most appropriate simulation program for conducting the intended experiment, contingent upon the accessible functionalities. This can be accomplished without acquiring knowledge of any specific configuration language or syntax of the utilized package, provided it is compatible with the MolComML program. It enables the description of components and functions absent in other languages [2].

MolComML emphasizes the modeling of various communication transmitting and receiving interfaces, communication channels, and communicative units, such as cells. Consequently, it offers an abstraction level appropriate for communication yet is excessively comprehensive for a systems perspective (systems biology) and overly simplistic for elucidating inner cellular mechanisms (synthetic biology) [2].

The design of MolComML seeks to integrate with existing markup languages. For example, it can incorporate existing information in Systems Biology Markup Language (SBML, examined in Chap. 2, Sect. 2.2.1) regarding certain biological entities (e.g., molecules or cells), without the need for complete redefinition. A more stringent integration can be attained between MolComML and Synthetic Biology Open Language (SBOL, examined in Chap. 2, Sect. 2.2.2). The combination of these two languages enables the building of applications that connect cellular behavior, represented by SBOL circuits, to extracellular signaling, which can be represented using MolComML. This technique would prevent the need to re-adapt these languages for use outside their original domains, a practice that generally yields unsuccessful results and results in complicated, often inflexible structures [2].

The current simulation tools for MCs systems, as noted in the previous chapter, employ various languages, configurations, and output files. Moreover, they incorporate characteristics that frequently render cross-validation and result replication impractical, particularly for intricate communication infrastructures. The development of MolComML has facilitated collaboration and knowledge sharing across various research domains [2].

As considered in the earlier chapter, the development of various tools for analysis, simulation, and experimentation in MCs, defined through particular and generally incompatible initiatives, necessitates the establishment of a standard or universal method. This would enable the independent sharing, refinement, and improvement of MCs research results, thereby facilitating the reuse of information with experimental projections and even lines of code in MCs scenarios. Therefore, the universal generic method known as MolComML is considered henceforth in this chapter.

MolComML provides a high level of abstraction suitable for the transfer of molecular information by employing systemic components in natural and/or synthetic biological communications. Additionally, MolComML enables convergence between numerical analysis and experimental synthesis through a markup language that facilitates cross-validation between experimentation and theoretical results, allowing for the combination of different software tools. This chapter will also analyze case studies of simulations developed using MolComML as it is indicated in Fig. 3.1 [6, 7].

The strength of MolComML lies in its integration of markup languages used across various scientific fields, such as [6, 7]:

- Mathematical Markup Language (MathML), which allows the inclusion of functions designed to represent equations and parameters [8].
- SBML [9] and SBOL [10], which facilitate the description of biological processes, among others.

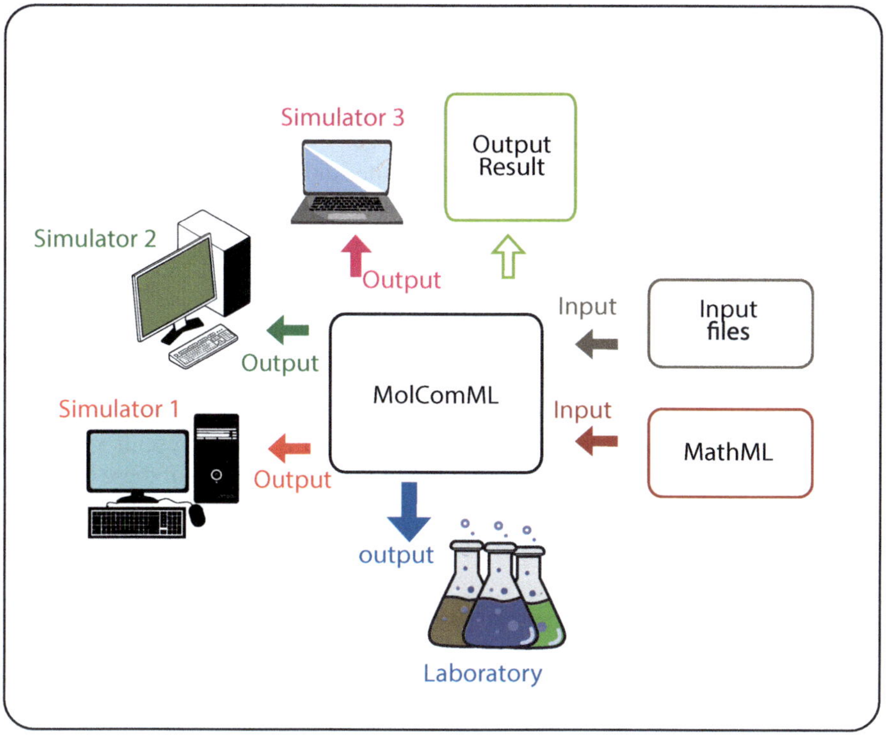

Fig. 3.1 MolComML aims to integrate diverse research efforts, ranging from experimental laboratory studies to various simulation platforms

MCs have rapidly developed in recent years (through numerous platforms mentioned in Chaps. 1 and 2), combining efforts across different disciplines. Thus, regardless of the environment in which nanocommunication information is analyzed, MolComML emerges as a common language to ensure precise consideration of such information. MolComML can be regarded as a parallel attempt to the IEEE 1906.1 and 1906.1.1 standards in standardizing schemes, programming languages, simulators, and experimental tools for MCs.

MolComML is deployed as an XML-based language that unifies both numerical analysis and experimental synthesis, ensuring flexibility [11]. Its design is oriented toward the convergence of all research fields through the integration of platforms and offering interoperability to represent elements, interactions, configurations, experimental results, and simulations in the MCs domain [12].

Regarding the guidelines of MolComML for handling nanocommunication systems at the molecular level, the following are noted [6]:

- Various MCs scenarios are represented at all levels of abstraction.
- Adjustments are made in different file formats to avoid rewriting models and consequently allow the use of various software tools.

- Models are preserved regardless of the software used.
- It is established as the sole language for analyzing MCs scenarios through software tools to synthesize experimental results.
- Availability for the use of software environments enabling access to models.
- It provides scalability of the markup language.
- Due to the constant evolution occurring in the study of MCs, it requires continuous updates to integrate new definitions and models.

The following sections analyze the constituent elements of MolComML.

3.2.1 Principal Components of MolComML

The Unified Modeling Language (UML) diagram in Fig. 3.2 represents some components that outline the structure of a general MCs system. Each of these contains a set of parameters that can be private and/or public and are defined according to the purpose of the simulation to be performed [7]. Figure 3.2 highlights several important aspects, such as [7]:

- A shaded blue area containing the main components that establish the fundamental structure of the language, including elements, topologies and network interfaces, communication protocols, states, interaction rules, channels, and modulation types.

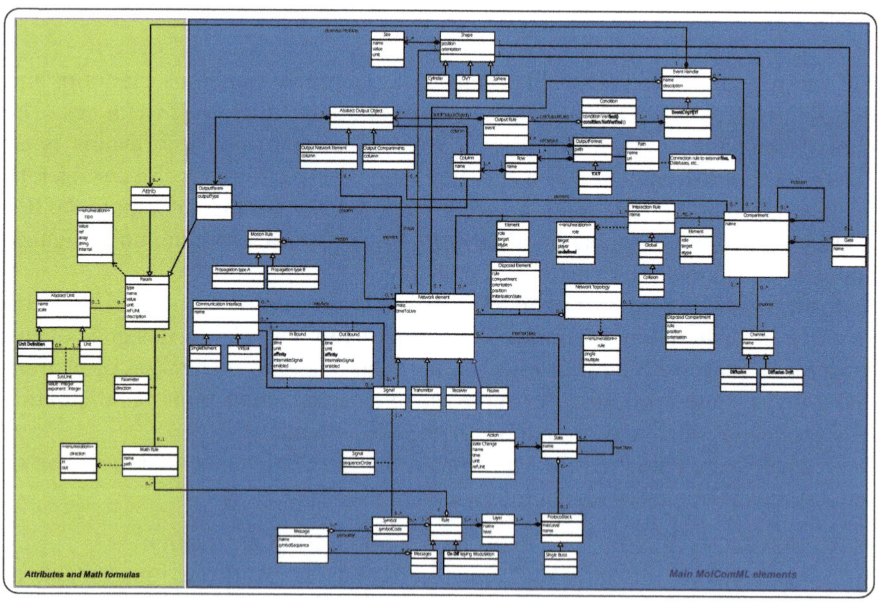

Fig. 3.2 UML Diagram in MolComML [7]

- And a shaded yellow area denoting attributes, rules, and mathematical formula parameters that are useful within the simulator.

3.2.1.1 Network Elements

The network elements class enables the construction of architecture by considering each network element with a set of general attributes that describe the element's shape and size, mass, lifespan, communication interface, motion rules, and the propagation process of the said element [7, 13].

The main network elements are defined as objects of the following types: transmitter, receiver, and signal to be transmitted. The following is a description of each element [11]:

- Transmitter Tx: The Transmitter object encodes the element that generates the information in the form of a molecular signal, including parameters, such as molecule creation rate, molecular emission rate, and the mechanism for molecule release [13].
- Receiver Rx: A receiver is the object where the information of the molecular signal arriving at the communication destination is decoded. This information may be altered due to fluctuations and the behavior of molecular channels. Additionally, it is possible to make multiple configurations, and different types of receivers can be assumed, such as absorbing receivers [13].
- Signal: A signal is the object that corresponds to the molecular information and will thus be emitted of the transmitters nanomachines toward the receivers nanomachines. Information can be represented by deoxyribonucleic acid (DNA), proteins, ions, and other mechanisms [13].

3.2.1.2 Communication Interface

The communication interface class is responsible for the exchange of information between network elements. For this to be possible, external communication interfaces must be defined between each of these elements. The interfaces have properties that describe the directionality of communication, affinity, and the types of signals transmitted and received [6, 7].

Figure 3.3 shows four network elements: transmitter Tx and receiver Rx (each with their respective biological receptors) and two types of signals (colored and uncolored circles), as well as the corresponding communication interfaces [6].

Figure 3.4 presents a code snippet used to define communication nodes, their addressing, delimitation, etc., which will act as simulation elements in MolComML [7].

Network elements can communicate over an interface, facilitating the characterization of each unit. A subset of parameters delineates the interface properties, and the relationships between the communication interface and the signaling element,

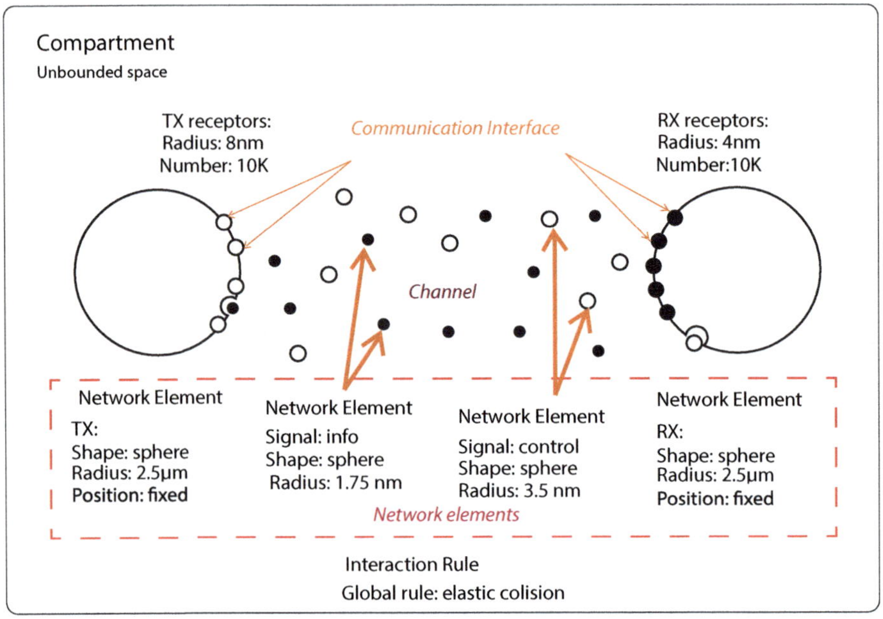

Fig. 3.3 Description of network elements in MolComML [7]

```
<listOfNetworkElement>
   <networkElement id="NE02"    name="RX_Node"   type="receiver">
        <shape      id="SH001"   type="sphere" />
        <motion     refID=""     name="none" />

        <child   id="NE02-1"    childOf="NE02"       name="RX" >
            <size     refShape="SH001"     name="radius" refUnit="U05"   unit="nm"
            value="5000" />
            <mass           value="100"     refUnit="U07" unit="pg"     />


            <timeToLive value="10"    refUnit="UD03" unit="hour" />


            <interface refID="CI01-02"    type="RX_FullSurface" value="1" />

          </child>
      </networkElement>
</listOfNetworkElement>
```

Fig. 3.4 Definition of network element in MolComML, in this case, is a spherical nanomachine of receptor type [6, 7]

inherited from the primary network element, are specified. To verify compatibility between the interface-signal element pair, two association classes mapped through data and rules are used accordingly, see Fig. 3.5 [7].

According to Fig. 3.3, it is possible to model the communication interface of the nodes (transmitter, receiver, and virtual elements) and two types of communication

```
<listOfCommunicationInterface>
   <CommunicationInterface  id="CI01"  name="Absorbing surface" type="virtual">

        <child id="CI01-01" childOf="CI01" name="TX_FullSurface" >
            <outbound enabled="yes">
            <signal refID="NE03-1" type="generic carrier"  time="0"    unit="s"
            refUnit="U03" affinity="1" internalizeSignal="no"/>
         </outbound>

         <inbound enabled="no">
         </inbound>
       </child>

       <child  id="CI01-02"  childOf="CI01"  name="RX_FullSurface" >
           <inbound enabled="yes" >
              <signal refID="NE03-1" type="generic carrier" time="0"   unit="s"
              refUnit="U03" affinity="1" internalizeSignal="yes" />
          </inbound>

          <outbound enabled="no">
          </outbound>
       </child>
    </CommunicationInterface>
</listOfCommunicationInterface>
```

Fig. 3.5 Definition of communication interfaces in MolComML [6, 7]

interfaces corresponding to two types of signals. Furthermore, both interfaces represent a distinct category, as each simulates an absorbing surface receptor, as illustrated in the code in Fig. 3.5. Consequently, for each category of communication interface, the MolComML file delineates the dimensions, affinity with signal kinds, quantity of receptors, and other relevant parameters [7].

3.2.1.3 Compartment

Compartment is somewhat considered a container characterized by a particular type and shape in which, for example, chemical substances, modeled cells, and network elements used in a simulation scenario are housed. This scenario can include multiple compartments of the same type, organized in a nested hierarchical architecture or close to each other. The interconnections among various compartments are illustrated by gates, which establish interaction protocols [7]. Every gate is distinguished by a distinct name, location, dimensions, form, and alignment. Each network piece in a simulation scenario is situated within a compartment [13].

Figure 3.6 specifies the space and boundaries of the compartment used in a simulation. Here, the compartment in use has a cubic shape with a side length of 1 mm and consists of absorbing walls, implying that a moving network element that comes into contact with one of the cube's walls will simply be removed from the simulation [6, 7].

```
<listOfCompartment>
     <Compartment id="C001" name="unbounded domain" >
          <shape       id="SH002" type="cubic" >
               <size        name="side"  value="1000"  unit="um"  refUnit="U04" />
          </shape>

          <channel     name="free diffusion"   refID="CH001" />

          <gate        name="" refID="" />
     </Compartment>
<listOfCompartment>
```

Fig. 3.6 Definition of compartment in MolComML [6, 7]

3.2.1.4 Gates

Gates determines a list of entries for the interconnection of adjacent or nested compartments. Each gate is identical and consists of a unique name, position, size, shape, and orientation. A gate facilitates the passage between surfaces that interconnect other compartments [6, 7].

3.2.1.5 Shape

Shape is established as a generic element whose attributes are *position and orientation*. It can be applied to any geometric object associated with network elements such as transmitters, receivers, or interconnection gates with other compartments. The shapes considered in this class include cylindrical, cubic, etc., inheriting the respective dimensions and position within the compartment. This is the case in Fig. 3.3, where spheres are used to represent the transmitter and receiver; this geometric shape is characterized by its radius, with values of 1000 ηm and 5000 ηm configured for each component, respectively. In the case of modeling a blood vessel, a cylinder could be used, with appropriate values for radius, length, and position [6, 7].

3.2.1.6 Interaction Rules

In *Interaction Rules*, rules are established through mathematical expressions imported from an external model, such as MathML. These rules are defined globally or specifically for their application throughout the communication environment or in a subset that is part of the simulation [6, 14].

Network elements and compartments will be associated with at least one interaction rule; the association will detail a role to describe the behavior of interactions between the objects involved. For example, in the case of molecular collisions, their behavior in the simulation can be determined through bonding, fusion, or bouncing.

```
<listOfInteractionRule>
    <globalRule      id="IR001" name="generalCollisionRule" type="global">
        <!-- This rule could be a set of math equations -->
        <mathRule id="MR001"       name="elasticCollision" path=
        "./mathEquations/resolveCollision.xml" >
            <!-- A list of <param> -->
        </mathRule>
        <!-- role  =      (1) target:  the target of the interaction
                          (2) undefined: the role is not important
                          (3) player: interaction only with the target -->
      <element type="NetworkElement" refID="*" etype="" role="undefined" targetID=
      "*"    />
    </globalRule>

    <specificRule  id="IR002"  name="transparentCollision"  type="collision">
       <mathRule  id="MR002"  name="transparentCollision"  path=
       "./mathEquations/transparent.xml" >
            <!-- A list of <param> -->
       </mathRule>
       <element type="NetworkElement" refID="NE02" etype="receiver" role="target"
           />
       <element type="NetworkElement" refID="NE01" etype="passive" role="player"
       targetID="NE02" />
       <element type="NetworkElement" refID="NE03" etype="signal"   role="player"
       targetID="NE02" />
    </specificRule>
</listOfInteractionRule>
```

Fig. 3.7 Definition of interaction rules [7]

Therefore, it is important to specify different behaviors for the interaction of the elements considered in the simulation [6, 14].

At the UML diagram in Fig. 3.2, three attributes have been included within the roles: *target*, *player*, and *undefined*, to specify the interaction between a set of components. The undefined role is used when the role of each element involved in the interaction is not essential, such as in an elastic collision. In Fig. 3.2, it can be observed that the Interaction Rule section depends on the role and its various options [7]:

- Target: It is used to identify or indicate the communication target in interaction.
- Player: Used when there is a single destination in the interaction.
- Undefined: Used when the role of each element involved in the interaction is unimportant.

Figure 3.7 indicates two types of interactions [7]:

- Global rule: To define an elastic collision, it must be considered that the entities involved in this event, such as network elements or compartments, are not significant and are thus set as undefined [7].
- Specific rule: This is conceptualized in the special case of a transparent biological receptor that acts as the communication destination *target* when interacting with network elements that will not be counted. The transparent receptor can only quantify the number of specific elements that arrive without any other specification [7].

Furthermore, an additional regulation is specified for activities to be executed in the molecular counting process while engaging with a transparent receptor target and the network elements to be monitored [7]. This phenomenon can be termed assimilation based on receptor-ligand compatibility principles via *affinity*. Figure 3.5 examines a simulated scenario including a complicated interaction between a molecule (the information signal of the network element), designated as the player, and a receptor element, identified as the target [7, 14]. Consequently, the network element designated as the target verifies if the network element operating as the player interacts with one of the communication interfaces compatible with the receptor surface. This methodology is illustrated in Fig. 3.3, as a particular rule governs the interaction of signals with the pertinent receptors positioned on the transmitter and receiver surfaces [7].

3.2.1.7 Communication Channels

At *Communication Channels*, the transmission channel responsible for conveying signals to the receiver is specified. There are various types of molecular propagation channels, including those based on diffusion principles (*Diffusion*) and diffusion in the presence of drift (*Diffusion Drift*). In this section, a communication channel element will be defined and connected to each compartment involved in the communication environment [7]. The local environmental conditions can be precisely described using mathematical rules defined in one or more MathML files (Fig. 3.8). Additionally, it is possible to initialize parameters used in imported equations. The molecular channel definition can be shared between one or more compartments, as exemplified in Figs. 3.3 and 3.8, which utilize a free diffusion model for both signals [7].

3.2.1.8 Network Topology

The relative movement of network elements within the simulation environment is described by *Network Topology*; it includes a list of network elements for each compartment, their starting position, and state [14]. References between topology and network elements or compartments are managed through the association of

```
<listOfChannel>
    <channel      id="CH001" name="free diffusion" type="Diffusion">
        <mathRule id="MR010"     name="brownian_diffusion"    path=
        "./mathEquations/brownian.xml" >
            <!-- A list of <param> -->
        </mathRule>
    </channel>
</listOfChannel>
```

Fig. 3.8 Communication channel definition [7]

```
<networkTopology>
 <disposedCompartment id="DC001" refID="C001" name="unbounded domain"rule="single"
 >
     <position refShape="SH002"  name="coords"  value="array(x=0; y=0; z=0)"
     unit=""    refUnit=""/>
         <disposedElement  id="DE001" refID="NE01-1" name="transmitter" rule=
         "single" >
             <position  refShape="SH001"    name="coords" value="array(x=5; y=0;
             z=0)"    unit="um" refUnit="U04" />
             <initialization      refID="IS01" />
         </disposedElement>

       <disposedElement  id="DE002" refID="NE02-1" name="receiver" rule="single">

           <position refShape="SH001" name="coords" value="array(x=0; y=0;
           z=0)"    unit="um" refUnit="U04" />
           <initialization      refID="IS02" />
       </disposedElement>
     disposedCompartment>
</networkTopology>
```

Fig. 3.9 Network topology definition [7]

classes in order, detailing each relationship. Additionally, a rule can be established manually or automatically for each element using mathematical formulas [7].

In Fig. 3.9, two network elements are highlighted: a transmitter node and a receiver node within a compartment (as referenced in earlier sections), as well as the initial state of each node. According to Fig. 3.3, the transmitter and receiver are fixed within the compartment because the network elements of a signal are generated during the execution of the simulation [7].

3.2.1.9 State

The *State* characterizes the internal state of each network element, as schematized in the UML diagram in Fig. 3.2, using the *internalState* relationship as a specification. Each state defines a list of actions for referencing the behavior of the element, as well as the time required to complete an action and proceed to the next state transition [7]. Each state can also be associated with an element of a protocol stack, as seen in Fig. 3.2, where both the transmitter and receiver possess internal states managed by the general communication protocol executed in both nodes [7, 15].

3.2.1.10 Protocol Stack

Network elements are associated with the *protocolStack* object through *internalState*, and the different actions related to each state are specified in the Finite State Machine (FSM) protocol, which fully details each communication rule executed by

```
<listOfProtocolStack>
  <protocolStack id="PS01" name="Emission of a Single Burst of carriers" type=
  "SingleBurst"  maxLevel="2">
       <layer level="1" name="Physical">
             <rule id="PR1-01"     type="on_off_keying modulation" >
                   <mathRule id="MRO20"    name="impulse"   path=
                        ". /mathEquations/impulse2.xml" >
                              <!-- A list of <param> -->
                        </mathRule>

                        <symbol id="SYM001"    symbolCode="1" >
                           <signal    id="MOLO01" type="carrier" refID="NE03-1"
                          sequenceOrder="1" >
                              <param type="value"    name="burst"    value="500"
                             unit=""       refUnit="" />
                             <param type="value" name="duration" value="0"
                              unit="s"   refUnit="U03" />
                              </signal>
                           </symbol>

                           <symbol id="SYM002"       symbolCode="0" >
                              <signal        1d="MOL002" type="carrier" refID="NE03-1"
                             sequenceOrder="2" >
                            <param type="value" name="burst" value="0"  unit=""
                                 refUnit=""  />
                            <param  type="value" name="duration" value="0"
                            unit="s"      refUnit="U03" />
                         </ signal>
                      </symbol>
             </rule>
         </layer>
         <layer level="2" name="signal_pattern">
            <rule id="PR2-01" type="messages" >
              <message id="MES001"  name="singleImpulse"        symbolSequence "[1 0]"
             symbolRef=" [SYM001 SYM0 02]" />
             <!-- e.g.    symbolSequence "[1 0 0 1 0]"-->
             </rule>
           </layer>
        </protocolstack>
</listOfProtocolStack>
```

Fig. 3.10 Protocol stack definition [7]

a network element. Furthermore, it is possible to define different protocol stacks, each with a customized organization, composed of a specific structure of level and identification [6, 13].

Furthermore, each protocol layer can be aligned with widely utilized protocol stacks in telecommunications, or entirely novel layers can be devised. The layers encompass a collection of regulations that dictate designated conduct. In the physical layer, each rule may encompass a compilation of related symbol elements. Each symbol is characterized by a collection of attributes, including burst type, carrier type, and burst duration, among others. Thus, each symbol element must be linked to a minimum of one message element to delineate the appropriate symbol sequence (Fig. 3.2) [6, 13].

To initialize all the required parameters in a customized way for the referenced element, the pertinent configuration is provided in standard XML format in Fig. 3.10, where two protocol layers will be used [7]:

- Layer 1: This layer is responsible for transmitting/receiving bursts to encode/detect symbols.

- Layer 2: This layer can decode messages, determine the signal level to limit receiver congestion, and send feedback signals.

In the example in Fig. 3.3, a bidirectional protocol is proposed, emulating the functionality of transmission control protocol (TCP) but operating in layer 2 according to the proposed architecture [7].

3.2.1.11 Event Scheduler

The *eventScheduler* element consists of an event controller that, through the *eventHandler* class, monitors a set of attributes established as *name* and *description*, aiming to detect any significant variations in values, thereby triggering the associated event accordingly. The event description is defined by the condition class through the Character DATA (CDATA[3]) section, with behavior generated through pseudocode. The event scheduler defines the initial state of each network element and the specific events that cause a state transition in a destination node because of an event or at scheduled times [12].

In Fig. 3.3, the scheduled event involves the Rx initiating a search process to find a Tx node. In this event, a process is initiated in a pair of biological nanomachines using a communication process between the Rx-Tx pair, like TCP feedback, though explicit acknowledgments are not used, as they would degrade communication performance due to the significant delay characteristic of MCs [16].

The number of events can cause state transitions in an FSM. For this reason, the information from both transmitter and receiver elements, as well as the events scheduled in a simulation, is periodically exported to a My Structured Query Language (MySQL) database [13].

3.2.1.12 Supports Elements

In the UML diagram in Fig. 3.2, on the left side, several support elements are visualized, describing attributes, parameters, units of measurement, and formulas that can be invoked by the MolComML elements [7, 13].

3.2.1.12.1 Param

Each element in a MolComML configuration file may include a set of user-defined attributes, allowing predefined properties to be extended or new ones to be introduced with high flexibility. The *Param* attribute defines custom parameters that

[3] This type of information is defined as blocks of text that XML recognizes.

function as predefined properties, typically employed for initializing biological entities [15, 17]. The attributes *type* and *name* are defined for these entities; *value* and *unit* refer to measurements, and *description* of the *param* is used to provide a brief description. The *type* attribute can assume fixed label values to specify in detail the type of information appearing in the *value* attribute [7].

The value attribute must be numeric and is classified as follows [17, 18]:

- *value*: It is employed for numeric data.
- *ref*: It is a reference to other elements in MolComML within the same configuration file (i.e., with a unique ID).
- *array*: It is employed for numerical arrays, to describe a position in a coordinate system.
- *string*: It is used for textual descriptions, like the *OutputParam* object, and serves to export specific information.
- *internal*: It refers to sub-attributes of some mathematical formulas described in the attached MathML files.

3.2.1.12.2 Unit

Unit is an object that can be defined as the unit of measurement related to the value attribute for each parameter and numerical attribute in MolComML. The *AbstractUnit* class inherits the *unit* and *unitDefinition* objects, which are established in simple or customized units provided by a combination of standard units depending on the *subUnit* object with the attributes *value* and *exponent*. For each unit, a unique ID, name, and scale must be defined [7, 13]. When custom units are used, a specific format is required that includes the value and an exponent to adequately combine each standard unit as indicated in Eq. (3.1) [17].

$$customUnit = \left(value \times 10^{scale}\right)^{esponent} \tag{3.1}$$

3.2.1.12.3 Math Rule

MathRule is conceptualized through an external file in MathML format. Subsequently, for each imported mathematical rule, a set of parameters can be defined to describe and initialize each related object, specifying the *direction* information for each one. These can be managed with the following attributes [6]:

- *direction* = *out*. Used to obtain the results of a formula.
- *direction* = *in*. Used to initialize a value.
- *direction* = *internal*. Can be defined in the same MathML file.

In some cases, it may be necessary to initialize a parameter with values defined in another section of the MolComML file, such as the size of a network element.

According to the notation in Eq. (3.2), the *value* attribute is established by referencing that value of the element as follows [7, 13]:

- *ID*: The first string considered is named and corresponds to a specific element in the MolComML file, defined as a network element.
- *Size*: The second string defined is the name of the attribute of some fixed elements in the subtree of the considered element, in this case, the size.
- *Name*: The third defined string is one of the attributes set for the extracted child elements, in this case, the name.
- [...]: Here, it is described that the value that the previously specified attribute string should take, in the case of Eq. (3.2), is radius.
- *Value*: If this element is found, the value related to the last attribute string is returned, in this case, the value [7].

$$ID.size.name\left[radius\right].value \tag{3.2}$$

In conclusion, the notation described returns the value of the "radius" attribute of the network element associated with the ID [7].

3.2.2 *Simulation of Case Studies of Biological System Scenarios with MolComML*

3.2.2.1 Case 1: Simulations of Diffusion-Based MCs Scenarios

To demonstrate the agility, feasibility, and ease of MolComML configuration, simulations were conducted with different characteristics in the propagation channel, including aqueous environments like blood, and gaseous environments, like carbon dioxide (CO_2) [19].

In each simulation, variations were made in the separation distance between the transmitter nanomachine (Tx) and the receiver nanomachine (Rx) and the size of the molecules emitted and released by Tx into the communication channel. The type of nanomachines used and characterized in the simulated MCs systems is spherical, with an absorbent nature specified at the molecular destination [19].

3.2.2.1.1 Simulation of MCs in a Gaseous Diffusion Environment: CO_2

To perform the simulations for this diffusion environment, the characteristics and parameters detailed in Table 3.1 were used. The most significant variation among the simulated propagation channels was the viscosity of the environment, where deposited molecules move, governed by free Brownian motion. The modulation type used in this case was on–off keying (OOK), representing bit-1 with 1000 molecules and no molecule for bit-0 [19].

Table 3.1 Simulation parameters for the gaseous diffusion environment: CO_2 [19]

Simulation parameters	Value	Unit
Distance, Tx-Rx	5 y 2.5	(μm)
Node radius, Tx	5	(μm)
Node radius, Rx	50	(ηm)
Molecule radius	1.75 and 0.85	(μm)
Number of molecules to represent bit-1 and bit-0	1000; 0	–
Transmission environment viscosity	1.5×10^{-5}	kg/ms

As shown in Table 3.1, the simulations modified the separation of distance between Tx and Rx nanomachines and the size of the molecules that Tx deposits into the propagation channel. These modifications were made to determine notable differences in information reception at the communication destination [19].

Figure 3.11 schematically presents the results obtained by performing the simulation in the BiNS2 simulator using the relevant MolComML configuration file, varying the separation distances between nanomachines with values of 5 μm and 2.5 μm, and setting the molecular radius at the emission end to 1.75 μm. In Fig. 3.11, a clear impact of the separation distance parameter on the number of molecules Rx captures compared to the total number of molecules Tx adds to the propagation channel is observed [19].

Additionally, Fig. 3.11 shows that when the molecular link distance is shorter (d2 = 2.5 μm, red dashed line), Rx nanomachine captures more information molecules, leading to adequate link quality, like conventional communication systems where shorter distances between communication ends are used [19].

Meanwhile, Fig. 3.12 presents the simulation results obtained with a molecular radius of 0.85 μm. In this case, the impact of the radius or size of the modulated molecules transmitted into the molecular propagation channel is reflected in a greater number of molecules detected by Rx when a distance of 5 μm was used. This situation is evident by comparing the results shown in Figs. 3.11 and 3.12, where an improvement in the number of molecules received is observed when a smaller molecular size (0.85 μm) is defined at a greater distance (5 μm). Concurrently, in Fig. 3.11 (blue line), approximately 490 molecules were received, while in Fig. 3.12, the number of captured molecules was approximately 550 (blue line) [19].

With the above, it is possible to identify that in a gaseous propagation environment, using smaller molecules can achieve better information molecule propagation even at greater distances [19].

3.2.2.1.2 Simulation of MCs in an Aqueous Diffusion Environment: Blood

This simulation scenario was performed in BiNS2 using the relevant MolComML configuration, employing an aqueous propagation channel, specifically corresponding to human blood, with characteristics and parameters detailed in Table 3.2 [19].

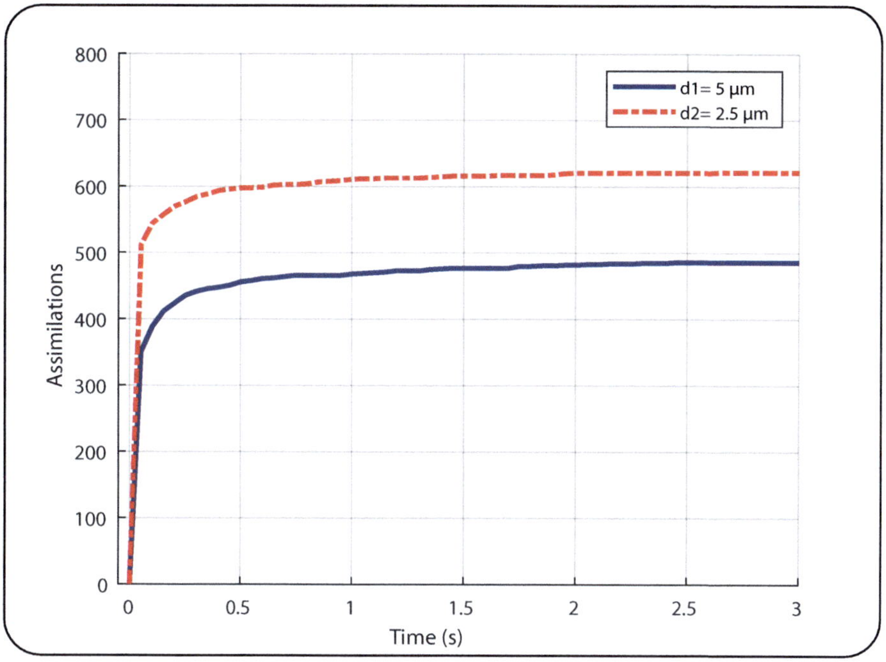

Fig. 3.11 Simulation results in the CO_2 gaseous diffusion transmission channel when the molecular radius corresponds to 1.75 μm [19]

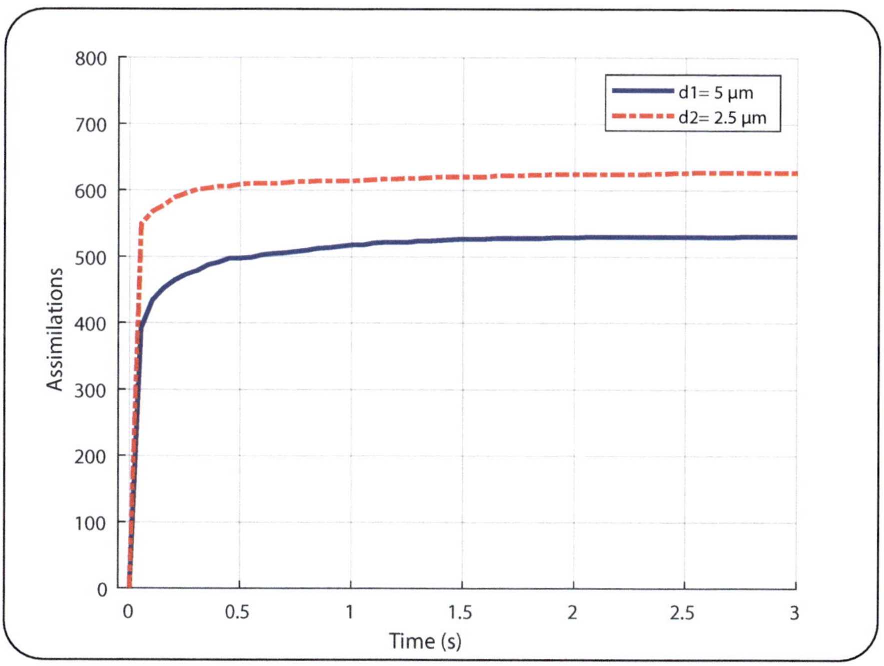

Fig. 3.12 Simulation results in the CO_2 gaseous diffusion transmission channel when the molecular radius was 0.85 μm [19]

Table 3.2 Simulation parameters for an aqueous propagation environment corresponding to human blood [19]

Simulation parameters	Value	Unit
Distance, Tx-Rx	5 y 2.5	μm
Node radius, Tx	5	μm
Node radius, Rx	50	ηm
Molecule radius	1.75 y 0.85	μm
Number of molecules represented by bit-1 and bit-0	1000; 0	–
Transmission environment viscosity	4×10^{-3}	kg/ms

Similar to the previous case, simulations were conducted varying the separation distance between nanomachines and the radius or size of the molecules that Tx emits into the propagation channel, using the same type of motion and modulation as in the previous case [19].

Figure 3.13 indicates that the results obtained show the interaction between the distance separating the nanomachines in the transmission environment and the number of molecules Rx detects and quantifies. This figure also shows that when a distance of d2 = 2.5 μm (the shortest distance in the simulation) was used, the number of molecules captured by Rx was higher compared to the greater distance [19].

Figure 3.14 summarizes the simulation results obtained by reducing the molecular radius parameter that Tx deposited into the propagation channel to 0.85 μm. This figure shows a lower incidence in the number of molecules captured by Rx compared to the results presented in Fig. 3.13 [19]. This occurs because the viscosity of an aqueous environment is much higher than the viscosity of a gas (CO_2) in the previous simulation, making molecular movement much more complex in the propagation channel. Additionally, the number of captured molecules tends to vary due to random motion and the diffusion effect occurring in a three-dimensional environment [19].

3.2.2.1.3 Results Comparison Using Different Simulators

To ensure appropriate reliability and validity of the results presented in this section, two different simulation programs, N3Sim and BiNS2, are used. Both can read MolComML input files, generate and compare similar results using the same configuration file. The transmission channel consists of a simple diffusion medium where molecular movement occurs in three dimensions [7, 19].

In this MCs system, Tx releases a burst of molecules (500 and 1000, respectively) toward Rx. The molecular burst sizes are set to test potential transmission signal dysfunctions [7, 19]. The MolComML configuration file initializes and configures the parameters in Table 3.3 [7, 19].

Figure 3.15 graphically demonstrates that using N3Sim and BiNS2 yields similar and therefore highly reliable results that align with the relevant theoretical support [14].

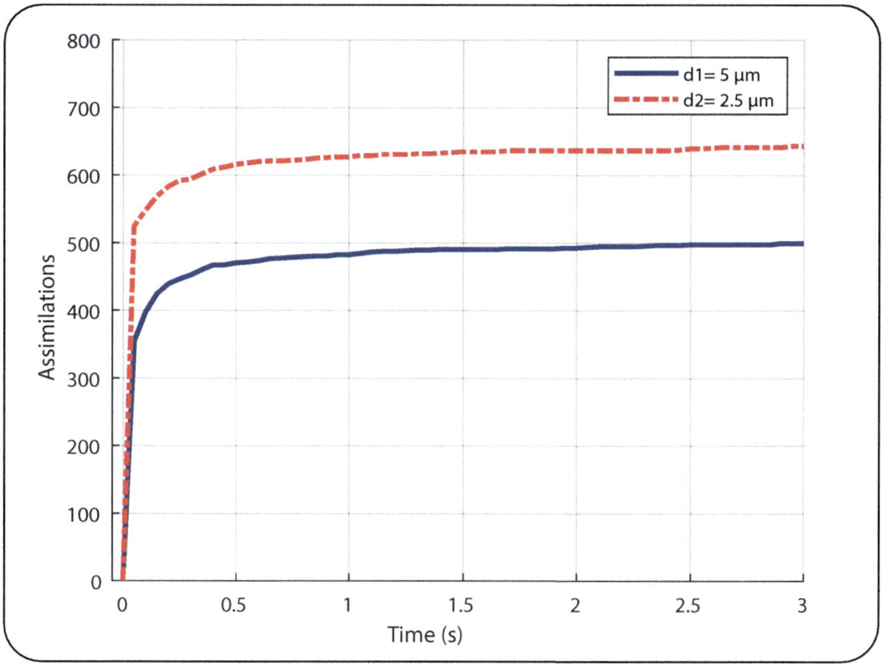

Fig. 3.13 Simulation results in an aqueous diffusion environment corresponding to blood when the molecular radius was 1.75 μm [19]

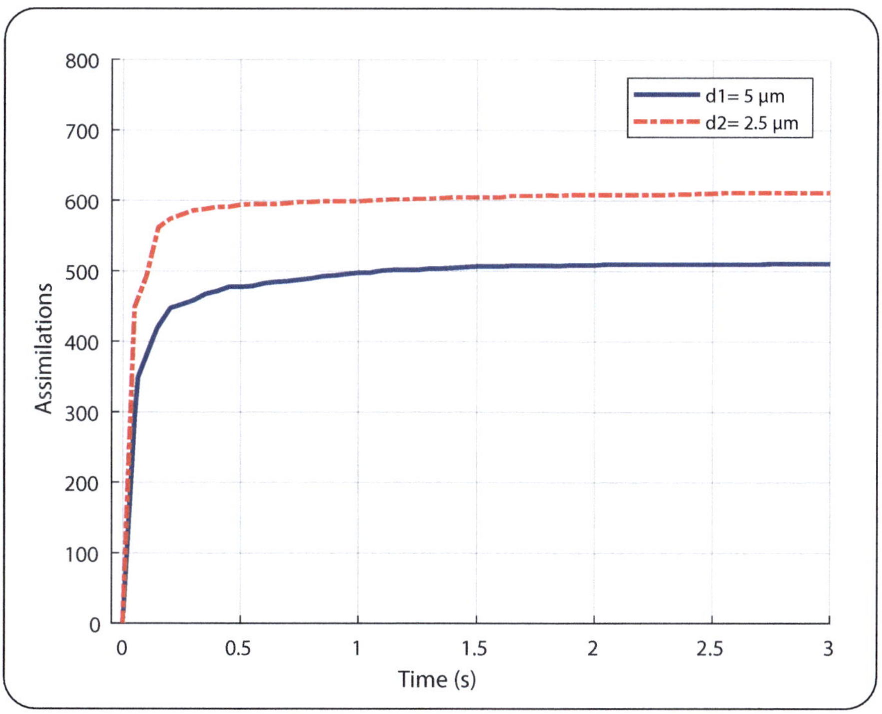

Fig. 3.14 Simulation results in an aqueous propagation environment corresponding to blood, using a molecular radius of 0.85 μm [19]

Table 3.3 Simulation parameters for comparison between N3Sim and BiNS2 simulators [7, 19]

Simulation parameters	Value	Unit
Distance, Tx-Rx	5	μm
Node radius, Tx	5	μm
Node radius, Rx	50	ηm
Molecule radius	1.75	μm
Number of molecules used to represent bit-1 and bit-0	(500; 1000); 0	–

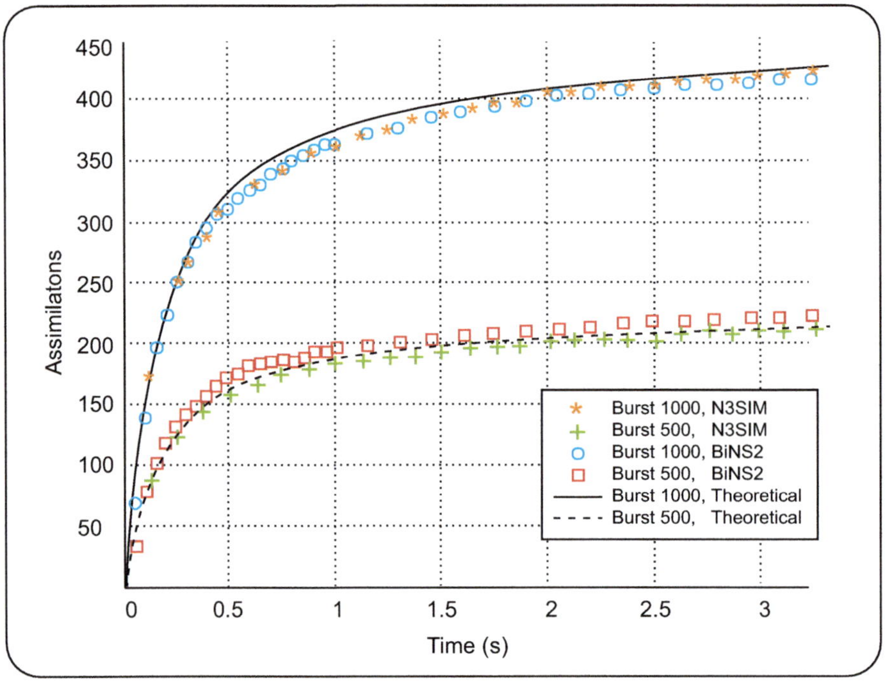

Fig. 3.15 Comparative simulation results between N3Sim and BiNS2

3.2.2.2 Case 2: MCs System Based on Internal Blood Vessel Flow

Complex environments involving nanocommunications within blood vessels can be developed in MolComML by programming multiple entities (cells and molecules) using tools provided by BiNS2. In this MCs scenario, the simulation is performed within a segment of a blood vessel containing white blood cells, red blood cells, and platelets. These components move due to collisions and according to

Hagen-Poiseuille's[4] law; the transmission channel is further characterized by Brownian motion [7, 19]. Additionally, circulating tumor cells (CTCs) are involved in simulation. At the communication destination, the receptors have fixed position and are transparent type, detecting the presence of CTCs via surface compatibility without affecting the movement of CTCs or surrounding blood cells. In BiNS2, a blood vessel section is modeled following particle propagation laws (primarily due to collisions and Hagen-Poiseuille's law), thus demonstrating the flexibility offered by MolComML [7, 19]. The biological parameters to be simulated correspond to those established by [20] and are specified in Table 3.4, considering relevant values for CTCs and red blood cells, which will have a higher concentration than other types of cells [7, 19].

Red blood cells are the most abundant component of blood, and their influence on the rheological qualities of this fluid is essential. The main effect is that these cells congregate along the central axis of a channel, where blood velocity is maximal, as dictated by Hagen-Poiseuille's equation, resulting in a cell-free layer adjacent to the vessel walls [7, 19]. The thickness of this layer is significant in smaller vessels, but it becomes insignificant in bigger vessels [19, 21]. In this simulation, an initial cell-free layer comparable to the size of red blood cells is assumed, though its

[4] Hagen-Poiseuille law: To state the Hagen-Poiseuille law, it is necessary to review some concepts. It should be noted that flow F is the displacement of volume (ΔV) per unit of time (ΔV), and they are related by Eq. (A) [20]:

$$F = \frac{\Delta V}{\Delta t} \tag{A}$$

Additionally, if a fluid moves inside a cylinder (and thus acquires the same cylindrical shape) with an area A and an average velocity $\overline{v}$, and if during a time interval Δt the fluid cylinder advances Δx so that a volume (of height or length L) is generated as the liquid moves from position P_1 to P_2, then Eq. (B) allows the determination of flow [20]:

$$F = \frac{\Delta V}{\Delta t} = \frac{A \cdot \Delta x}{\Delta t} = A \cdot \frac{\Delta x}{\Delta t} = A \cdot \overline{v} \tag{B}$$

Equation B holds at any point along the cylinder, regardless of the irregularity of the cross-section.

In a well-defined physical system, it is also possible to predict flow based on the cylinder's geometry and the fluid properties. Jean Poiseuille observed the flow of liquids in small-diameter cylinders and derived the law that bears his name. In a straight and rigid cylinder, the flow is determined by Eq. (C), known as the Hagen-Poiseuille law [20]:

$$F = \Delta P \cdot \frac{\pi \cdot r^4}{8 \cdot \eta \cdot l} = \frac{\Delta P}{R} \tag{C}$$

where F is the flow, ΔP is the driving pressure, r and l are the internal radius and the length of the cylinder, respectively, and η is the viscosity. The Hagen-Poiseuille equation requires that the driving pressure and the resulting flow remain constant and it implies three key points:

- Flow is directly proportional to the axial pressure difference ΔP. The proportionality constant is $\frac{\pi \cdot r^4}{8 \cdot \eta \cdot l}$, defined as the inverse of resistance R.
- Flow is directly proportional to the fourth power of the cylinder's radius.
- Flow is inversely proportional to the length of the cylinder and the fluid viscosity [20].

Table 3.4 Simulation parameters for Case 2 [7, 19]

General parameters		
Blood vessel length	6.0	mm
Blood vessel radius	30	μm
Mean flow velocity	0.5	mm/s
Viscosity	1.3	mPa.s
Temperature	310	K°
Time step	100	μs
Red blood cells		
Concentration	5×10^6	cell/mm^3
Radius	2.9	μm
White blood cells		
Concentration	7.5×10^3	cell/mm^3
Radius	3.8	μm
Platelets		
Concentration	2.5×10^5	cell/mm^3
Radius	1.0	μm
CTC and receiver node		
CTC concentration	10^{-4}	cell/mm^3
CTC and receiver radius	5.0	μm
CTC initial distance from axis	[0; 12.5; 25]	μm
Initial distance between the CTC and the receiver	6	mm

size can change in space and time according to the previously mentioned particle movement law [7, 19].

In these simulations, the CTCs are positioned at the beginning of the blood vessel, and both the receptor and the CTCs are placed at varying distances from the vessel axis, ranging from 0 to 25 μm. The initial longitudinal distance from the CTCs to the receptor node is equal to the total length of the vessel, which is 6 mm. Additionally, it is assumed that the receptor and the CTCs are of the same size [7, 19].

The goal of the simulation is to maximize the probability of CTC detection using MolComML, which allows monitoring the positions of the CTCs along the blood vessel through the output information. Extensive simulations have been performed using BiNS2 to visualize the various propagation patterns of the CTCs. Each different configuration has been simulated (with different colors) at least 10 times (Fig. 3.16) [7, 19].

In Fig. 3.16, the following aspects are highly relevant [17, 19]:

- In the top-view figures, a black dot represents the different initial positions of the CTCs. The dashed circular line symbolizes the closest position to the blood vessel walls.
- The side-view illustrations depict the distance of the CTCs from the central axis as a function of time. The vertical red dashed line approximately marks the moment when the cells cross the midpoint of the blood vessel (3 mm). The dashed lines represent the point at which the cells are closest to the vessel walls.
- The duration of each execution in the same simulation can vary, and the total propagation time depends on the cell's speed, which, in turn, is a function of the

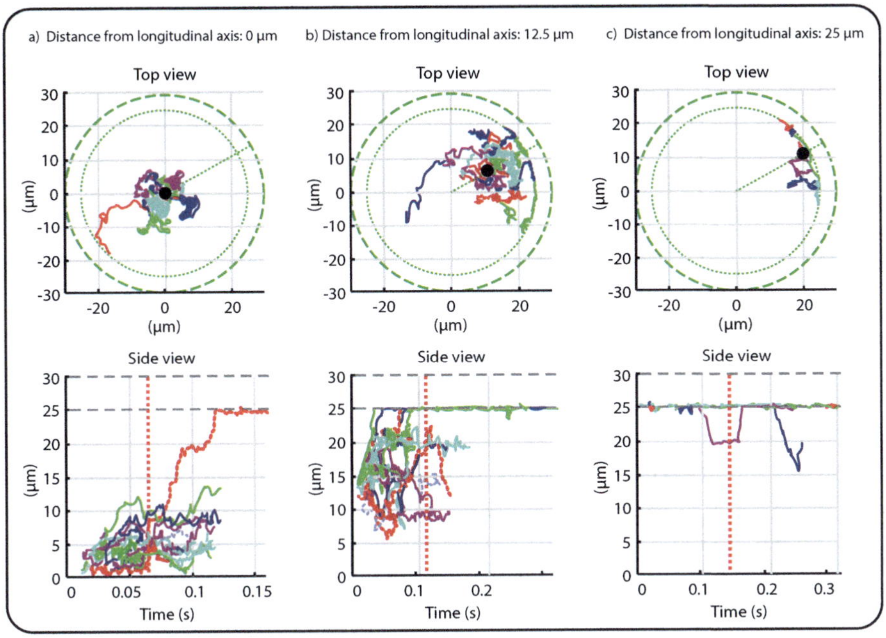

Fig. 3.16 Propagation profile of CTCs at different distances. (**a**) $d = 0$ μm. (**b**) $d = 12.5$ μm. (**c**) $d = 25$ μm along the length of the blood vessel 6 mm [7, 19]

actual distance from the central axis (according to Hagen-Poiseuille's law). Therefore, crossing time increases as the CTCs approach the blood vessel walls.

- In Fig. 3.16a, in the top view the CTCs propagate randomly around their initial position; however, their distance from the axis increases. This occurs because the highest velocity profile and collisions affect the central position, causing lighter cells (with lower density) like CTCs to bounce off heavier cells toward the endothelium, where the blood density is lower.
- The top view in Fig. 3.16b reveals that the CTCs propagate randomly around their initial position. A similar behavior is observed in the final simulation of this case study.
- Figure 3.16c shows the results when the CTCs are positioned near the endothelium and tend to remain near it, with a slight angular displacement from the initial position of approximately 30 degrees ($\pi/6$ rad) in both directions.

Figure 3.17 presents the mass probabilities versus the distance from the vessel axis for the three simulated distances, highlighting the following specifications [17, 19]:

- The assumed positions of the CTCs approximate a Gaussian distribution centered between 0.5 and 0.75 μm, as the mean value is shifted to the right, i.e., toward the endothelium (Fig. 3.17a).
- Figure 3.17b determines the intermediate initial position of the CTCs, assuming the position is closer to the endothelium, and for the second section, the position is set at an axis distance corresponding to 12.5 μm. The side view indicates that in

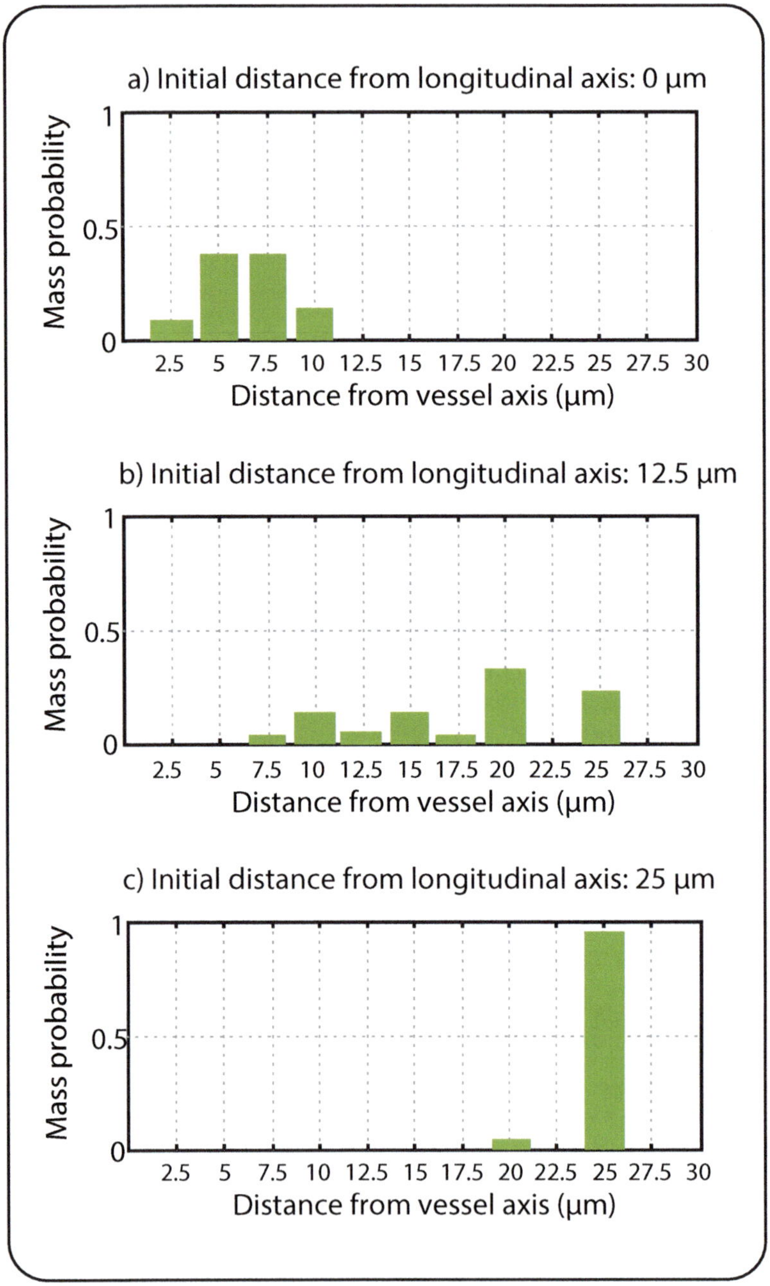

Fig. 3.17 Propagation profile of CTCs at different distances from the blood vessel axis: (**a**) d = 0 μm. (**b**) d = 12.5 μm. (**c**) d = 25 μm [19]

most cases, the CTCs tend to move toward the endothelium. The CTCs remained for a shorter time inside the simulated vessel section compared to the first half, due to the higher propagation speed. Nonetheless, in over 90% of cases, the CTCs reached a farther position than the initial position, suggesting that in long vessels, the general tendency of CTC movement is directed toward the vessel walls.

- In Fig. 3.17c, almost all CTCs maintained their initial position (near the endothelium) [7, 19].

The simulation results indicate that CTCs migrate toward the endothelium irrespective of their initial location, suggesting that CTC receivers should be positioned adjacent to the endothelium to enhance the likelihood of capturing CTCs within blood vessels [7].

3.2.2.3 Case 3: Combination of Simulators

MolComML allows for specifying configuration files and output formats of configurations and for interacting jointly with simulation tools such as BiNS2 and N3SIM, enabling the exchange of functions and the complementarity of different tools that can read/write MolComML configuration files, thus enhancing integration [7]. This new simulation scenario consists of neurotransmitter diffusion directed from the axon to the dendrite [22]. This communication occurs via messenger molecules with a radius of 0.2 ηm. The distance between the transmitter and the receptor is 20 ηm in the synaptic cleft [22], the transmitter radius is 20 ηm, and the receptor radius (receptors on the dendrites) is 1 ηm, as shown in Fig. 3.18 [7].

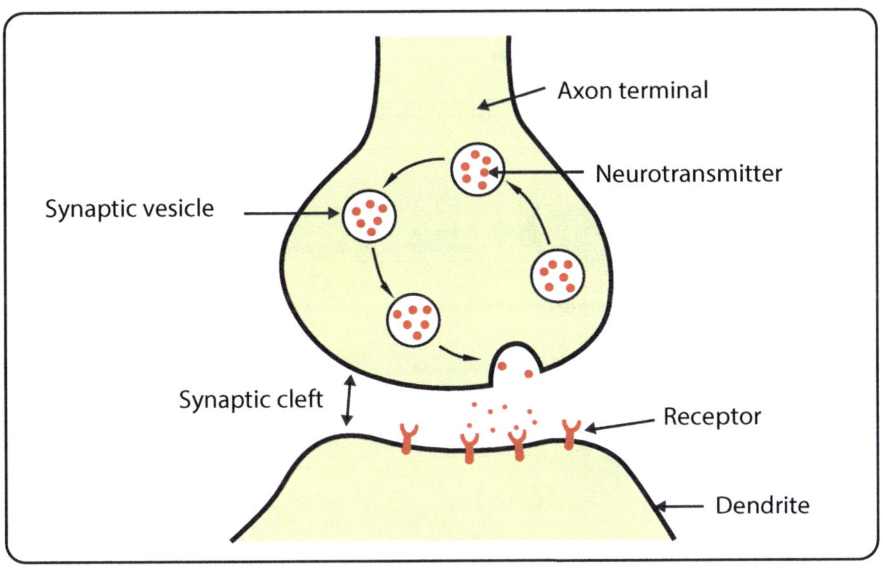

Fig. 3.18 Communications via diffusion in a synaptic cleft

In the simulations, the number of transmitter and receptor nodes used is set to $N = 1$, 20, 50, and 100 for each type of node, and the time interval is defined as 0.1 ηs. The transmitter releases 1000 molecules via diffusion in the synaptic cleft, leading to a number of absorptions by the receptor [7].

N3SIM is used to accelerate simulations by neglecting molecular collisions, thus reducing simulation time, with minimal impact on the output information in the scenario, although it has the drawback of only simulating absorbing receptors without implementing functions associated with individual receptors [7]. BiNS2 enables modeling the general scenario with absorbing receptors in specific positions on the receptor surface (dendrite). Using both simulators together facilitates faster simulations [7].

MolComML is a common language for describing the simulation, which is initialized in BiNS2, where nodes are placed according to the geometry of the scenario to be simulated. The result obtained is a simulation status report that serves as the configuration input for the N3SIM simulator (Fig. 3.19) [7].

Molecular emission is carried out by multiple nodes (neurotransmitter carriers), and molecule capture is performed by multiple absorbing receptors (receptors on the dendrite). Thus, a simulation behavior resembling an MIMO (multiple-input/multiple-output) system is implemented using N3SIM. At the same time, a set of information is loaded into the MolComML file produced by the BiNS2 simulator to transfer the position of each node to N3SIM through the MolComML configuration file, where the simulation is configured using a list of node coordinates [7]. BiNS2 uses an algorithm to calculate the positions of the transmitter and receptor nodes, considering a specific mathematical rule for the network topology in the MolComML configuration file [7].

Figure 3.20 demonstrates that the simulation durations achieved through the integrated use of the N3SIM and BiNS2 simulators are around six times shorter than those derived solely from BiNS2 [7].

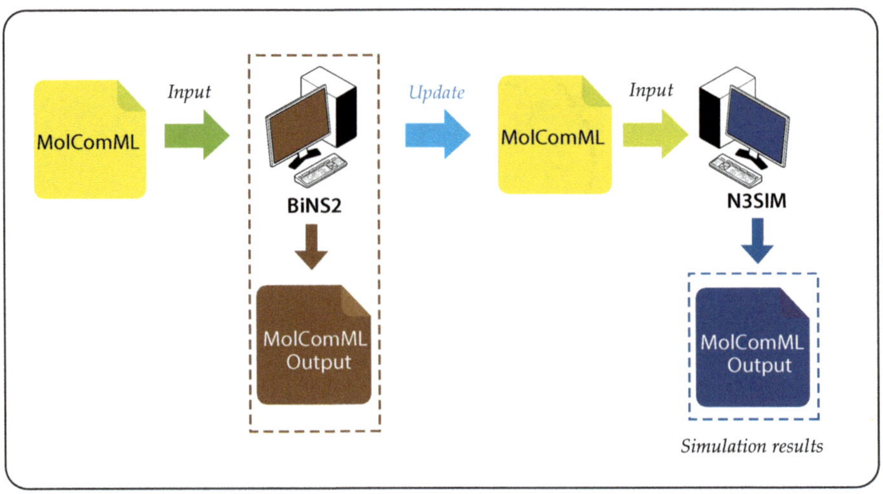

Fig. 3.19 Simultaneous use of two simulators for MCs integrated into MolComML

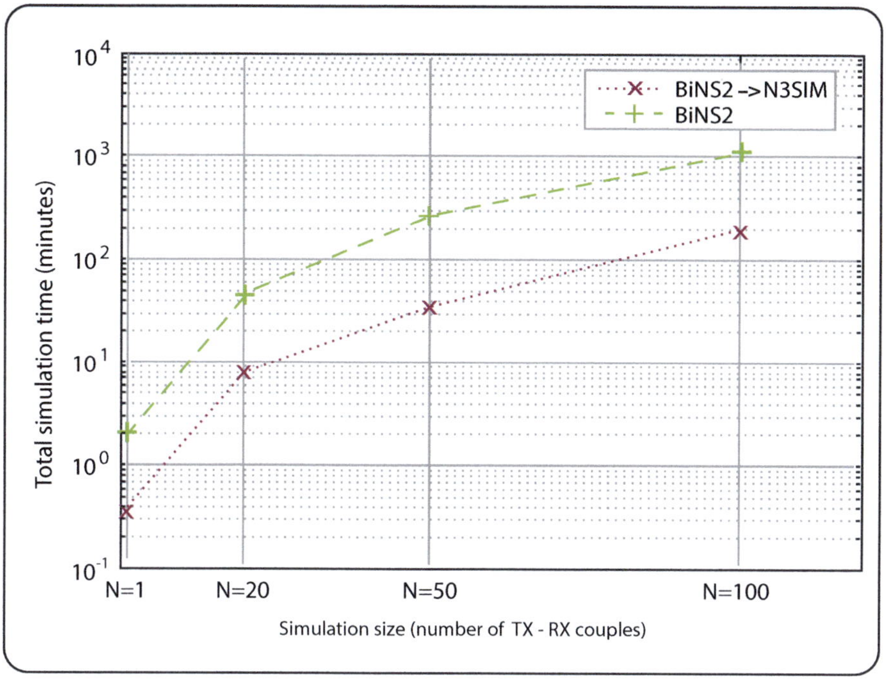

Fig. 3.20 Total simulation time comparison

3.3 IEEE Standard Data Model for Nanoscale Communication Systems

IEEE advances the 1906.1 and 1906.1.1 nanocommunications standards. These standards delineate concepts, nomenclature, a conceptual model, and standardized measurements for ad hoc networks and MCs at the nanoscale, namely within dimensions ranging from 1 nm to 100 nm. A nanometer is designated as the lower threshold to exclude individual atoms or small clusters from being classified as nanoscale entities; for context, the diameter of a hydrogen atom is 0.5 nm. The upper limit is the dimension at which material properties undergo substantial alterations from their macroscopic scale, the dimension at which an individual can directly observe them without magnification assistance. A fundamental prerequisite of a nanoscale communication system is the presence of a critical component that exists at the nanoscale in at least one dimension [3, 4].

In its first version, the IEEE 1906.1 recommended practice (approved in 2015) introduced a conceptual framework along with standardized terminology for ad hoc nanoscale communication networks and MCs. More specifically, this recommended practice encompasses the following elements [3, 23, 24]:

- The definition of nanoscale communication networks.
- The theoretical framework for ad hoc nanoscale communication networks.

- The standardized terminology for nanoscale communication networking includes: (1) defining a nanoscale communication channel and highlighting its key differences from macroscale channels; (2) specifying the abstract interfaces between nanoscale communication channels and nanoscale systems; (3) identifying performance metrics relevant to ad hoc nanoscale communication networks; and (4) describing the relationship between nanoscale and traditional communication networks, including essential high-level components, such as coding and packet structures, addressing, routing, localization, layering, and reliability mechanisms.

A new version of the standard, 1906.1.1, was developed in [4] and adopted in 2020 by the Internet Engineering Task Force, Request for Comment 7950, wherein IEEE develops a unified Yet Another Next Generation (YANG) data model for IEEE 1906.1 nanoscale communication systems. This data model comprises a set of YANG modules that delineate nanoscale communication systems and their corresponding physical quantities, adhering to IEEE Std 1906.1-201, a unified foundation for all nanoscale communication technologies. The model represents physics specific to the nanoscale. The physics are classified as non-standard, as mandated by IEEE Std 1906.1-2015. The model delineates remote configuration and management for the operation and analysis of nanoscale communication networks. A self-describing data structure is characterized by the framework for datastores and repositories of nanoscale communication experimental data, facilitating a unified comprehension of the data across diverse nanoscale communication media and technologies [4, 23, 24].

The YANG data model establishes a standardized framework for network administration and configuration data in nanoscale communication systems. This accomplishes multiple objectives: It mandates compliance with IEEE Std 1906.1(TM)-2015. It delineates nanoscale communication systems. It delineates the essential physics influencing IEEE 1906.1 systems. It delineates the configuration and administration for simulation and analysis. It delineates a self-descriptive data structure utilized in archives of nanoscale communication experimental data. An established network, management, and configuration data model facilitates the effective comprehension and use of IEEE 1906.1 systems and simulations [4, 23, 24].

A standardized data model is required to guarantee that systems and simulations adhere to IEEE Std 1906.1-2015. A standardized data model is required to function as documentation that is comprehensible to both humans and machines for IEEE 1906.1 systems. A data model is required to capture fundamental physics, as small-scale communication systems directly interact with nanoscale physics. A standardized data model is essential for the correct and equitable comparison of IEEE 1906.1 systems. Repositories of experimental data from small-scale communication systems necessitate precise and comprehensive documentation for the data to hold significance. This prevalent data model offers a self-describing framework that fulfills this objective [4, 23, 24].

The IEEE 1906 framework can be correlated or aligned with communication protocols. Numerous efforts have been made to correlate nanoscale communication

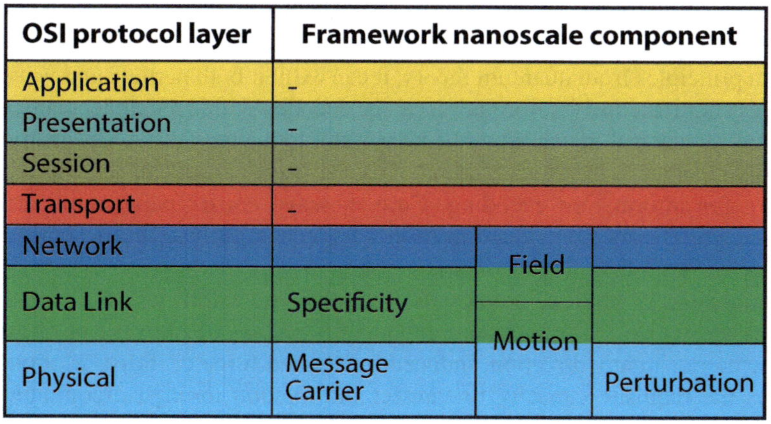

Fig. 3.21 Example mapping between the OSI model and nanoscale communication network architecture

networks with the Open Systems Interconnection (OSI) model, defined in ISO/IEC Standard 7498-1:1994 in various manners, as illustrated in the example (Fig. 3.21). Nanoscale communication systems are less complex and more challenging to program than macroscale systems due to their size. The IEEE 1906 framework exemplifies those prevalent in natural, small-scale environments, such as biological systems. The outcome is diminished control over the implementation of the specifics of OSI logical levels. Perturbation, for instance, is more straightforward and generally exhibits a more mechanical execution. Motion exhibits greater randomness, for instance, influenced by Brownian forces. A gradient or field is necessary to enhance mobility. There is diminished control over specificity (addressing) [3, 4, 23, 24].

The OSI model (ISO/IEC 7498-1) delineates the conventional seven-layer protocol architecture. The IEEE 1906 framework is positioned among the lowest levels of the OSI model. Specifically [3, 4, 23, 24]:

- Message pertains to a classical frame, packet, or Protocol Data Unit (PDU).
- Message carrier (Component 0) pertains to a wave, which is a property that encodes information.
- Motion (Component 1) pertains closely to the classical physical layer (wave propagation).
- Field (Component 2) corresponds closely to the traditional data connection and network layers, facilitating node-to-node information transfer.
- Perturbation (Component 3) is approximately analogous to classical modulation at the physical layer.
- Specificity (Component 4) pertains closely to classical addressing within the data connection layer.

In the next paragraphs, these components will be explained briefly.

- Component 0, message carrier, is the carrier that facilitates the transmission of the message. It may take the form of either a particle or a wave, and consistent with principles from quantum theory, it can exhibit both particle and wave properties simultaneously. Information can be encoded within the molecular structure of the carrier as it travels from the transmitter to the receiver. Additionally, oscillatory changes in the concentration of the message carrier can serve as a mechanism for information encoding. Core message carrier parameters, including molecular motors, are defined in 1906.1.1 (Sect. 3.3.3.1) with a designated label name [3, 4, 23, 24].
- Component 1, Motion, defines the mobility characteristics of Component 0 (message carrier). The motion component facilitates the movement of the message carrier in any direction, induced by applied force or thrust. Motion facilitates the requisite capacity to transfer information through a communication channel. Message carriers can be either active, generating their own movement or passive being transported by the medium. Examples include molecular diffusion in fluids, Brownian motion, and self-propelled movement. Motion (Component 1) is closely associated with the traditional physical layer, particularly wave propagation. Motion specifications, including diffusion movement, are defined in 1906.1.1 (Sect. 3.3.3.3) with a designated label [3, 4, 23, 24].
- Component 2, Field, defines organized movement of Component 1 (Motion). The field component provides the service of organized motion for message carriers. It can be thought of as a virtual waveguide in communications. The field may be implemented internally or externally relative to the medium. Examples include an internal implementation, including swarm motion or flocking behavior; external implementations are nonturbulent fluid flow, electromagnetic (EM) field, chemical gradient released to guide movement of bacteria, and molecular motors guided by microtubules. Field (Component 2) relates approximately to the classical data link and network layers (ensuring node-to-node information flow). In 1906.1.1 (Sect. 3.3.3.3) are defined some field specifications (with a label name) include the concentration gradient in this component [3, 4, 23, 24].
- Component 3, Perturbation, specifies the signal conveyed by Component 0 (message carrier). The perturbation component provides the service of varying message carriers as needed to represent a signal. This may be thought of as modulation (signal impression) [3, 4, 23, 24].

  Examples include signals based on the number of received message carriers, controlled dense-versus-sparse concentrations of molecules, the simple on-versus-off flow of signal molecules, using different types of message carriers, and modifying the conformation of molecules (e.g., DNA) to represent multiple states. In 1906.1.1 (Sect. 3.3.3.1) are defined some perturbation specifications (with a label name) which include the molecular structure in this component. Perturbation (Component 3) relates approximately to classical modulation at the physical level [3, 4, 23, 24].
- Component 4, Specificity, determines the selective reception of Component 3 (Perturbation). The specificity component provides the service of sensing or reception of a message carrier by a target. This can be mapped to addressing in

Table 3.5 IEEE 1906.1 core component specification for 1906.1.1 standard for nanoscale communication systems [23]

Label (name)	Parent component	Component
Bacterium	Messagecarrier	Component 0
Calcium-ion		
Charge		
Electromagnetic-wave-variation		
Ligand		
Motor		
Diffusion	Motion	Component 1
Motion-type		
Potential-difference		
Walking		
Wave-guide		
Compartmentalized	Field	Component 2
Concentration-gradient		
Directional-antenna		
Microtubules		
Nanostructure-orientation		
Concentration-change	Perturbation	Component 3
Electrical-current-variation		
Molecular-structure		
Transmission-rate		
Antenna-aperture	Specificity	Component 4
Receptor-sensitivity		
Electrical-charge		

classical communication systems. Examples include the shape or affinity of a molecule to a particular target and complementary DNA for hybridization. In 1906.1.1 (Sect. 3.3.3.2) are defined some specificity specifications (with a label name) including receptor sensitivity in this component. Specificity (Component 4) relates approximately to classical addressing at the data link layer [3, 4, 23, 24] (Table 3.5).

To facilitate research convergence toward common objectives in the field of nanoscale communication, the availability of a modular and freely accessible simulation platform is crucial. In this regard, a novel, open-source, and extensible simulation tool has been developed and selected as a reference model within the IEEE P1906.1 working group (and recognized in 1906.1.1 standard). This simulation tool, built on top of NS-3,[5] is readily accessible online [3, 4, 25, 26].

The core component of the simulator is responsible for representing the elements defined within the communication framework. Its models implement essential parameters and shared functionalities applicable across different communication

[5] ns-3 | a discrete-event network simulator for Internet systems (nsnam.org).

schemes, thereby enabling consistent interaction and coordinated operation among system components during message exchange [3, 4, 25, 26].

In NS-3, the *P1906MessageCarrier* class defines the message entity that can be created and transmitted between devices. By default, it contains a single data member, a pointer to an *ns3:Packet* object which represents the bit sequence encapsulated within the message. This structure can be extended with additional attributes to tailor the message carrier to specific modeling requirements and application scenarios. The *P1906NetDevice* class represents the network device participating in the communication process. It includes the *m_interface* at a member, which corresponds to the communication interface implemented by the *P1906CommunicationInterface* class. The communication interface is composed of the *P1906TransmitterCommunicationInterface* and *P1906ReceiverCommunication Interface* classes, which manage the transmission and reception processes, respectively. The *P1906Medium* class plays a central role in the communication process by coordinating message delivery among connected devices. It leverages the *P1906Motion* class to model propagation dynamics and transmission delay effects. Additionally, *P1906Perturbation*, *P1906Specificity*, and *P1906Field* classes are used to model the perturbation, specificity, and field components, respectively [3, 4, 25, 26]. The operation of nanoscale communication according to IEEE standards is illustrated in Fig. 3.22 and described in the following section [3, 4, 25, 26].

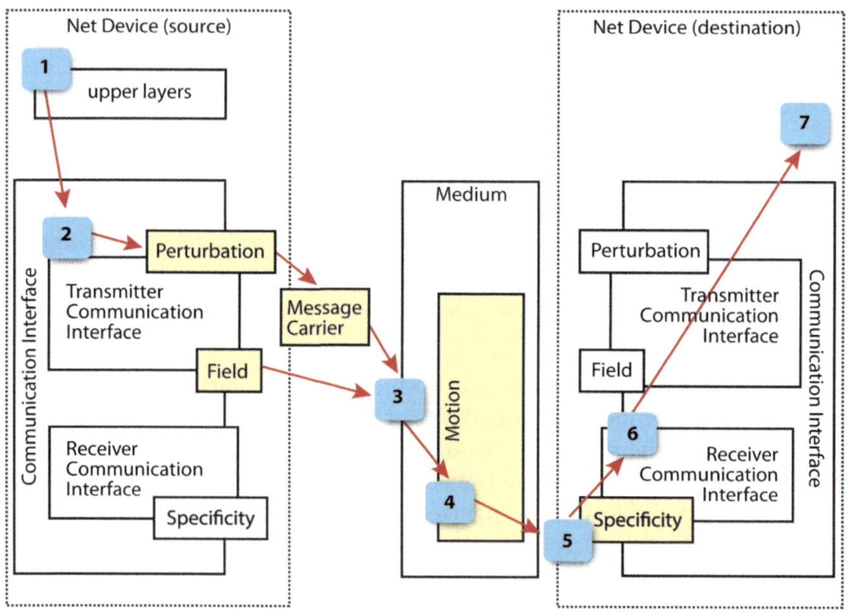

Fig. 3.22 Overview of the procedures involved in communication as represented by the simulation software tool within the established simulation framework. Source: Adapted from IEEE P1906.1/Draft 1.0 Recommended Practice for Nanoscale and MCs and IEEE Standard Data Model for Nanoscale Communication Systems Framework [3, 4, 25, 26]

1. The *NetDevice* receives a message from the upper protocol layers and forward it to the transmitter communication interface.
2. The perturbation component is used to generate the message carrier.
3. The transmitter communication interface initiates the propagation within the medium by releasing the message carrier along with the corresponding perturbation and field components.
4. The motion component alters the characteristics of the message carrier, including parameters, such as propagation attenuation and transmission delay.
5. The message carrier is received by the receiver, and the specificity component checks for compatibility.
6. If compatible, the message is delivered to the higher layers.
7. Finally, the message is received by the higher layer's dimension.

3.3.1 Example of Open-Source Implementation of NS-3 Defined in IEEE Nanostandards for the Case of MCs

The generic example of simulation provided by the IEEE 1906.1 and 1906.1.1 standards allows to understand what are the parameters and the general programming environment for MCs; with such fundamentals in [3, 4, 27, 28] the compilation of the aforementioned example is performed based on the open-source code defined in NS-3 provided by such standards.

To evaluate the behavior of a communication at the nanolevel, in [3, 4, 27, 28] the variables that intervene and influence each of the elements that constitute an MC system are considered. These results are presented based on specific values and characteristics of information that must be established to execute the source code provided by the IEEE 1906 protocol family standards. Consequently, parameters such as the following should be specified [3, 4, 27, 28]:

- The number of molecules that are transmitted, which in this case is $Q = 50{,}000$.
- The distance between the transmitter and receiver $r = 0.001$ (m).
- The diffusion coefficient $D = 1$ ($\eta m^2/ns$).
- The molecular pulse interval equals 1 (ms).

These values are entered to gain insight into simulation behavior and can be modified for simulation analysis. In addition, simulation characteristics with different parameters are evaluated to make comparisons between various simulation scenarios. In this way, (programmed) classes are implemented to simulate the components of the IEEE nanocommunications standards, based on which the performance of the link is measured, according to the modification of the variables indicated in the preceding paragraph, through which the communication performance is determined by means of [3, 4, 27, 28]:

- Molecular pulse
- The delay pulse in the medium

- The width of the molecular pulse
- The speed of transmission of information
- The ratio of molecular signals obtained
- The capacity of the molecular channel.

3.3.1.1 Implementation of Molecular Bonding

- **Creation of the P1906 helper.**
  The P1906 framework helper of the IEEE 1906.1 and 1906.1.1.1 standards is responsible for managing and creating nanomachine objects and enabling all register components related to the P1906 framework. The most important function of this class is to connect the components and attributes of the communicational endpoints. To visualize the communication process, logs were implemented in the method *EnableLogComponents* (with its corresponding functions), which allows the display of program debugging messages [3, 4, 27, 28].
  The method *connect* of the P1906 helper makes it possible to connect the components, tributes, and devices (nodes) involved in the communication; the main objective of the method is to facilitate the implementation and interaction of all the components of the molecular link. Other *helpers* defined by the NS-3 software allow the execution of the same functions as the PointToPointHelper classes and CsmaHelper, among others [3, 4, 27, 28].
- **Creation of nodes.**
  Nodes can be created using the classes *NetdeviceContainer* and *NodeContainer*. The class *NetDeviceContainer* allows the establishment of several communicational endpoints with the same characteristics, in addition to providing support when a large number of network components need to be defined. The class *NodeContainer* is used to install a certain number of nodes that share the same characteristics; this class contains multiple nodes defined in the parameter *Ptr < Node >*, which is used to refer to each of the nodes that it contains [3, 4, 27, 28].
- **Creation of the propagation medium and the motion component.**
  To create the components and entities of the module *P1906*, pointers are used, which specify a memory address. These pointers contain the type of variable to which they point [3, 4, 27, 28].
- **Creation of components and their devices (nodes).**
  As in the previous case, the creation of new variables is done by means of pointers. The variables that are created in this section of the code refer to the network communication endpoints, the communication interface, the specificity component, field, and disturbance [3, 4, 27, 28].
- **Establishment of the position and mobility model of the communicational extremes (transmitter and receiver).**
  Choosing a mobility model for the nodes will depend on the application for which the simulation is intended. These mobility models are usually added to the nodes to provide them with motion characteristics [3, 4, 27, 28].

The positions of the nodes are set by the *PositionAllocator parameter* and are generally only used at the beginning of the information configuration in the simulation, to determine the initial position of the nodes. In the case of the molecular simulation example specified in the IEEE 1906 protocol family standards, there is a list of options with two positions corresponding to the transmitting/receiving entities, which are assigned a constant mobility model; consequently, the nodes will remain in the initial position [3, 4, 27, 28].

- **Connection of the communicational ends, nodes, media, components and entities.**
  In order to connect the elements of the nanonet and their characteristics, there are helpers that allow this task to be carried out in a simplified way; thus, the class *helper* by means of its *connect* makes it possible to perform this function through which connection information relevant to the node, transmitter, receiver, communication medium, and components used in the topology to be simulated is specified [3, 4, 27, 28].
- **Creation of the PDU (*Protocol Data Unit*) package.**
  The information to be transferred on the link is implemented by means of the class *Packet*, which allows the creation of packets to be transmitted between nodes. In the case of the molecular simulation example specified in the IEEE 1906 protocol family standards, a string of characters composed of logical zeros (equivalent to one byte of information) is created; this string will be sent through the transmitter's interface [3, 4, 27, 28].

3.3.1.2 Messages Transfer

The sending of messages is carried out through the communication interface of the transmitter; this interface initiates the communication process, which will access all the components and methods involved in the communication [3, 4, 27, 28].

3.3.2 Communication Process Between Components

In the frame of reference of the IEEE 1906 protocol family standards, the communication process between the different components that integrate the nanoscale communication network is made effective by executing the following procedure [3, 4, 27, 28]:

- The class *NetDevice* receives a message from the upper layers and forward it to the transmitter's communication interface.
- The message carrier is created with the help of the disturbance component.
- The transmitter communication interface activates the propagation in the transmission medium and transfers the message carrier, disturbance, and field components.

- The motion component modifies the properties of the message carrier.
- The message carrier arrives at the receiver and the specificity component checks for compatibility.
- If the message is compatible, it is forwarded to the upper layers.
- The upper layers receive the message.

3.3.3 Implementation Process of the Classes that Execute the Components of the IEEE Nanoscale Framework

3.3.3.1 Message Carrier Class

In a nanoscale communication system, the message carrier can be a simple particle whose movement can be defined, for example, by free diffusion and thus provides the service to contain and transport the information of a message. The class (extended from the core module of the NS-3 code) for the previously mentioned functions corresponds to *MolMessagecarrier* in which the carrier itself and the duration and interval of the pulse in which the molecules are released are added. Additionally, the start time at which the information is transmitted and the number of molecules per pulse interval are configured by the same class. The methods *SetMessage* and *GetMessage* are functions inherited from the core model of the NS-3 code that capture the message and store it in the carrier [3, 4, 27, 28].

3.3.3.2 Perturbation Class

This class implements the disturbance component of the *P1906 module* and specifies the signal transmitted by the message carrier. The perturbation component can be considered as the one that performs the modulation of the information signal and produces the variation of the concentration or movement for the signal, as required. To generate the component relevant to the variation in the message carrier, the perturbation component uses the method *CreateMessageCarrier* which encodes the information to be transmitted and associates it with the carrier. In the case of the molecular simulation example specified in the IEEE 1906 protocol family standards, functionalities are added that make it possible to establish the message carrier with the characteristics relevant to those of a molecular signal. This component facilitates the storage of data on the quantity of molecules delivered per pulse and the temporal delay between successive pulses, with this information encoded on the message carrier at the beginning of communication [3, 4, 27, 28].

Figure 3.23 illustrates how a one-byte information packet is transmitted. It consists of 8 pulses in which molecule transfer occurs or does not occur, to forward the value specified in the simulation (message value equal to 136 in Fig. 3.23); two active pulses in which the molecules are released and six inactive pulses [3, 4, 27, 28].

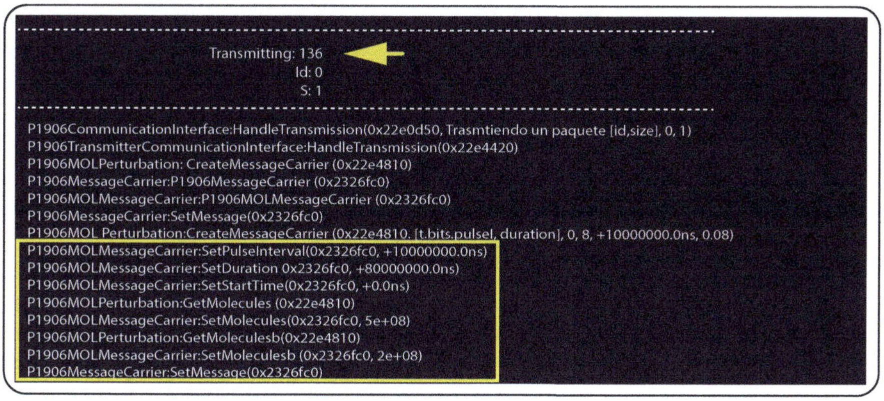

Fig. 3.23 Flowchart of the *Molecular Simulation.cc* example file from the IEEE 1906 protocol family standards framework [3, 4, 27, 28]

The simulation in reference has a pulse-based OOK modulation scheme, where the transmitter sends a pulse of molecules to represent bit-1 and no molecules to represent bit-0. This type of modulation is specified by Eq. (3.3):

$$f(t) = \begin{cases} Q_1 \prod(t), \text{bit } 1 \\ 0, \text{bit } 0 \end{cases} \tag{3.3}$$

where Q_1 represents the molecular concentration and $\prod(t)$ denotes a rectangular pulse with unit amplitude; the pulse amplitude is the unit of time from the end of one pulse to the beginning of the next [3, 4, 27, 28].

3.3.3.3 Motion Class

This class implements the motion component of the *P1906 module* and supports a propagation model using molecular sensing techniques in terms of an ideal diffusion process. This component provides the motion service for the message carrier in any direction and allows us to obtain the necessary potential to transport the information through a communication channel. The motion of each of the molecules released by the transmitter is modeled by Brownian motion, which represents free motion in space. In the simulation, it is considered that there are no obstacles between the communicating ends, and, consequently, it is an ideal motion in which other entities are not affected. The movement in question can be modeled by means of a Wiener notation process of $x \sim N(m, \sigma^2)$. When the transmitter must emit information, a pulse of molecules is released, which generates peaks of molecular concentration at the transmitter location that then propagate throughout space; the propagation of the pulse in reference is modeled analytically by the equations governing Fick's diffusion laws punctuated by Eq. (3.4) [3, 4, 27, 28]:

$$c(r,t) = \frac{Q}{(4\pi Dt)^{3/2}} e^{-\frac{r^2}{4Dt}} \tag{3.4}$$

where D is the diffusion coefficient of the medium, t is the time, r is the distance from the transmitter, and Q is the number of molecules. This equation is known as the pulse equation and allows modeling metrics of MCs in the *P1906 module*, such as delay, minimum pulse width, and channel capacity [3, 4, 27, 28].

The motion component of the *P1906 module* uses Fick's law to evaluate the propagation delay in the medium, which is calculated at the time instant at which the pulse reaches its maximum value and is expressed in Eq. (3.5) [3, 4, 27, 28]:

$$t_d = \frac{r^2}{6D} \tag{3.5}$$

where r corresponds to the distance from the transmitter to the receiver, and D is the diffusion coefficient.

With the help of the *Mobility* of NS-3, a motion is obtained based on the stochastic properties of motion and Fick's laws, given a concentration of molecules, the potential of the module also enables the calculation of the propagation delay by means of mathematical models and causes the message to propagate through the medium [3, 4, 27, 28].

3.3.3.4 Class

This component organizes the motion component; that is, it guides the motion in one direction. The field component of the *P1906 module* implemented in [3, 4, 27, 28] is null; therefore, the molecular motion will not have a particular direction and, consequently, the diffusion of information molecules becomes omnidirectional.

3.3.3.5 Specificity Class

This class implements the specificity component of the *P1906 module* and provides the service of detecting or receiving a message carrier; additionally, it verifies that the physical interface can receive and correctly interpret the transmitted message, by using the method *CheckRxCompatibility* returning as result a Boolean data (true or false), in case the corresponding compatibility exists between the transmitter and the receiver. In the case of such compatibility, the specificity component verifies that the channel capacity is greater than or equal to the transmission rate. The transmission rate (v_t) is calculated as the inverse of the molecular pulse interval (l_p), as indicated in Eq. (3.6) [3, 4, 27, 28].

$$v_t = \frac{1}{l_p} \tag{3.6}$$

By calculating the transmission rate, the specificity component determines the channel capacity, assuming the use of a pulse-width detection scheme, which has a key impact on the detection of molecular signals. To calculate the channel capacity, it is important to know the molecular pulse width, because this will be the main limitation on the achievable transmission rate. The inverse of the pulse width will be equal to the maximum capacity that the channel supports. This value must be taken into consideration for checking the Fick limits in the communication medium. The molecular model of the IEEE 1906.1 and 1906.1.1 standards produces a Boolean value corresponding to true if the Fick limit is met, and a Boolean value corresponding to false if the capacity of the molecular channel is less than the transmission rate, so that the Fick limit is not respected, and, consequently, the link fails. This check allows the receiver to determine the compatibility of the message with its interface [3, 4, 27, 28].

3.3.4 Communication Topology for Simulation

The topology in Fig. 3.24 [27, 28] denotes a point-to-point connection, in which the communication medium, nanotransmitters, and nanoreceivers do not present any obstacles or external agents interfering in the link, in which the field component establishes an omnidirectional movement of the molecular carriers. This topology enables the analysis of performance metrics such as the propagation of the molecular pulse, the delay pulse in the propagation medium, the width of the molecular pulse, the information transmission rate, the ratio of molecular signals obtained, and the channel capacity. In the topology, certain values such as the number of molecules to be transferred, the pulse interval, the distance between nodes, and the diffusion coefficient vary in the different components of the IEEE 1906 protocol family standards and their consequent variability of values depending on the specification of the initial values of the parameters in question [3, 4, 27, 28].

Conversely, in [27, 28], the *Matlab* and *Mathematica* programs are used to analyze the results from the simulation of the molecular example established in the reference framework of the IEEE 1906.1 and 1906.1.1, and by means of the scientific libraries (GSL-GNU *Scientific Library*) enabled in NS-3, it is possible to export files with the results of the simulation of the molecular example, thus making it possible to plot them. In addition, files with extension. *sh* declare a certain number of commands to allow inserting default values in the simulation and, consequently, comparing the results.

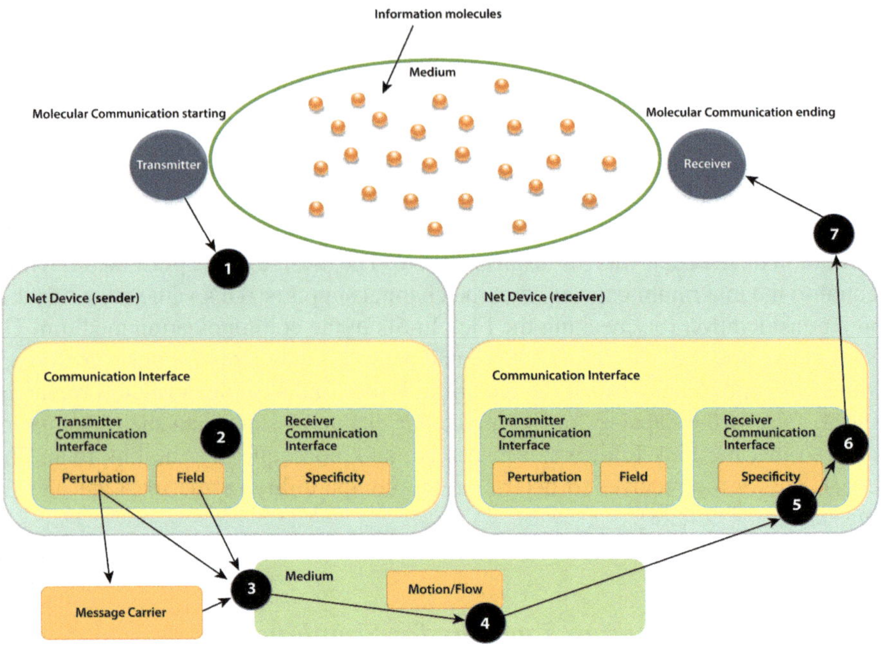

Fig. 3.24 Topological distribution of the communication elements used in the simulation of the molecular example of the IEEE 1906 protocol family standards

3.3.5 Communication Performance

When analyzing the performance of a communication link, it is necessary to examine each of the elements involved in the link, which defines the way in which information is transferred in the communication system. We then discuss the results obtained by performing the simulation of the MCs example established in the IEEE 1906 protocol family standards for nanocommunications, considering all the communicational aspects of the previous sections [3, 4, 27, 28].

3.3.5.1 Molecular Pulse Propagation

The molecular pulse propagation obtained from Fick's laws produces an important impact on the communication performance, as shown in Eq. (3.4); such propagation depends on several parameters, the alteration of one of them generates an important change in the binding performance [3, 4, 27, 28].

One of the most important parameters is precisely the distance between the transmitter and receiver (both considered as communication nodes or simply nodes), because at a greater distance between these nodes, the communication performance could be inherently affected by diffusion noise. To analyze the propagation behavior in [27, 28], the distance between the nodes has been modified to determine the maximum distance at which MCs can be successful through the proper control of parameters such as the pulse interval, the number of molecules, and the diffusion coefficient.

Figure 3.25 [27, 28] shows the results of the simulation of the molecular example of the IEEE nanoscale standards, in terms of the molecular pulse obtained in the transmission. It can be noticed that there are losses in the propagation of such pulse; the reason for these losses is found in the presence of diffusion noise that occurs as a consequence of the stochastic nature of Brownian motion; this contaminating affectation in the information signal becomes more evident as the distance between nodes is greater, severely degrading the performance of the communication [3, 4, 27, 28].

Fick's laws provide the maximum value at which a transmitter node can be distanced from its receiver without the latter failing to detect molecular signals. If, for example, a receiver node has a molecular concentration detection scheme, the detection level will depend on the concentration density (which is associated with the distance) and, consequently, these parameters will establish the effectiveness or not of the communication. In this sense, if a receiving node is located at a great distance from the transmitting node, the molecular concentration will not exceed a certain threshold level necessary for proper molecular detection and, consequently, the message will not be received. Figure 3.26 [27, 28] shows the molecular concentration obtained in the simulation of the molecular example of the IEEE 1906 protocol family standards, which is based on ideal diffusion conditions in the propagation medium; therefore, each of the released molecules has the same probability of being directed to the right or to the left from its previous position [3, 4, 27, 28].

In the simulation of the IEEE nanoscale example, OOK modulation is used for information transmission, its implementation is achieved through the perturbation component that encapsulates in the carrier signal the characteristics of an amplitude-modulated signal; thus, the receiver's detection threshold will determine the molecular concentration and the consequent decision of whether the information that has arrived corresponds to a bit-1 or bit-0 according to the interval times of each bit [27, 28].

Figure 3.27 [27, 28] allows the visualization of the transmission of molecular information using different distances of the link; in part (a) of the figure it can be seen that the communication was successful because the limitations established by the molecular concentration threshold have been overcome; while in part (b) of the figure, the amount of molecules captured by the receiver is not enough to overcome the concentration threshold; consequently, the transmission fails and the message is not received.

3.3.5.2 Delay Pulse

In the simulation of the molecular example of the IEEE 1906 protocol family standards, there are important metrics that have been implemented to know the performance of the communication. One of these (which is based on the transfer of molecular signals) is the delay pulse between the transmitter and receiver.

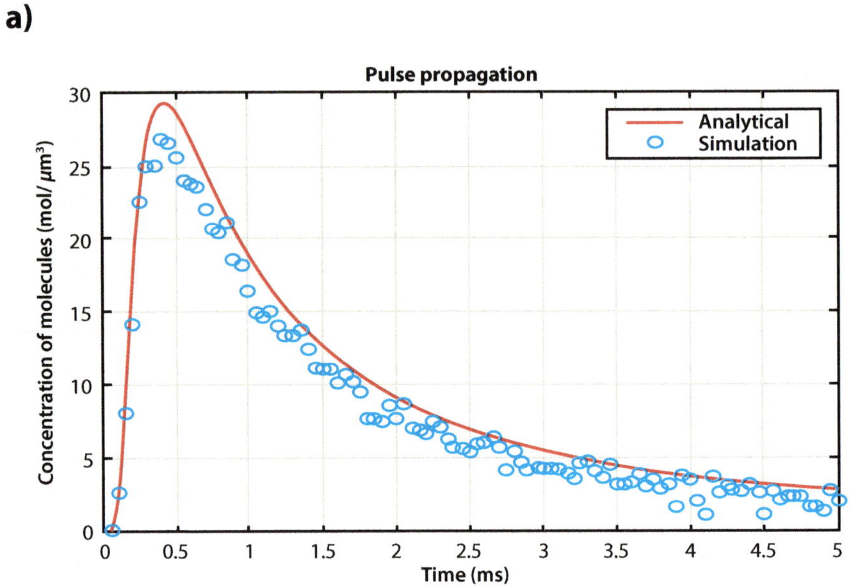

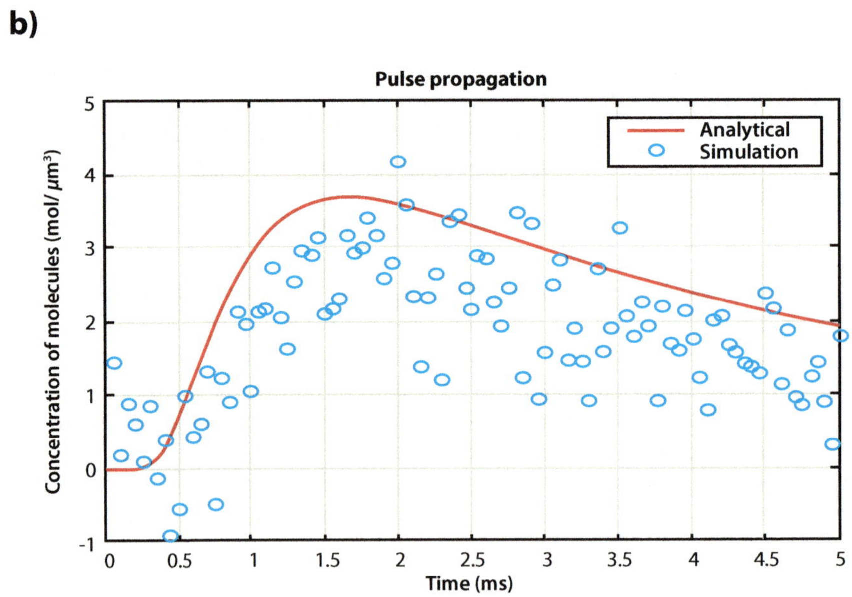

Fig. 3.25 Resulting molecular pulse as a function of time obtained from the NS-3 simulation of the molecular example of the IEEE 1906 protocol family standards

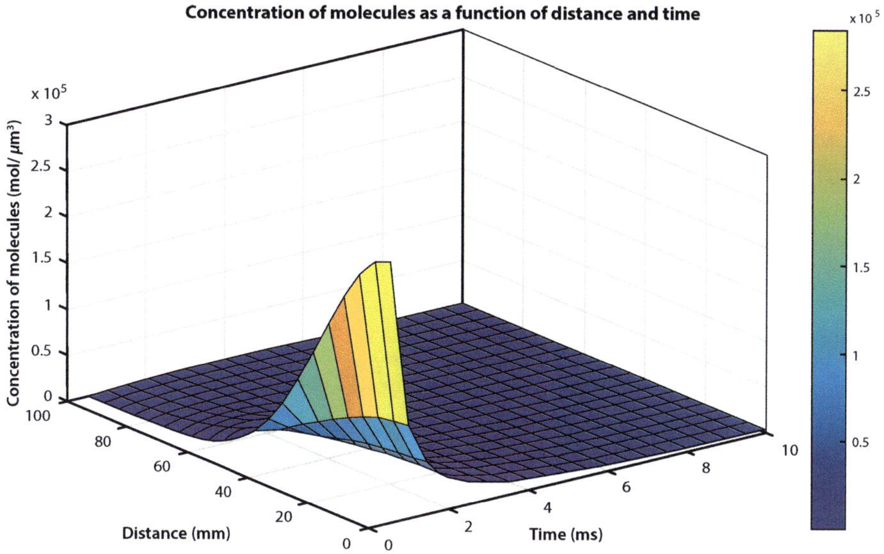

Fig. 3.26 Molecular concentration as a function of distance and time derived from the NS-3 simulation of the molecular example of the IEEE 1906 protocol family standards

Equation (3.7) is used to study the delay, which is implemented by the motion component of the IEEE 1906.1 and 1906.1.1 standards framework [27, 28].

$$d_S\left(bw\right)=\frac{S_{pn}\left(bw\right)}{S_p} \tag{3.7}$$

where d_S is the degradation of the nanocommunication supersystem defined in [3, 4], S_{pn} is the performance of the supersystem with the embedded nanoscale communication network, *bw* is the bandwidth of the embedded nanoscale communication network, and S_p is the performance of the original supersystem (without the embedded nanoscale communication network) [3, 4, 27, 28].

The delay in the transmission pulse is measured for each distance at which the receiving node is placed, so the result of each of the distances allows us to know how this delay affects the communication. In order to obtain these results, files are executed whose extension is.*sh*, which facilitates the performance of several simulations and the capture of the desired values, in this case the delay pulse [3, 4, 27, 28].

Figure 3.28 [3, 4, 27, 28] exhibits the pulse delay result after simulating the molecular example of the IEEE 1906 protocol family standards, where a pulse of 50,000 molecules with a diffusion coefficient equal to 1 (ηm²/ns) [3, 4, 27, 28].

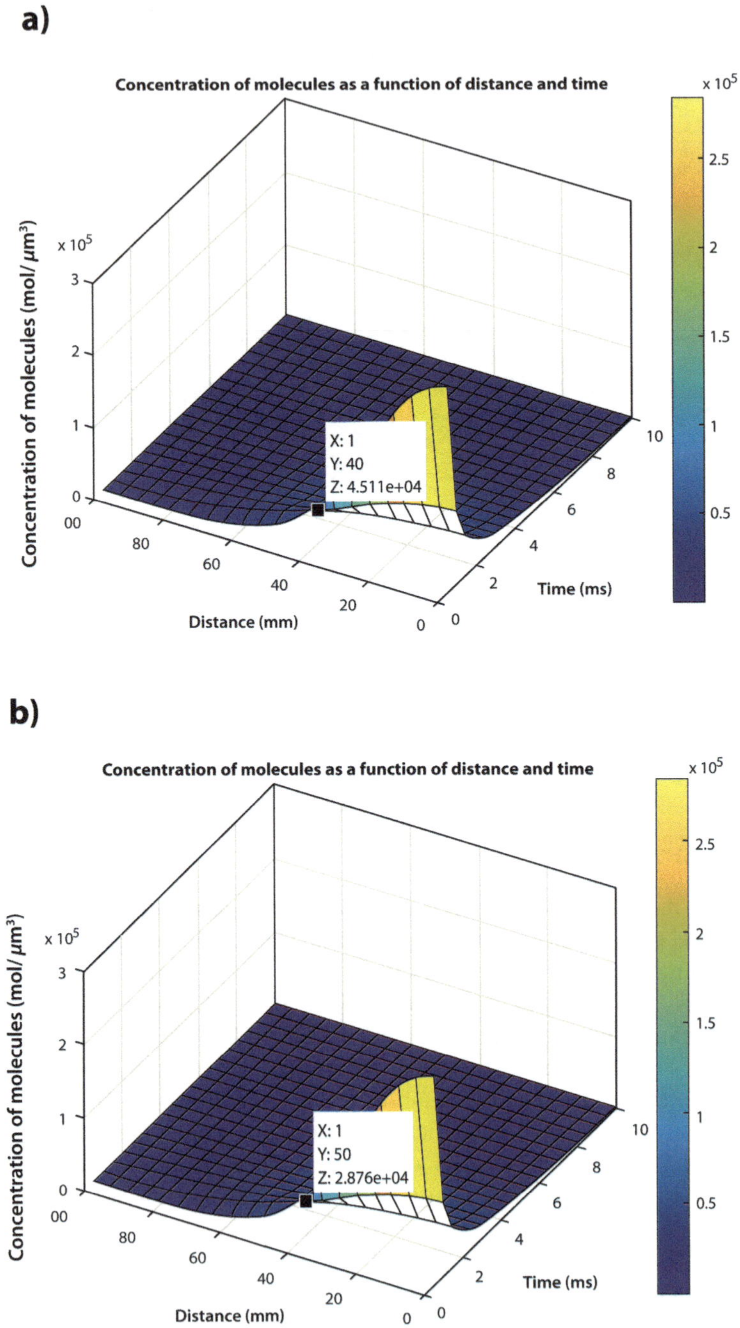

Fig. 3.27 Molecular communication graphs obtained from the simulation in NS-3 of the molecular example of the IEEE 1906 protocol family standards, set at different distances: (**a**) 0.04 (m) and (**b**) 0.05 (m)

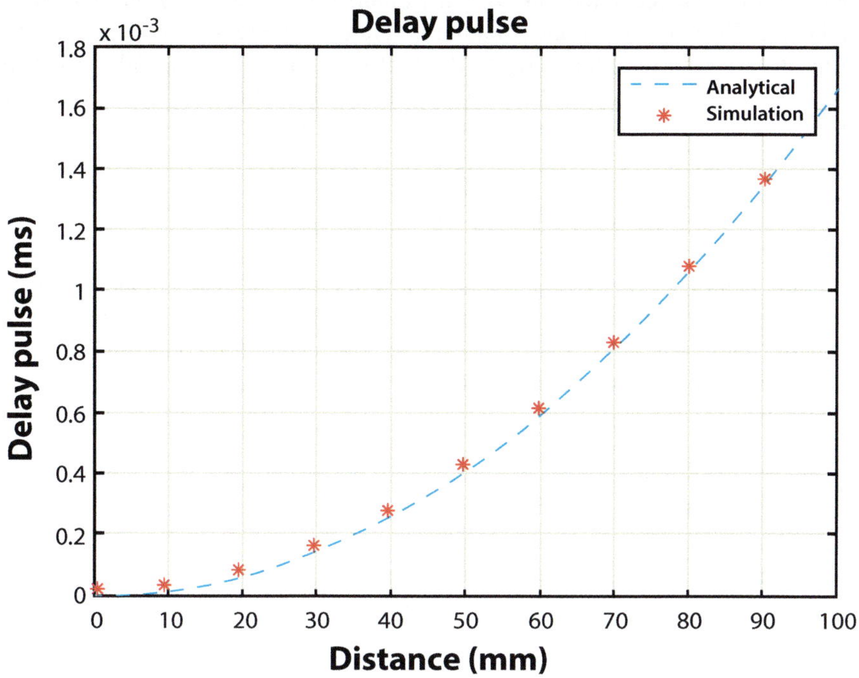

Fig. 3.28 Delay pulse obtained from the simulation of the molecular example of the IEEE 1906 protocol family standards based on a communication range contemplated between 0 (m) and 0.1 (m)

3.3.5.3 Molecular Pulse Width and Transmission Rate

The molecular pulse width is a highly relevant parameter in the simulation of the molecular example of the IEEE 1906 protocol family standards, because it determines the transmission speed that can be achieved in communication. One of the objectives of the molecular link is to achieve the highest possible bit transfer rate between the nanotransmitter and the nanoreceiver, and this rate is related to the duration of the symbols that are sent. The pulse width is obtained from the specificity component and allows us to know the molecular channel capacity; for the analysis of this metric, the Lambert function W must be discussed, which is defined as the inverse of the function $f(w) = We^{w}$, where e^{w} is a natural exponential function and W is any complex number. Therefore, the time variable is expressed as [3, 4, 27, 28]:

$$t = -\frac{r^2}{6DW\left(-\frac{1}{2^{\frac{2}{3}}e}\right)} \tag{3.8}$$

where r is the distance in the link, W is the Lambert function, and D is the diffusion coefficient. The solutions of the equation in reference are two, in which time values are taken at which the amplitude of the molecular pulse corresponds to half the maximum value and are expressed in Eq. (3.9). The difference of the two times punctuated in Eq. (3.9) specifies the pulse width [3, 4, 27, 28]:

$$t_1 = \frac{0.0728\,r^2}{D}, \quad t_2 = \frac{0.5229\,r^2}{D} \tag{3.9}$$

Figure 3.29 [3, 4, 27, 28] denotes the results of pulse width as a function of transmission distance acquired from the molecular example simulation of IEEE nanoscale standards, while Fig. 3.30 [3, 4, 27, 28] shows the result of the transmission rate in the molecular channel when using various values of pulse intervals ($Q(t)$) [3, 4, 27, 28].

3.3.5.4 Signal Ratio

In the simulation of the molecular example of the IEEE nanoscale standards, a diffusion environment is established in the communications channel; therefore, the modeling of this channel will be based on the equations that govern the diffusion

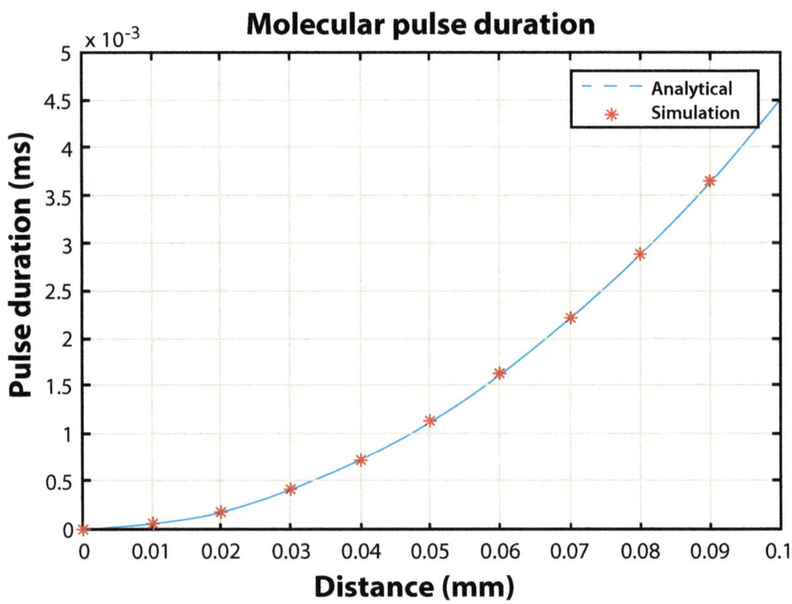

Fig. 3.29 Molecular pulse width obtained in the simulation of the molecular example of the IEEE 1906 protocol family standards for 100 realizations at different distances

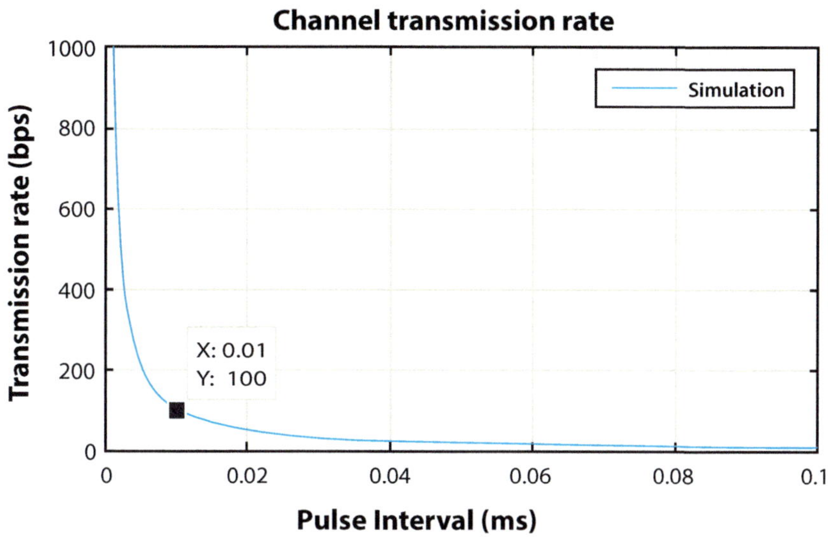

Fig. 3.30 Transmission rate results on the molecular channel in the simulation of the molecular example of the IEEE 1906 protocol family standards

process. In these channels, the molecules move randomly, an event that could generate ISI (Intersymbol Interference). In order to reduce the effect of signal contamination, certain time parameters are used to determine the duration of each symbol. In the case of the simulation carried out in [3, 4, 27, 28], pulse intervals are added to indicate the time in which a symbol is transmitted, in order to ensure that the following symbol does not present reading errors. Thus, a message of 1 byte of information will consist of 8 pulses of 0.001 (s) and therefore to transmit the message, the message carrier component will indicate the message duration time referred to in 0.008 (s) [3, 4, 27, 28].

One of the most important difficulties for the processing of the perturbation component when using OOK modulation is the presence of residual molecules that remain in the medium for longer than the time corresponding to a single bit, this problem is evident when the diffusion coefficient is low and the remaining molecules interfere in the detection of the following bit [28].

As shown in Fig. 3.31 [27, 28], the transmitter releases molecules at a fixed rate of 50,000 molecules per 0.001 (s) based on the bit sequence that has been broadcast. The disturbance component implements an OOK modulation on the message carrier signal; therefore, to measure the performance of this signal, the desired signal strength ratio (E_S) and the received interference intensity (E_I) are computed as follows [3, 4, 27, 28]:

$$E_S = \int_0^{T_{\text{obs}}} S_{\text{out}}(t)\,dt \tag{3.10}$$

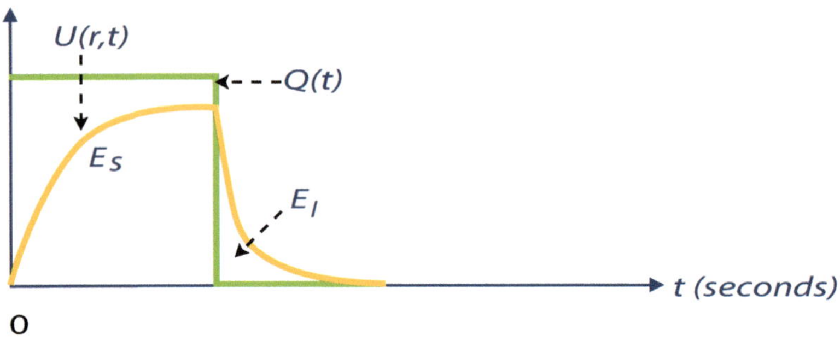

Fig. 3.31 Overview of the signal ratio in a molecular link of the molecular example of the IEEE 1906 protocol family standards

$$E_I = \int_0^{T_{obs}} i(t)\,dt \tag{3.11}$$

Signals $S_{out}(t)$, e $i(t)$ correspond to the transmitter output signal and the interference signal picked up by the receiver, respectively [3, 4, 27, 28].

The observation time (T_{obs}) corresponds to the product of the total number of bits to be transmitted and the bit duration. To obtain Fig. 3.31, the observation time has been set in the message carrier to the relevant value of 8,000,000 (ns) [3, 4, 27, 28].

3.3.5.5 Channel Capacity

The last communicational parameter analyzed in [3, 4, 27, 28] corresponds to the capacity of the communication channel, which can vary depending on the physical characteristics established in the implementation of the link, for example, through the configuration of the number of molecules or the diffusion coefficient. Figure 3.32 [3, 4, 27, 28] exposes the channel capacity as a result of the simulation of the molecular example of the IEEE nanoscale standards specifying different distances in the transmission. As can be seen in this figure, the greater the link distance, the channel is not shown to be appropriate to achieve adequate communication, completely analogous to what happens in traditional communication systems [3, 4, 27, 28].

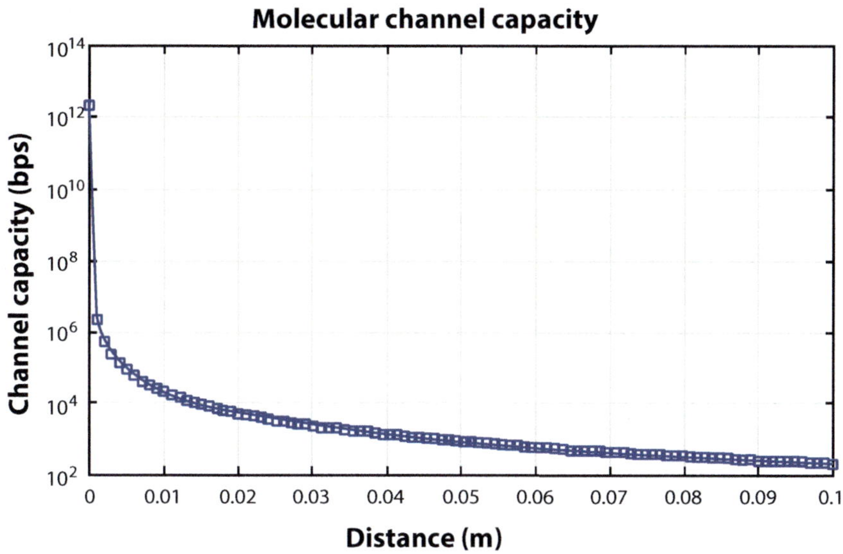

Fig. 3.32 Channel capacity obtained from the simulation of the molecular example of the IEEE 1906 protocol family standards

References

1. T. Suda, T. Nakano, Molecular communication: a personal perspective. IEEE Trans. Nanobiosci. **17**(4), 424–432 (2018)
2. L. Felicetti, S.S. Assaf, M. Femminella, G. Reali, E. Alarcon, J. Sole-Pareta, The molecular communications markup language (MolComML). Nano Commun. Netw. **16**, 12–25 (2018)
3. IEEE recommended practice for nanoscale and molecular communication framework, IEEE Std 1906.1-2015 (2016), pp. 1–64. https://doi.org/10.1109/IEEESTD.2016.7378262
4. IEEE standard data model for nanoscale communication systems, IEEE Std 1906.1.1-2020 (2020), pp. 1–142. https://doi.org/10.1109/IEEESTD.2020.9285373
5. S.S. Assaf, *Synthetic Biology Guidelines for Diffusion Based Molecular Communication* (2019)
6. E. Alarcon et al., MolComML: the molecular communication markup language, in *Proceedings of the 3rd ACM International Conference on Nanoscale Computing and Communication*, (2016), pp. 1–6
7. L. Felicetti, M. Femminella, T. Ivanov, P. Lio, G. Reali, A big-data layered architecture for analyzing molecular communications systems in blood vessels, in *Proceedings of the 4th ACM International Conference on Nanoscale Computing and Communication*, (2017), pp. 1–2
8. G. Reali et al., Simulation tools for molecular communications. TCSIM Newsl. **22**, 8–12 (2016)
9. M. Galdzicki et al., The Synthetic Biology Open Language (SBOL) provides a community standard for communicating designs in synthetic biology. Nat. Biotechnol. **32**(6), 545–550 (2014)
10. S. Neymotin, *Integrating Systems Biology Markup Language (SBML) with NEURON*
11. F. Achard, G. Vaysseix, E. Barillot, XML, bioinformatics and data integration. Bioinformatics **17**(2), 115–125 (2001)
12. H.B. Yilmaz, A.C. Heren, T. Tugcu, C.-B. Chae, Three-dimensional channel characteristics for molecular communications with an absorbing receiver. IEEE Commun. Lett. **18**(6), 929–932 (2014)

13. H. Yan et al., Molecular communication in nanonetworks. Nano Biomed. Eng. **8**(4), 274–287 (2016)
14. N. Farsad, H.B. Yilmaz, A. Eckford, C.-B. Chae, W. Guo, A comprehensive survey of recent advancements in molecular communication. IEEE Commun. Surv. Tutor. **18**(3), 1887–1919 (2016)
15. I. Llatser et al., Exploring the physical channel of diffusion-based molecular communication by simulation, in *2011 IEEE Global Telecommunications Conference-GLOBECOM 2011*, (2011), pp. 1–5
16. L. Felicetti, M. Femminella, G. Reali, T. Nakano, A.V. Vasilakos, TCP-like molecular communications. IEEE J Sel Areas Commun **32**(12), 2354–2367 (2014)
17. C. Schölzel, V. Blesius, G. Ernst, A. Dominik, Characteristics of mathematical modeling languages that facilitate model reuse in systems biology: A software engineering perspective. NPJ Syst. Biol. Appl. **7**(1), 1–20 (2021)
18. A. Akkaya, G. Genc, T. Tugcu, HLA based architecture for molecular communication simulation. Simul. Model. Pract. Theory **42**, 163–177 (2014)
19. E.A. Pilligua Vélez, *Simulación de sistemas de comunicación biológicos en entornos de difusión y de células tumorales en vasos sanguíneos mediante la herramienta de nanocomunicaciones a nivel molecular MolComML* (Universidad Nacional de Chimborazo, Riobamba, 2024)
20. W.F. Boron, E.L. Boulpaep, *Fisiología médica* (Elsevier Health Sciences, 2017) https://books.google.com.ec/books?id=1J_HDgAAQBAJ
21. W. Pan, B. Caswell, G.E. Karniadakis, A low-dimensional model for the red blood cell. Soft Matter **6**(18), 4366–4376 (2010)
22. E.R. Kandel, J.H. Schwartz, T.M. Jessell, S. Siegelbaum, A.J. Hudspeth, *Principles of neural science*, vol 4 (McGraw-Hill, New York, 2000)
23. Y. Cevallos et al., *Molecular Communications: An Analysis from Networking Theories Perspective* (Springer Nature, 2023)
24. Y. Cevallos et al., Modeling gene expression and protein delivery as an end-to-end digital communication system. Open Bioinform. J. **14**(1), 21 (2021)
25. S.F. Bush, Interoperable nanoscale communication [future directions]. IEEE Consum. Electron. Mag. **6**(2), 39–47 (2017)
26. S.F. Bush, J.L. Paluh, G. Piro, V. Rao, R.V. Prasad, A. Eckford, Defining communication at the bottom. IEEE Trans. Mol. Biol. Multi-Scale Commun. **1**(1), 90–96 (2015)
27. A.G. Rivera Gaibor, *Simulación de nano sistemas de comunicaciones en NS3 en función a los componentes comunicacionales establecidos por el estándar IEEE 1906.1* (Universidad Nacional de Chimborazo, Riobamba, 2021)
28. Y. Cevallos et al., *Comunicaciones Moleculares Un análisis desde el paradigma protocolario de IEEE* (2023)

GPSR Compliance

The European Union's (EU) General Product Safety Regulation (GPSR) is a set of rules that requires consumer products to be safe and our obligations to ensure this.

If you have any concerns about our products, you can contact us on ProductSafety@springernature.com

In case Publisher is established outside the EU, the EU authorized representative is:

Springer Nature Customer Service Center GmbH
Europaplatz 3
69115 Heidelberg, Germany

Batch number: 10746987

Printed by Printforce, the Netherlands